# Homotopical Algebraic Geometry II: Geometric Stacks and Applications

MEMOIRS
of the
American Mathematical Society

Number 902

# Homotopical Algebraic Geometry II: Geometric Stacks and Applications

Bertrand Toën
Gabriele Vezzosi

May 2008 • Volume 193 • Number 902 (second of 5 numbers) • ISSN 0065-9266

**American Mathematical Society**
Providence, Rhode Island

2000 *Mathematics Subject Classification.*
Primary 14A20, 18G55, 18F10, 55U40, 55P42, 55P43, 18F20, 18D10, 18E30, 18G35, 18G30, 13D10, 55N34.

**Library of Congress Cataloging-in-Publication Data**

Toën, Bertrand, 1973–
Homotopical algebraic geometry II : geometric stacks and applications / Bertrand Toën, Gabriele Vezzosi.
p. cm. — (Memoirs of the American Mathematical Society, ISSN 0065-9266 ; no. 902)
"May 2008, volume 193, number 902 (second of 5 numbers)."
Includes bibliographical references and index.
ISBN 978-0-8218-4099-3 (alk. paper)
1. Algebraic stacks. 2. Algebra, Homological. 3. Geometry, Algebraic. 4. Categories (Mathematics) I. Vezzosi, Gabriele, 1966– II. Title.

QA564.T638 2008
516.3′5—dc22 2008060003

## Memoirs of the American Mathematical Society

This journal is devoted entirely to research in pure and applied mathematics.

**Subscription information.** The 2008 subscription begins with volume 191 and consists of six mailings, each containing one or more numbers. Subscription prices for 2008 are US$675 list, US$540 institutional member. A late charge of 10% of the subscription price will be imposed on orders received from nonmembers after January 1 of the subscription year. Subscribers outside the United States and India must pay a postage surcharge of US$38; subscribers in India must pay a postage surcharge of US$43. Expedited delivery to destinations in North America US$53; elsewhere US$130. Each number may be ordered separately; *please specify number* when ordering an individual number. For prices and titles of recently released numbers, see the New Publications sections of the *Notices of the American Mathematical Society.*

**Back number information.** For back issues see the *AMS Catalog of Publications.*

Subscriptions and orders should be addressed to the American Mathematical Society, P. O. Box 845904, Boston, MA 02284-5904, USA. *All orders must be accompanied by payment.* Other correspondence should be addressed to 201 Charles Street, Providence, RI 02904-2294, USA.



*Memoirs of the American Mathematical Society* (ISSN 0065-9266) is published bimonthly (each volume consisting usually of more than one number) by the American Mathematical Society at 201 Charles Street, Providence, RI 02904-2294, USA. Periodicals postage paid at Providence, RI. Postmaster: Send address changes to Memoirs, American Mathematical Society, 201 Charles Street, Providence, RI 02904-2294, USA.


This publication is indexed in *Science Citation Index*®, *SciSearch*®, *Research Alert*®, *CompuMath Citation Index*®, *Current Contents*®*/Physical, Chemical & Earth Sciences.*
Printed in the United States of America.

∞ The paper used in this book is acid-free and falls within the guidelines established to ensure permanence and durability.
Visit the AMS home page at http://www.ams.org/

10 9 8 7 6 5 4 3 2 1 13 12 11 10 09 08

This work is dedicated to Alexandre Grothendieck

# Contents

# Abstract

[1] This is the second part of a series of papers called "HAG", and devoted to develop the foundations of *homotopical algebraic geometry*. We start by defining and studying generalizations of standard notions of linear algebra in an abstract monoidal model category, such as derivations, étale and smooth morphisms, flat and projective modules, etc. We then use our theory of stacks over model categories, introduced in [**HAGI**], in order to define a general notion of geometric stack over a base symmetric monoidal model category $C$, and prove that this notion satisfies the expected properties.

The rest of the paper consists in specializing $C$ in order to give various examples of applications in several different contexts. First of all, when $C = k - Mod$ is the category of $k$-modules with the trivial model structure, we show that our notion gives back the algebraic $n$-stacks of C. Simpson. Then we set $C = sk - Mod$, the model category of simplicial $k$-modules, and obtain this way a notion of geometric $D^-$*-stack* which are the main geometric objects of *derived algebraic geometry*. We give several examples of derived version of classical moduli stacks, as the $D^-$-stack of local systems on a space, the $D^-$-stack of algebra structures over an operad, the $D^-$-stack of flat bundles on a projective complex manifold, etc. We also present the cases where $\mathcal{C} = C(k)$ is the model category of unbounded complexes of $k$-modules, and $C = Sp^\Sigma$ the model category of symmetric spectra. In these two contexts we give some examples of geometric stacks such as the stack of associative dg-algebras, the stack of dg-categories, and a geometric stack constructed using topological modular forms.

`There are more things in heaven and earth, Horatio,`
`than are dreamt of in our philosophy. But come...`

W. Shakespeare, *Hamlet*, Act 1, Sc. 5.

`Mon cher Cato, il faut en convenir, les forces de l'ether nous pénètrent,`
`et ce fait déliquescent il nous faut l'appréhendcr coute que coute.`

P. Sellers, *Quand la panthère rose s'en mêle.*

[1]2000 Mathematics Subject Classification: 14A20, 18G55, 18F10, 55U40, 55P42, 55P43, 18F20, 18D10, 18E30 18G35, 18G30, 13D10, 55N34.
Keywords: Algebraic stacks, higher stacks, derived algebraic geometry
Received by the Editor June 14 2005

# Introduction

This is the second part of a series of papers called "HAG", devoted to start the development of *homotopical algebraic geometry*. The first part [**HAGI**] was concerned with the homotopical generalization of sheaf theory, and contains the notions of model topologies, model sites, stacks over model sites and model topoi, all of these being homotopical versions of the usual notions of Grothendieck topologies, sites, sheaves and topoi. The purpose of the present work is to use these new concepts in some specific situations, in order to introduce a very general notion of *geometric stacks*, a far reaching homotopical generalization of the notion of algebraic stacks introduced by P. Deligne, D. Mumford and M. Artin. This paper includes the general study and the standard properties of geometric stacks, as well as various examples of applications in the contexts of algebraic geometry and algebraic topology.

**Reminders on abstract algebraic geometry.** A modern point of view on algebraic geometry consists of viewing algebraic varieties and schemes through their functors of points. In this functorial point of view, schemes are certain sheaves of sets on the category of commutative rings endowed with a suitable topology (e.g. the Zariski topology). Keeping this in mind, it turns out that the whole theory of schemes can be completely reconstructed starting from the symmetric monoidal category $\mathbb{Z} - Mod$ of $\mathbb{Z}$-modules alone. Indeed, the category of commutative rings is reconstructed by taking the category of commutative monoids in $\mathbb{Z} - Mod$. Flat morphisms can be recognized via the exactness property of the base change functor on the category of modules. Finitely presented morphisms are recognized via the usual categorical characterization in terms of commutation of mapping sets with respect to filtered colimits. Finally, Zariski open immersion can be defined as flat morphisms of finite presentation $A \longrightarrow B$ such that $B \simeq B \otimes_A B$. Schemes are then reconstructed as certain Zariski sheaves on the opposite category of commutative rings, which are obtained by gluing affine schemes via Zariski open immersions (see for example the first chapters of [**Dem-Gab**]).

The fact that the notion of schemes has such a purely categorical interpretation has naturally lead to the theory of *relative algebraic geometry*, in which the base symmetric monoidal category $\mathbb{Z} - Mod$ is replaced by an abstract base symmetric monoidal category $\mathcal{C}$, and under reasonable assumptions on $\mathcal{C}$ the notion of *schemes over* $\mathcal{C}$ can be made meaningful as well as useful (see for example [**Del1, Ha**] for some applications).

The key observation of this work is that one can generalize further the theory of relative algebraic geometry by requiring $\mathcal{C}$ to be endowed with an additional *model category* structure, compatible with its monoidal structure (relative algebraic geometry is then recovered by taking the trivial model structure), in such a way that the notions of *schemes* and more generally of *algebraic spaces* or *algebraic stacks* still have a natural and useful meaning, compatible with the homotopy theory carried by $\mathcal{C}$. In this work, we present this general theory, and show how this enlarges the field of applicability by investigating several examples not covered by the standard theory of relative algebraic geometry. The most important of these applications is the existence of foundations for *derived algebraic geometry*, a global counter part of the derived deformation theory of V. Drinfel'd, M. Kontsevich and al.

**The setting.** Our basic datum is a symmetric monoidal model category $\mathcal{C}$ (in the sense of [**Ho1**]), on which certain conditions are imposed (see assumptions 1.1.0.1, 1.1.0.2, 1.1.0.3 and 1.1.0.4). We briefly discuss these requirements here. The model category $\mathcal{C}$ is assumed to satisfy some reasonable additional properties (as for example being proper, or that cofibrant objects are flat for the monoidal structure). These assumptions are only made for the convenience of certain constructions, and may clearly be omitted. The model category $\mathcal{C}$ is also assumed to be *combinatorial* (see e.g. [**Du2**]), making it reasonably behaved with respect to localization techniques. The first really important assumption on $\mathcal{C}$ states that it is pointed (i.e. that the final and initial object coincide) and that its homotopy category $\mathrm{Ho}(\mathcal{C})$ is additive. This makes the model category $\mathcal{C}$ *homotopically additive*, which is a rather strong condition, but is used all along this work and seems difficult to avoid (see however [**To-Va2**]). Finally, the last condition we make on $\mathcal{C}$ is also rather strong, and states that the theory of commutative monoids in $\mathcal{C}$, and the theory of modules over them, both possess reasonable model category structures. This last condition is of course far from being satisfied in general (as for example it is not satisfied when $\mathcal{C}$ is the model category of complexes over some ring which is not of characteristic zero), but all the examples we have in mind can be treated in this setting[2]. The model categories of simplicial modules, of complexes over a ring of characteristic zero, and of symmetric spectra are three important examples of symmetric monoidal model category satisfying all our assumptions. More generally, the model categories of sheaves with values in any of these three fundamental categories provide additional examples.

**Linear and commutative algebra in a symmetric monoidal model category.** An important consequence of our assumptions on the base symmetric monoidal model category $\mathcal{C}$ is the existence of reasonable generalizations of general constructions and results from standard linear and commutative algebra. We have gathered some of these notions (we do not claim to be exhaustive) in §1.1. For example, we give definitions of derivations as well as of cotangent complexes representing them, we define formally étale morphisms, flat morphisms, open Zariski immersions, formally unramified morphisms, finitely presented morphism of commutative monoids and modules, projective and flat modules, Hochschild cohomology, etc. They are all generalizations of the well known notions in the sense that when applied to the case where $\mathcal{C} = \mathbb{Z} - Mod$ with the trivial model structure we find back the usual notions. However, there are sometimes several nonequivalent generalizations, as for example there exist at least two, nonequivalent reasonable generalizations of smooth morphisms which restrict to the usual one when $\mathcal{C} = \mathbb{Z} - Mod$. This is why we have tried to give an overview of several possible generalizations, as we think all definitions could have their own interest depending both on the context and on what one wants to do with them. Also we wish to mention that all these notions depend heavily on the base model category $\mathcal{C}$, in the sense that the same object in $\mathcal{C}$, when considered

[2] Alternatively, one could switch to $E_\infty$-algebras (and modules over them) for which useful model and semi-model structures are known to exist, thanks to the work of M. Spitzweck [**Sp**], in much more general situations than for the case of commutative monoids (and modules over them).

in different model categories structures on $\mathcal{C}$, might not behave the same way. For example, a commutative ring can also be considered as a simplicial commutative ring, and the notion of finitely presented morphisms is not the same in the two cases. We think that keeping track of the base model category $\mathcal{C}$ is rather important, since playing with the change of base categories might be very useful, and is also an interesting feature of the theory.

The reader will immediately notice that several notions behave in a much better way when the base model category satisfies certain stability assumptions (e.g. is a stable model category, or when the suspension functor is fully faithful, see for example Prop. 1.2.6.5, Cor. 1.2.6.6). We think this is one of the main features of homotopical algebraic geometry: linear and commutative algebra notions tend to be better behaved as the base model category tend to be "more" stable. We do not claim that everything becomes simpler in the stable situation, but that certain difficulties encountered can be highly simplified by enlarging the base model category to a more stable one.

**Geometric stacks.** In §1.3 we present the general notions of geometric stacks relative to our base model category $\mathcal{C}$. Of course, we start by defining $Aff_{\mathcal{C}}$, the model category of affine objects over $\mathcal{C}$, as the opposite of the model category $Comm(\mathcal{C})$ of commutative monoids in $\mathcal{C}$. We assume we are given a *model (pre-)topology* $\tau$ on $Aff_{\mathcal{C}}$, in the sense we have given to this expression in [**HAGI**, Def. 4.3.1] (see also Def. 1.3.1.1). We also assume that this model topology satisfies certain natural assumptions, as quasi-compactness and the descent property for modules. The model category $Aff_{\mathcal{C}}$ together with its model topology $\tau$ is a model site in the sense of [**HAGI**, Def. 4.3.1] or Def. 1.3.1.1, and it gives rise to a model category of stacks $Aff_{\mathcal{C}}^{\sim,\tau}$. The homotopy category of $Aff_{\mathcal{C}}^{\sim,\tau}$ will simply be denoted by $\mathrm{St}(\mathcal{C},\tau)$. The Yoneda embedding for model categories allows us to embed the homotopy category $\mathrm{Ho}(Aff_{\mathcal{C}})$ into $\mathrm{St}(\mathcal{C},\tau)$, and this gives a notion of representable stack, our analog of the notion of affine scheme. Geometric stacks will result from a certain kind of gluing representable stacks.

Our notion of geometric stack is relative to a class of morphisms $\mathbf{P}$ in $Aff_{\mathcal{C}}$, satisfying some compatibility conditions with respect to the topology $\tau$, essentially stating that the notion of morphisms in $\mathbf{P}$ is local for the topology $\tau$. With these two notions, $\tau$ and $\mathbf{P}$, we define by induction on $n$ a notion of $n$-geometric stack (see 1.3.3.1). The precise definition is unfortunately too long to be reproduced here, but one can roughly say that $n$-geometric stacks are stacks $F$ whose diagonal is $(n-1)$-representable (i.e. its fibers over representable stacks are $(n-1)$-geometric stacks), and which admits a covering by representable stacks $\coprod_i U_i \longrightarrow F$, such that all morphisms $U_i \longrightarrow F$ are in $\mathbf{P}$.

The notion of $n$-geometric stack satisfies all the expected basic properties. For example, geometric stacks are stable by (homotopy) fiber products and disjoint unions, and being an $n$-geometric stack is a local property (see Prop. 1.3.3.3, 1.3.3.4). We also present a way to produce $n$-geometric stacks as certain quotients of groupoid actions, in the same way that algebraic stacks (in groupoids) can always be presented as quotients of a scheme by a smooth groupoid action (see Prop. 1.3.4.2). When a property $\mathbf{Q}$ of morphisms in $Aff_{\mathcal{C}}$ satisfies a certain compatibility with both $\mathbf{P}$ and $\tau$, there exists a natural notion of $\mathbf{Q}$-morphism between $n$-geometric stacks, satisfying all the expected properties (see Def. 1.3.6.4, Prop. 1.3.6.3). We define the stack of quasi-coherent modules, as well its sub-stacks of vector bundles and of perfect modules (see Thm. 1.3.7.2, Cor. 1.3.7.3). These also behave as expected, and for example the stack of vector bundles is shown to be 1-geometric as soon as the class **P** contains the class of smooth morphisms (see Cor. 1.3.7.12).

**Infinitesimal theory.** In §1.4, we investigate the infinitesimal properties of geometric stacks. For this we define a notion of derivation of a stack $F$ with coefficients in a module, and the notion of cotangent complex is defined via the representability of the functor of derivations (see Def. 1.4.1.4, 1.4.1.5, 1.4.1.7). The object representing the derivations, the cotangent complex, is not in general an object in the base model category $\mathcal{C}$, but belongs to the stabilization of $\mathcal{C}$ (this is of course related to the well known fact that cotangent spaces of algebraic stacks are not vector spaces but rather complexes of vector spaces). This is why these notions will be only defined when the suspension functor of $\mathcal{C}$ is fully faithful, or equivalently when the stabilization functor from $\mathcal{C}$ to its stabilization is fully faithful (this is again an incarnation of the fact, mentioned above, that homotopical algebraic geometry seems to prefer stable situations). In a way, this explains from a conceptual point of view the fact that the infinitesimal study of usual algebraic stacks in the sense of Artin is already part of homotopical algebraic geometry, and does not really belong to standard algebraic geometry. We also define stacks having an *obstruction theory* (see Def. 1.4.2.1), a notion which controls obstruction to lifting morphisms along a first order deformation in terms of the cotangent complex. Despite its name, having an obstruction theory is a property of a stack and not an additional structure. Again, this notion is really well behaved when the suspension functor of $\mathcal{C}$ is fully faithful, and this once again explains the relevance of derived algebraic geometry with respect to infinitesimal deformation theory. Finally, in the last section we give sufficient conditions (that we called *Artin's conditions*) insuring that any $n$-geometric stack has an obstruction theory (Thm. 1.4.3.2). This last result can be considered as a far reaching generalization of the exixtence of cotangent complexes for algebraic stacks as presented in [**La-Mo**].

**Higher Artin stacks (after C. Simpson).** As a first example of application, we show how our general notion of geometric stacks specializes to C. Simpson's algebraic $n$-stacks introduced in [**S3**]. For this, we let $\mathcal{C} = k - Mod$, be the symmetric monoidal category of $k$-modules (for some fixed commutative ring $k$), endowed with its trivial model structure. The topology $\tau$ is chosen to be the étale (ét) topology, and **P** is chosen to be the class of smooth morphisms. We denote by $\mathrm{St}(k)$ the corresponding homotopy category of ét-stacks. Then, our definition of $n$-geometric stack gives back the notion of algebraic $n$-stack introduced in [**S3**] (except that the two $n$'s might differ); these stacks will be called *Artin $n$-stacks* as they contain the usual algebraic stacks in the sense of Artin as particular cases (see Prop. 2.1.2.1). However, all our infinitesimal study (cotangent complexes and obstruction theory) does not apply here as the suspension functor on $k - Mod$ is the zero functor. This should not be viewed as a drawback of the theory; on the contrary we rather think this explains why deformation theory and obstruction theory in fact already belong to the realm of derived algebraic geometry, which is our next application.

**Derived algebraic geometry: $D^-$-stacks.** Our second application is the so-called *derived algebraic geometry*. The base model category $\mathcal{C}$ is chosen to be $sk{-}Mod$, the symmetric monoidal model category of simplicial commutative $k$-modules, $k$ being some fixed base ring. The category of affine objects is $k - D^-Aff$, the opposite model category of the category of commutative simplicial $k$-algebras. In this setting, our general notions of flat, étale, smooth morphisms and Zariski open immersions all have explicit descriptions in terms of standard notions (see Thm. 2.2.2.6). More precisely, we prove that a morphism of simplicial commutative $k$-algebras $A \longrightarrow B$ is flat (resp. smooth, resp. étale, resp. a Zariski open immersion) in the general sense we have given to these notions in §1.2, if and only if it satisfies the following two conditions

- The induced morphism of affine schemes
$$Spec\, \pi_0(B) \longrightarrow Spec\, \pi_0(A)$$
is flat (resp. smooth, resp. étale, resp. a Zariski open immersion) in the usual sense.
- The natural morphism
$$\pi_*(A) \otimes_{\pi_0(A)} \pi_0(B) \longrightarrow \pi_*(B)$$
is an isomorphism.

We endow $k - D^-Aff$ with the étale model topology, a natural extension of the étale topology for affine schemes; the corresponding homotopy category of $D^-$-stacks is simply denoted by $D^-\mathrm{St}(k)$. The class $\mathbf{P}$ is taken to be the class of smooth morphisms. The $n$-geometric stacks in this context will be called *$n$-geometric $D^-$-stacks*, where the notation $D^-$ is meant to bring to mind the negative bounded derived category[3]. An important consequence of the above descriptions of étale and smooth morphisms is that the natural inclusion functor from the category of $k$-modules to the category of simplicial $k$-modules, induces a full embedding of the category of Artin $n$-stacks into the category of $n$-geometric $D^-$-stack. This inclusion functor $i$ has furthermore a right adjoint, called the truncation functor $t_0$ (see Def. 2.2.4.3), and the adjunction morphism $it_0(F) \longrightarrow F$ provides a closed embedding of the classical Artin $n$-stack $it_0(F)$ to its derived version $F$, which behaves like a formal thickening (see Prop. 2.2.4.7). This is a global counterpart of the common picture of derived deformation theory of a formal classical moduli space sitting as a closed sub-space in the corresponding formal derived moduli space.

We also prove that our general conditions for the existence of an obstruction theory are satisfied, and so any $n$-geometric $D^-$-stack has an obstruction theory (see Prop. 2.2.3.3). An important particular case is when this result is applied to the image of an Artin $n$-stack via the natural inclusion functor $i$; we obtain in this way the existence of an obstruction theory for *any* Artin $n$-stack, and in particular the existence of a cotangent complex. This is a very good instance of our principle that things simplifies when the base model category $\mathcal{C}$ becomes more stable: the infinitesimal study of classical objects of algebraic geometry (such as schemes, algebraic stacks or Artin $n$-stacks) becomes conceptually clearer and behaves much better when we consider these objects as geometric $D^-$-stacks.

Finally, we give several examples of $D^-$-stacks being derived versions of some well known classical moduli problems. First of all the $D^-$-stack of *local systems* on a topological space, and the $D^-$-stack of *algebra structures* over a given operad are shown to be 1-geometric (see Lem. 2.2.6.3, Prop. 2.2.6.8). We also present derived versions of the *scheme of morphisms* between two projective schemes, and of the moduli stack of *flat bundles* on a projective complex manifold (Cor. 2.2.6.14 and Cor. 2.2.6.15). The proofs that these last two stacks are geometric rely on a special version of J. Lurie's

[3]Recall that the homotopy theory of simplicial $k$-modules is equivalent to the homotopy theory of negatively graded cochain complexes of $k$-modules. Therefore, derived algebraic geometry can also be considered as algebraic geometry over the category of negatively graded complexes.

representability theorem (see [**Lu1**] and Appendix C).

**Complicial algebraic geometry: $D$-stacks.** What we call *complicial algebraic geometry* is an unbounded version of derived algebraic geometry in which the base model category is $C(k)$ the category of complexes over some commutative ring $k$ (of characteristic zero), and is presented in §2.3. It turns out that linear algebra over $C(k)$ behaves rather differently than over the category of simplicial $k$-modules (corresponding to complexes in non-positive degrees). Indeed, the smooth, étale and Zariski open immersion can not be described using a simple description on homotopy groups anymore. For example, a usual ring $A$ may have Zariski open localizations $A \longrightarrow A'$ in the context of complicial algebraic geometry such that $A'$ is not cohomologically concentrated in degree 0 anymore. Also, a usual non affine scheme might be affine when considered as a scheme over $C(k)$: for example any quasi-compact open subscheme of a usual affine scheme is representable by an affine scheme over $C(k)$ (see example 2.3.1.5).

This makes the complicial theory rather different from derived algebraic geometry for which the geometric intuition was instead quite close to the usual one, and constitutes a very interesting new feature of complicial algebraic geometry. The category $Aff_{\mathcal{C}}$ is here the opposite of the category of unbounded commutative differential graded algebras over $k$. It is endowed with a *strong étale topology*, and the corresponding homotopy category of $D$-stacks is simply denoted by $D\mathrm{St}(k)$. A new feature is here the existence of several interesting choices for the class $\mathbf{P}$. We will present two of them, one for which $\mathbf{P}$ is taken to be the class of perfect morphisms, a rather weak notion of smoothness, and a second one for which $\mathbf{P}$ is taken to be the class of *fip-smooth* morphism, a definitely stronger notion behaving similarly to usual smooth morphisms with respect to lifting properties. We check that such choices satisfy the required properties in order for $n$-geometric stacks (called *weakly $n$-geometric $D$-stacks* and *$n$-geometric $D$-stacks*, according to the choice of $\mathbf{P}$) to make sense. Furthermore, for our second choice, we prove that Artin's conditions are satisfied, and thus that $n$-geometric $D$-stacks have a good infinitesimal theory. We give several examples of weakly geometric $D$-stacks, the first one being the $D$-stack of *perfect modules* **Perf**. We also show that the $D$-stack of *associative algebra structures* **Ass** is a weakly 1-geometric $D$-stack. Finally, the $D$-stack of *connected dg-categories* $\mathbf{Cat}_*$ is shown to be weakly 2-geometric. It is important to note that these $D$-stacks can not be reasonably described as geometric $D^-$-stacks, and provide examples of truly "exotic" geometric objects.

Suitable slight modifications of the $D$-stacks **Perf**, **Ass** and $\mathbf{Cat}_*$ are given and shown to be geometric. This allows us to study their tangent complexes, and show in particular that the infinitesimal theory of a certain class of dg-algebras and dg-categories is controlled by derivations and Hochschild cohomology, respectively (see Cor. 2.3.5.9 and Cor. 2.3.5.12). We also show that Hochschild cohomology does not control deformations of general dg-categories in any reasonable sense (see Cor. 2.3.5.13 and Rem. 2.3.5.14). This has been a true surprise, as it contradicts some of the statements one finds in the existing literature, including some made by the authors themselves (see e.g. [**To-Ve2**, Thm. 5.6]).

**Brave new algebraic geometry: $S$-stacks.** Our last context of application, briefly presented in §2.4, is the one where the base symmetric monoidal model category is $\mathcal{C} = Sp^{\Sigma}$, the model category of symmetric spectra ([**HSS, Shi**]), and gives rise to what we call, after F. Waldhausen, *brave new algebraic geometry*. Like in the complicial case, the existence of negative homotopy groups makes the general theory of flat, smooth, étale morphisms and of Zariski open immersions rather different from the corresponding one in derived algebraic geometry. Moreover, typical phenomena coming from the existence of Steenrod operations makes the notion of smooth morphism even more exotic and rather different from algebraic geometry; to give just a striking example, $\mathbb{Z}[T]$ is *not* smooth over $\mathbb{Z}$ in the context of brave new algebraic geometry. Once again, we do not think this is a drawback of the theory, but rather an interesting new feature one should contemplate and try to understand, as it might reveal interesting new insights also on classical objects. In brave new algebraic geometry, we also check that the strong étale topology and the class **P** of fip-smooth morphisms satisfy our general assumptions, so that $n$-geometric stacks exists in this context. We call them $n$-geometric *$S$-stacks*, while the homotopy category of $S$-stacks for the strong étale topology is simply denoted by $\mathrm{St}(S)$. As an example, we give a construction of a 1-geometric $S$-stack starting from the "sheaf" of topological modular forms (Thm. 2.4.2.1).

**Relations with other works.** It would be rather long to present all related works, and we apologize in advance for not mentioning all of them.

The general fact that the notion of geometric stack only depends on a topology and a choice of the class of morphisms **P** has already been stressed by Carlos Simpson in [**S3**], who attributes this idea to C. Walter. Our general definition of geometric $n$-stacks is a straightforward generalization to our abstract context of the definitions found in [**S3**].

Originally, derived algebraic geometry have been approached using the notion of *dg-schemes*, as introduced by M. Kontsevich, and developed by I. Ciocan-Fontanine and M. Kapranov. We have not tried to make a full comparison with our theory. Let us only mention that there exists a functor from dg-schemes to our category of 1-geometric $D^{-}$-stacks (see [**To-Ve2**, §3.3]). Essentially nothing is known about this functor: we tend to believe that it is not fully faithful, though this question does not seem very relevant. On the contrary, the examples of *dg*-schemes constructed in [**Ci-Ka1, Ci-Ka2**] do provide examples of $D^{-}$-stacks and we think it is interesting to look for derived moduli-theoretic interpretations of these (i.e. describe their functors of points).

We would like to mention that an approach to formal derived algebraic geometry has been settled down by V. Hinich in [**Hin2**]. As far as we know, this is the first functorial point of view on derived algebraic geometry that appeared in the literature.

There is a big overlap between our Chapter 2.2 and Jacob Lurie's thesis [**Lu1**]. The approach to derived algebraic geometry used by J. Lurie is different from ours as it is based on a notion of $\infty$-category, whereas we are working with model categories. The simplicial localization techniques of Dwyer and Kan provide a way to pass from model categories to $\infty$-categories, and the "strictification" theorem of [**To-Ve1**, Thm. 4.2.1] can be used to see that our approach and Lurie's approach are in fact equivalent (up to some slight differences, for instance concerning the notion of descent). We think that the present work and [**Lu1**] can not be reasonably considered as totally independent, as their authors have been frequently communicating on the subject since the spring 2002. It seems rather clear that we all have been influenced by these communications and that we have greatly benefited from the reading of the first drafts of [**Lu1**]. We have to mention however that a huge part of the material of the present paper had been already announced in earlier papers (see e.g. [**To-Ve4, To-Ve2**]), and have been worked out since the summer 2000 at the time were our project has started. The two works are also rather disjoint and complementary, as [**Lu1**] contains much more materials on derived algebraic geometry than what we have included in §2.2 (e.g. a wonderful generalization of Artin's representability theorem, to state only the most striking result). On the other hand, our "HAG" project has also been motivated by rather exotic contexts of applications, as the ones exposed for example in §2.3 and §2.4 , and which are not covered by the framework of [**Lu1**].

The work of K. Behrend on *dg*-schemes and *dg*-stacks [**Be1, Be2**] has been done while we were working on our project, and therefore §2.2 also has some overlaps with his work. However, the two approaches are rather different and nonequivalent, as K. Behrend uses a 2-truncated version of our notions of stacks, with the effect of killing some higher homotopical information. We have not investigated a precise comparison between these two approaches in this work, but we would like to mention that there exists a functor from our category of $D^-$-stacks to K. Behrend's category of *dg*-sheaves. This functor is extremely badly behaved: it is not full, nor faithful, nor essentially surjective, nor even injective on isomorphism classes of objects. The only good property is that it sends 1-geometric Deligne-Mumford $D^-$-stacks to Deligne-Mumford *dg*-stack in Behrend's sense. However, there are non geometric $D^-$-stacks that become geometric objects in Behrend's category of *dg*-sheaves.

Some notions of étale and smooth morphisms of commutative $S$-algebras have been introduced in [**MCM**], and they seem to be related to the general notions we present in §1.2. However a precise comparison is not so easy. Moreover, [**MCM**] contains some wrong statements like the fact that *thh*-smoothness generalizes smoothness for discrete algebras (right after Definition 4.2) or like Lemma 4.2 (2). The proof of Theorem 6.1 also contains an important gap, since the local equivalences at the end of the proof are not checked to glue together.

Very recentely, J. Rognes has proposed a brave new version of Galois theory, including brave new notions of étaleness which are very close to our notions (see [**Ro**]).

A construction of the moduli of dg-algebras and dg-categories appears in [**Ko-So**]. These moduli are only formal moduli by construction, and we propose our $D$-stacks **Ass** and $\mathbf{Cat}_*$ as their global geometrical counterparts.

We wish to mention the work of M. Spitzweck [**Sp**], in which he proves the existence of model category structures for $E_\infty$-algebras and modules in a rather general context. This work can therefore be used in order to suppress our assumptions on the existence of model category of commutative monoids. Also, a nice symmetric monoidal model category of motivic complexes is defined in [**Sp**], providing a new interesting context to investigate. It has been suggested to us to consider this example of *algebraic geometry over motives* by Yu. Manin, already during spring 2000, but we do not have at the moment interesting things to say on the subject.

Finally, J. Gorski has recently constructed a $D^-$-stack version of the Quot functor (see [**Go**]), providing this way a functorial interpretation of the derived Quot scheme of [**Ci-Ka1**]. A geometric $D^-$-stack classifying objects in a dg-category has been recently constructed by the first author and M. Vaquié in [**To-Va1**].

**Acknowledgments.** First of all, we are very grateful to C. Simpson for sharing with us all his ideas on higher stacks, and for encouraging us to pursue our project. We owe him a lot.

We are very grateful to J. Lurie for various communications on the subject, and for sharing with us his work [**Lu1**]. We have learned a lot about derived algebraic geometry from him.

We thank M. Vaquié for reading a former version of the present work, and for his comments and suggestions.

For various conversations on the subject we thank M. Anel, J. Gorski, A. Hirschowitz, A. Joyal, M. Kontsevich, L. Katzarkov, H. Miller, T. Pantev, C. Rezk, J. Rognes, S. Schwede, B. Shipley, M. Spitzweck, N. Strickland and J. Tapia.

Finally, we thank both referees for their careful reading of the manuscript and for their interesting and useful remarks and suggestions.

**Notations and conventions.** We will use the word *universe* in the sense of [**SGA4-I**, Exp.I, Appendice]. Universes will be denoted by $\mathbb{U} \in \mathbb{V} \in \mathbb{W} \ldots$. For any universe $\mathbb{U}$ we will assume that $\mathbb{N} \in \mathbb{U}$. The category of sets (resp. simplicial sets, resp. ...) belonging to a universe $\mathbb{U}$ will be denoted by $Set_{\mathbb{U}}$ (resp. $SSet_{\mathbb{U}}$, resp. ...). The objects of $Set_{\mathbb{U}}$ (resp. $SSet_{\mathbb{U}}$, resp. ...) will be called $\mathbb{U}$-sets (resp. $\mathbb{U}$-simplicial sets, resp. ...). We will use the expression $\mathbb{U}$-*small set* (resp. $\mathbb{U}$-*small simplicial set*, resp. ...) to mean *a set isomorphic to a set in* $\mathbb{U}$ (resp. *a simplicial set isomorphic to a simplicial set in* $\mathbb{U}$, resp. ...). A unique exception concerns categories. The expression $\mathbb{U}$-*category* refers to the usual notion of [**SGA4-I**, $IDef.1.2$], and denotes a category $\mathcal{C}$ such that for any two objects $x$ and $y$ in $\mathcal{C}$ the set $Hom_{\mathcal{C}}(x,y)$ is $\mathbb{U}$-small. In the same way, a category $\mathcal{C}$ is $\mathbb{U}$-*small* is it is isomorphic to some element in $\mathbb{U}$.

Our references for model categories are [**Ho1**] and [**Hi**]. By definition, our model categories will always be *closed* model categories, will have all *small* limits and colimits and the functorial factorization property. The word *equivalence* will always mean *weak equivalence* and will refer to a model category structure.

The homotopy category of a model category $M$ is $W^{-1}M$ (see [**Ho1**, Def. 1.2.1]), where $W$ is the subcategory of equivalences in $M$, and it will be denoted as $\mathrm{Ho}(M)$. The sets of morphisms in $\mathrm{Ho}(M)$ will be denoted by $[-,-]_M$, or simply by $[-,-]$ when the reference to the model category $M$ is clear. We will say that two objects in a model category $M$ are equivalent if they are isomorphic in $\mathrm{Ho}(M)$. We say that two model categories are *Quillen equivalent* if they can be connected by a finite string of Quillen adjunctions each one being a Quillen equivalence. The mapping space of morphisms between two objects $x$ and $y$ in a model category $M$ is denoted by $Map_M(x,y)$ (see [**Ho1**, §5]), or simply $Map(x,y)$ if the reference to $M$ is clear. The simplicial set depends on the choice of cofibrant and fibrant resolution functors, but is well defined as an object in the homotopy category of simplicial sets $\mathrm{Ho}(SSet)$. If $M$ is a $\mathbb{U}$-category, then $Map_M(x,y)$ is a $\mathbb{U}$-small simplicial set.

The homotopy fiber product (see [**Hi**, 13.3, 19.5], [**DHK**, Ch. XIV] or [**DS**, 10] ) of a diagram $x \longrightarrow z \longleftarrow y$ in a model category $M$ will be denoted by $x \times^h_z y$. In the same way, the homotopy push-out of a diagram $x \longleftarrow z \longrightarrow y$ will be denoted by $x \coprod^{\mathbb{L}}_z y$. For a pointed model category $M$, the suspension and loop functors functor will be denoted by

$$S : \mathrm{Ho}(M) \longrightarrow \mathrm{Ho}(M) \qquad \mathrm{Ho}(M) \longleftarrow \mathrm{Ho}(M) : \Omega.$$

Recall that $S(x) := * \coprod^{\mathbb{L}}_x *$, and $\Omega(x) := * \times^h_x *$.

When a model category $M$ is a simplicial model category, its simplicial sets of morphisms will be denoted by $\underline{Hom}_M(-,-)$, and their derived functors by $\mathbb{R}\underline{Hom}_M$ (see [**Ho1**, 1.3.2]), or simply $\underline{Hom}(-,-)$ and $\mathbb{R}\underline{Hom}(-,-)$ when the reference to $M$ is clear. When $M$ is a symmetric monoidal model category in the sense of [**Ho1**, §4], the derived monoidal structure will be denoted by $\otimes^{\mathbb{L}}$.

For the notions of $\mathbb{U}$-cofibrantly generated, $\mathbb{U}$-combinatorial and $\mathbb{U}$-cellular model category, we refer to [**Ho1, Hi, Du2**], or to [**HAGI**, Appendix], where the basic definitions and crucial properties are recalled in a way that is suitable for our needs.

As usual, the standard simplicial category will be denoted by $\Delta$. The category of simplicial objects in a category $\mathcal{C}$ will be denoted by $s\mathcal{C} := \mathcal{C}^{\Delta^{op}}$. In the same way, the category of co-simplicial objects in $\mathcal{C}$ will be denoted by $cs\mathcal{C}$. For any simplicial object $F \in s\mathcal{C}$ in a category $\mathcal{C}$, we will use the notation $F_n := F([n])$. Similarly, for any co-simplicial object $F \in \mathcal{C}^{\Delta}$, we will use the notation $F_n := F([n])$. Moreover, when $\mathcal{C}$ is a model category, we will use the notation

$$|X_*| := Hocolim_{[n]\in\Delta^{op}} X_n$$

for any $X_* \in s\mathcal{C}$.

A sub-simplicial set $K \subset L$ will be called *full* is $K$ is a union of connected components of $L$. We will also say that a morphism $f : K \longrightarrow L$ of simplicial sets is *full* if it induces an equivalence bewteen $K$ and a full sub-simplicial set of $L$. In the same way, we will use the expressions *full sub-simplicial presheaf*, and *full morphisms of simplicial presheaves* for the levelwise extension of the above notions to presheaves of simplicial sets.

For a Grothendieck site $(\mathcal{C}, \tau)$ in a universe $\mathbb{U}$, we will denote by $Pr(\mathcal{C})$ the category of presheaves of $\mathbb{U}$-sets on $\mathcal{C}$, $Pr(\mathcal{C}) := \mathcal{C}^{Set_{\mathbb{U}}^{op}}$. The subcategory of sheaves on $(\mathcal{C}, \tau)$ will be denoted by $Sh_{\tau}(\mathcal{C})$, or simply by $Sh(\mathcal{C})$ if the topology $\tau$ is unambiguous.

All complexes will be cochain complexes (i.e. with differential increasing the degree by one) and therefore will look like

$$\cdots \longrightarrow E^n \xrightarrow{d_n} E^{n+1} \longrightarrow \cdots \longrightarrow E^0 \longrightarrow E^1 \longrightarrow \cdots.$$

The following notations concerning various homotopy categories of stacks are defined in the main text, and recalled here for readers' convenience (see also the Index at the end of the book).

$$\mathrm{St}(\mathcal{C}, \tau) := \mathrm{Ho}(Aff_{\mathcal{C}}^{\sim,\tau})$$

$$\mathrm{St}(k) := \mathrm{Ho}(k - Aff^{\sim,\text{ét}})$$

$$D^-\mathrm{St}(k) := \mathrm{Ho}(k - D^-Aff^{\sim,\text{ét}})$$

$$D\mathrm{St}(k) := \mathrm{Ho}(k - DAff^{\sim,\text{s-ét}})$$

$$\mathrm{St}(S) := \mathrm{Ho}(SAff^{\sim,\text{s-ét}})$$

# Part 1

# General theory of geometric stacks

# Introduction to Part 1

In this first part we will study the general theory of stacks and geometric stacks over a base symmetric monoidal model category $\mathcal{C}$. For this, we will start in §1.1 by introducing the notion of a *homotopical algebraic context* (HA *context* for short), which consists of a triple $(\mathcal{C}, \mathcal{C}_0, \mathcal{A})$ where $\mathcal{C}$ is our base monoidal model category, $\mathcal{C}_0$ is a sub-category of $\mathcal{C}$, and $\mathcal{A}$ is a sub-category of the category $Comm(\mathcal{C})$ of commutative monoids in $\mathcal{C}$; we also require that the triple $(\mathcal{C}, \mathcal{C}_0, \mathcal{A})$ satisfies certain compatibility conditions. Although this might look like a rather unnatural and complicated definition, it will be shown in §1.2 that this data precisely allows us to define abstract versions of standard notions such as derivations, unramified, étale, smooth and flat morphisms. In other words a HA context describes an abstract context in which the basic notions of linear and commutative algebra can be developed.

The first two chapters are only concerned with purely algebraic notions and the geometry only starts in the third one, §1.3. We start by some reminders on the notions of model topology and of model topos (developed in [**HAGI**]), which are homotopical versions of the notions of Grothendieck topology and of Grothendieck topos and which will be used all along this work. Next, we introduce a notion of a *homotopical algebraic geometry context* (HAG *context* for short), consisting of a HA context together with two additional data, $\tau$ and $\mathbf{P}$, satsfying some compatiblity conditions. The first datum $\tau$ is a model topology on $Aff_{\mathcal{C}}$, the opposite model category of commutative monoids in $\mathcal{C}$. The second datum $\mathbf{P}$ consists of a class of morphisms in $Aff_{\mathcal{C}}$ which behaves well with respect to $\tau$. The model topology $\tau$ gives a category of stacks over $Aff_{\mathcal{C}}$ (a homotopical generalization of the category of sheaves on affine schemes) in which everything is going to be embedded by means of a Yoneda lemma. The class of morphisms $\mathbf{P}$ will then be used in order to define *geometric stacks* and more generally *n-geometric stacks*, by considering successive quotient stacks of objects of $Aff_{\mathcal{C}}$ by action of groupoids whose structural morphisms are in $\mathbf{P}$. The compatibility axioms between $\tau$ and $\mathbf{P}$ will insure that this notion of geometricity behaves well, and satisfies the basic expected properties (stability by homotopy pullbacks, gluing and certain quotients).

In §1.4, the last chapter of part I, we will go more deeply into the study of geometric stacks by introducing infinitesimal constructions such as derivations, cotangent complexes and obstruction theories. The main result of this last chapter states that any geometric stack has an obstruction theory (including a cotangent complex) as soon as the HAG context satisfies suitable additional conditions.

CHAPTER 1.1

# Homotopical algebraic context

The purpose of this chapter is to fix once for all our base model category as well as several general assumptions it should satisfy.

All along this chapter, we refer to [**Ho1**] for the general definition of monoidal model categories, and to [**Schw-Shi**] for general results about monoids and modules in monoidal model categories.

From now on, and all along this work, we fix three universes $\mathbb{U} \in \mathbb{V} \in \mathbb{W}$ (see, e.g. [**SGA4-I**, Exp.I, Appendice]). We also let $(\mathcal{C}, \otimes, \mathbf{1})$ be a symmetric monoidal model category in the sense of [**Ho1**, §4]. We assume that $\mathcal{C}$ is a $\mathbb{V}$-small category, and that it is $\mathbb{U}$-combinatorial in the sense of [**HAGI**, Appendix].

We make a first assumption on the base model category $\mathcal{C}$, making it closer to an additive category. Recall that we denote by $Q$ a cofibrant replacement functor and by $R$ a fibrant replacement functor in $M$.

ASSUMPTION 1.1.0.1. (1) *The model category $\mathcal{C}$ is proper, pointed (i.e. the final object is also an initial object) and for any two object $X$ and $Y$ in $\mathcal{C}$ the natural morphisms*

$$QX \coprod QY \longrightarrow X \coprod Y \longrightarrow RX \times RY$$

*are all equivalences.*

(2) *The homotopy category* $\mathrm{Ho}(\mathcal{C})$ *is an additive category.*

Assumption 1.1.0.1 implies in particular that finite homotopy coproducts are also finite homotopy products in $\mathcal{C}$. It is always satisfied when $\mathcal{C}$ is furthermore a stable model category in the sense of [**Ho1**, §7]. Note that 1.1.0.1 implies in particular that for any two objects $x$ and $y$ in $\mathcal{C}$, the set $[x, y]$ has a natural abelian group structure.

As $(\mathcal{C}, \otimes, \mathbf{1})$ is a symmetric monoidal category, which is closed and has $\mathbb{U}$-small limits and colimits, all the standard notions and constructions of linear algebra makes sense in $\mathcal{C}$ (e.g. monoids, modules over monoids, operads, algebra over an operad ...). The category of all associative, commutative and unital monoids in $\mathcal{C}$ will be denoted by $Comm(\mathcal{C})$. Objects of $Comm(\mathcal{C})$ will simply be called *commutative monoids in* $\mathcal{C}$, or *commutative monoids* if $\mathcal{C}$ is clear. In the same way, one defines $Comm_{nu}(\mathcal{C})$ to be the category of non-unital commutative monoids in $\mathcal{C}$. Therefore, our convention will be that monoids are unital unless the contrary is specified.

The categories $Comm(\mathcal{C})$ and $Comm_{nu}(\mathcal{C})$ are again $\mathbb{U}$-categories which are $\mathbb{V}$-small categories, and possess all $\mathbb{U}$-small limits and colimits. They come equipped with natural forgetful functors

$$Comm(\mathcal{C}) \longrightarrow \mathcal{C} \qquad Comm_{nu}(\mathcal{C}) \longrightarrow \mathcal{C},$$

possessing left adjoints

$$F : \mathcal{C} \longrightarrow Comm(\mathcal{C}) \qquad F_{nu} : \mathcal{C} \longrightarrow Comm_{nu}(\mathcal{C})$$

sending an object of $\mathcal{C}$ to the free commutative monoid it generates. We recall that for $X \in \mathcal{C}$ one has

$$F(X) = \coprod_{n\in\mathbb{N}} X^{\otimes n}/\Sigma_n$$

$$F_{nu}(X) = \coprod_{n\in\mathbb{N}-\{0\}} X^{\otimes n}/\Sigma_n,$$

where $X^{\otimes n}$ is the $n$-tensor power of $X$, $\Sigma_n$ acts on it by permuting the factors and $X^{\otimes n}/\Sigma_n$ denotes the quotient of this action in $\mathcal{C}$.

Let $A \in Comm(\mathcal{C})$ be a commutative monoid. We will denote by $A - Mod$ the category of unital left $A$-modules in $\mathcal{C}$. The category $A - Mod$ is again a $\mathbb{U}$-category which is a $\mathbb{V}$-small category, and has all $\mathbb{U}$-small limits and colimits. The objects in $A - Mod$ will simply be called *$A$-modules*. It comes equiped with a natural forgetful functor

$$A - Mod \longrightarrow \mathcal{C},$$

possessing a left adjoint

$$A \otimes - : \mathcal{C} \longrightarrow A - Mod$$

sending an object of $\mathcal{C}$ to the free $A$-module it generates. We also recall that the category $A - Mod$ has a natural symmetric monoidal structure $- \otimes_A -$. For two $A$-modules $X$ and $Y$, the object $X \otimes_A Y$ is defined as the coequalizer in $\mathcal{C}$ of the two natural morphisms

$$X \otimes A \otimes Y \longrightarrow X \otimes Y \qquad X \otimes A \otimes Y \longrightarrow X \otimes Y.$$

This symmetric monoidal structure is furthermore closed, and for two $A$-modules $X$ and $Y$ we will denoted by $\underline{Hom}_A(X,Y)$ the $A$-module of morphisms. One has the usual adjunction isomorphisms

$$Hom(X \otimes_A Y, Z) \simeq Hom(X, \underline{Hom}_A(Y,Z)),$$

as well as isomorphisms of $A$-modules

$$\underline{Hom}_A(X \otimes_A Y, Z) \simeq \underline{Hom}(X, \underline{Hom}_A(Y,Z)).$$

We define a morphism in $A - Mod$ to be a *fibration* or an *equivalence* if it is so on the underlying objects in $\mathcal{C}$.

ASSUMPTION 1.1.0.2. *Let $A \in Comm(\mathcal{C})$ be any commutative monoid in $\mathcal{C}$. Then, the above notions of equivalences and fibrations makes $A - Mod$ into a $\mathbb{U}$-combinatorial proper model category. The monoidal structure $- \otimes_A -$ makes furthermore $A - Mod$ into a symmetric monoidal model category in the sense of* [**Ho1**, §4].

Using the assumption 1.1.0.2 one sees that the homotopy category $\mathrm{Ho}(A - Mod)$ has a natural symmetric monoidal structure $\otimes^{\mathbb{L}}_A$, and derived internal $Hom$'s associated to it $\mathbb{R}\underline{Hom}_A$, satisfying the usual adjunction rule

$$[X \otimes^{\mathbb{L}}_A Y, Z] \simeq [X, \mathbb{R}\underline{Hom}_A(Y,Z)].$$

ASSUMPTION 1.1.0.3. *Let $A$ be a commutative monoid in $\mathcal{C}$. For any cofibrant object $M \in A - Mod$, the functor*

$$- \otimes_A M : A - Mod \longrightarrow A - Mod$$

*preserves equivalences.*

Let us still denote by $A$ a commutative monoid in $\mathcal{C}$. We have categories of commutative monoids in $A-Mod$, and non-unital commutative monoids in $A-Mod$, denoted respectively by $A-Comm(\mathcal{C})$ and $A-Comm_{nu}(\mathcal{C})$, and whose objects will be called *commutative A-algebras* and *non-unital commutative A-algebras*. They come equipped with natural forgetful functors

$$A - Comm(\mathcal{C}) \longrightarrow A - Mod \qquad A - Comm_{nu}(\mathcal{C}) \longrightarrow A - Mod,$$

possessing left adjoints

$$F_A : A - Mod \longrightarrow A - Comm(\mathcal{C}) \qquad F_A^{nu} : A - Mod \longrightarrow A - Comm_{nu}(\mathcal{C})$$

sending an object of $A-Mod$ to the free commutative monoid it generates. We recall that for $X \in A - Mod$ one has

$$F_A(X) = \coprod_{n\in\mathbb{N}} X^{\otimes_A n}/\Sigma_n$$

$$F_A^{nu}(X) = \coprod_{n\in\mathbb{N}-\{0\}} X^{\otimes_A n}/\Sigma_n,$$

where $X^{\otimes_A n}$ is the $n$-tensor power of $X$ in $A - Mod$, $\Sigma_n$ acts on it by permuting the factors and $X^{\otimes_A n}/\Sigma_n$ denotes the quotient of this action in $A - Mod$.

Finally, we define a morphism in $A - Comm(\mathcal{C})$ or in $A - Comm_{nu}(\mathcal{C})$ to be a *fibration* (resp. an *equivalence*) if it is so as a morphism in the category $\mathcal{C}$ (or equivalently as a morphism in $A - Mod$).

ASSUMPTION 1.1.0.4. *Let $A$ be any commutative monoid in $\mathcal{C}$.*

(1) *The above classes of equivalences and fibrations make the categories $A - Comm(\mathcal{C})$ and $A - Comm_{nu}(\mathcal{C})$ into $\mathbb{U}$-combinatorial proper model categories.*
(2) *If $B$ is a cofibrant object in $A - Comm(\mathcal{C})$, then the functor*

$$B \otimes_A - : A - Mod \longrightarrow B - Mod$$

*preserves equivalences.*

REMARK 1.1.0.5. One word concerning non-unital algebras. We will not really use this notion in the sequel, except at one point in order to prove the existence of a cotangent complex (so the reader is essentially allowed to forget about this unfrequently used notion). In fact, by our assumptions, the model category of non-unital commutative $A$-algebras is Quillen equivalent to the model category of augmented commutative $A$-algebra. However, the categories themselves are not equivalent, since the category $\mathcal{C}$ is not assumed to be strictly speaking additive, but only additive up to homotopy (e.g. it could be the model category of symmetric spectra of [**HSS**]). Therefore, we do not think that the existence of the model structure on $A-Comm(\mathcal{C})$ implies the existence of the model structure on $A - Comm_{nu}(\mathcal{C})$; this explains why we had to add condition (1) on $A-Comm_{nu}(\mathcal{C})$ in Assumption 1.1.0.4. Furthermore, the passage from augmented $A$-algebras to non-unital $A$-algebras will be in any case necessary to construct a certain Quillen adjunction during the proof of Prop. 1.2.1.2, because such a Quillen adjunction does not exist from the model category of augmented $A$-algebras (as it is a composition of a left Quillen functor by a right Quillen equivalence).

An important consequence of assumption 1.1.0.4 (2) is that for $A$ a commutative monoid in $\mathcal{C}$, and $B$, $B'$ two commutative $A$-algebras, the natural morphism in $\mathrm{Ho}(A-$

$Mod$)

$$B \coprod_A^{\mathbb{L}} B' \longrightarrow B \otimes_A^{\mathbb{L}} B'$$

is an isomorphism (here the object on the left is the homotopy coproduct in $A-Comm(\mathcal{C})$, and the one on the right is the derived tensor product in $A-Mod$).

An important remark we will use implicitly very often in this paper is that the category $A-Comm(\mathcal{C})$ is naturally equivalent to the comma category $A/Comm(\mathcal{C})$, of objects under $A$. Moreover, the model structure on $A-Comm(\mathcal{C})$ coincides through this equivalence with the comma model category $A/Comm(\mathcal{C})$.

We will also fix a full subcategory $\mathcal{C}_0$ of $\mathcal{C}$, playing essentially the role of a kind of "$t$-structure" on $\mathcal{C}$ (i.e. essentially defining which are the "non-positively graded objects", keeping in mind that in this work we use the cohomological grading when concerned with complexes). More precisely, we will fix a subcategory $\mathcal{C}_0 \subseteq \mathcal{C}$ satisfying the following conditions.

ASSUMPTION 1.1.0.6. (1) $\mathbf{1} \in \mathcal{C}_0$.
(2) *The full subcategory $\mathcal{C}_0$ of $\mathcal{C}$ is stable by equivalences and by $\mathbb{U}$-small homotopy colimits.*
(3) *The full subcategory $\mathrm{Ho}(\mathcal{C}_0)$ of $\mathrm{Ho}(\mathcal{C})$ is stable by the monoidal structure $-\otimes^{\mathbb{L}}-$ (i.e. for $X$ and $Y$ in $\mathrm{Ho}(\mathcal{C}_0)$ we have $X \otimes^{\mathbb{L}} Y \in \mathrm{Ho}(\mathcal{C}_0)$).*

Recall that as $\mathcal{C}$ is a pointed model category one can define its *suspension functor*

$$\begin{array}{cccc} S: & \mathrm{Ho}(\mathcal{C}) & \longrightarrow & \mathrm{Ho}(\mathcal{C}) \\ & x & \mapsto & * \coprod_x^{\mathbb{L}} * \end{array}$$

left adjoint to the *loop functor*

$$\begin{array}{cccc} \Omega: & \mathrm{Ho}(\mathcal{C}) & \longrightarrow & \mathrm{Ho}(\mathcal{C}) \\ & x & \mapsto & := * \times_x^h *. \end{array}$$

We set $\mathcal{C}_1$ to be the full subcategory of $\mathcal{C}$ consisting of all objects equivalent to the suspension of some object in $\mathcal{C}_0$. The full subcategory $\mathcal{C}_1$ of $\mathcal{C}$ is also closed by equivalences, homotopy colimits and the derived tensor structure. We will denote by $Comm(\mathcal{C})_0$ the full subcategory of $Comm(\mathcal{C})$ consisting of commutative monoids whose underlying $\mathcal{C}$-object lies in $\mathcal{C}_0$. In the same way, for $A \in Comm(\mathcal{C})$ we denote by $A-Mod_0$ (resp. $A-Mod_1$, resp. $A-Comm(\mathcal{C})_0$) the full subcategory of $A-Mod$ consisting of $A$-modules whose underlying $\mathcal{C}$-object lies in $\mathcal{C}_0$ (resp. of $A-Mod$ consisting of $A$-modules whose underlying $\mathcal{C}$-object lies in $\mathcal{C}_1$, resp. of $A-Comm(\mathcal{C})$ consisting of commutative $A$-algebras whose underlying $\mathcal{C}$-object lies in $\mathcal{C}_0$).

An important consequence of Assumption 1.1.0.6 is that for any morphism $A \longrightarrow B$ in $Comm(\mathcal{C})_0$, and any $M \in A-Mod_0$, we have $B \otimes_A^{\mathbb{L}} M \in \mathrm{Ho}(B-Mod_0)$. Indeed, any such $A$-module can be written as a homotopy colimit of $A$-modules of the form $A^{\otimes^{\mathbb{L}} n} \otimes^{\mathbb{L}} M$, for which we have

$$B \otimes_A^{\mathbb{L}} A^{\otimes^{\mathbb{L}} n} \otimes^{\mathbb{L}} M \simeq A^{\otimes^{\mathbb{L}}(n-1)} \otimes^{\mathbb{L}} M.$$

In particular the full subcategory $Comm(\mathcal{C})_0$ is closed by homotopy push-outs in $Comm(\mathcal{C})$. Passing to the suspension we also see that for any $M \in A-Mod_1$, one also has $B \otimes_A^{\mathbb{L}} M \in \mathrm{Ho}(B-Mod_1)$.

REMARK 1.1.0.7. (1) The reason for introducing the subcategory $\mathcal{C}_0$ is to be able to consider reasonable infinitesimal lifting properties; these infinitesimal lifting properties will be used to develop the abstract obstruction theory of geometric stacks in §1.4. It is useful to keep in mind that $\mathcal{C}_0$ plays a role analogous to a kind of $t$-structure on $\mathcal{C}$, in that it morally defines which are the non-positively graded objects (a typical example will appear in §2.3 where $\mathcal{C}$ will be the model category of unbounded complexes and $\mathcal{C}_0$ the subcategory of complexes with vanishing positive cohomology). Different choices of $\mathcal{C}_0$ will then give different notions of formal smoothness (see §1.2.8), and thus possibly different notions of geometric stacks. We think that playing with $\mathcal{C}_0$ as a degree of freedom is an interesting feature of our abstract infinitesimal theory.

(2) Assumptions 1.1.0.1, 1.1.0.2 and 1.1.0.3 are not really serious, and are only useful to avoid taking too many fibrant and cofibrant replacements. With some care, they can be omitted. On the other hand, the careful reader will probably be surprised by assumption 1.1.0.4, as it is known not to be satisfied in several interesting examples (e.g. when $\mathcal{C}$ is the model category of complexes over some commutative ring $k$, not of characteristic zero). Also, it is well known that in some situations the notion of commutative monoid is too strict and it is often preferable to use the weaker notion of $E_\infty$-monoid. Two reasons has led us to assume 1.1.0.4. First of all, for all contexts of application of the general theory we will present in this work, there is always a base model category $\mathcal{C}$ for which this condition is satisfied and gives rise to the correct theory. Moreover, if one replaces commutative monoids by $E_\infty$-monoids then assumption 1.1.0.4 is almost always satisfied, as shown in [**Sp**], and we think that translating our general constructions should then be a rather academic exercise. Working with commutative monoids instead of $E_\infty$-monoids simplifies a lot the notations and certain constructions, and in our opinion this theory already captures the real essence of the subject.

Finally, in partial defense of our choice, let us also mention that contrary to what one could think at first sight, working with $E_\infty$-monoids would not strictly speaking increase the degree of generality of the theory. Indeed, one of our major application is to the category of simplicial $k$-modules, whose category of commutative monoids is the category of simplicial commutative $k$-algebras. However, if $k$ has non-zero characteristic, the homotopy theory of simplicial commutative $k$-algebras is *not* equivalent to the homotopy theory of $E_\infty$-monoids in simplicial $k$-modules (the latter is equivalent to the homotopy theory of $E_\infty$-$k$-algebras in non positive degrees). Therefore, using $E_\infty$-monoids throughout would prevent us from developing derived algebraic geometry as presented in §2.2.

We list below some important examples of symmetric monoidal model categories $\mathcal{C}$ satisfying the four above assumptions, and of crucial importance for our applications.

(1) Let $k$ be any commutative ring, and $\mathcal{C} = k - Mod$ be the category of $\mathbb{U}$-$k$-modules, symmetric monoidal for the tensor product $\otimes_k$, and endowed with its trivial model structure (i.e. equivalences are isomorphisms and all morphisms are fibrations and cofibrations). Then assumptions 1.1.0.1, 1.1.0.2, 1.1.0.3 and 1.1.0.4 are satisfied. For $\mathcal{C}_0$ one can take the whole $\mathcal{C}$ for which Assumption 1.1.0.6 is clearly satisfied.

(2) Let $k$ be a commutative ring of characteristic 0, and $\mathcal{C} = C(k)$ be the category of (unbounded) complexes of $\mathbb{U}$-$k$-modules, symmetric monoidal for the tensor product of complexes $\otimes_k$, and endowed with its projective model structure for which equivalences are quasi-isomorphisms and fibrations are epimorphisms (see [**Ho1**]). Then, the category $Comm(\mathcal{C})$ is the

category of commutative differential graded $k$-algebras belonging to $\mathbb{U}$. For $A \in Comm(\mathcal{C})$, the category $A - Mod$ is then the category of differential graded $A$-modules. It is well known that as $k$ is of characteristic zero then assumptions 1.1.0.1, 1.1.0.2, 1.1.0.3 and 1.1.0.4 are satisfied (see e.g; [**Hin1**]). For $\mathcal{C}_0$ one can take either the whole $\mathcal{C}$, or the full subcategory of $\mathcal{C}$ consisting of complexes $E$ such that $H^i(E) = 0$ for any $i > 0$, for which Assumption 1.1.0.6 is satisfied.

A similar example is given by non-positively graded complexes $\mathcal{C} = C^-(k)$.

(3) Let $k$ be any commutative ring, and $\mathcal{C} = sMod_k$ be the category of $\mathbb{U}$-simplicial $k$-modules, endowed with the levelwise tensor product and the usual model structure for which equivalences and fibrations are defined on the underlying simplicial sets (see e.g. [**Goe-Ja**]). The category $Comm(\mathcal{C})$ is then the category of simplicial commutative $k$-algebras, and for $A \in sMod_k$, $A - Mod$ is the category of simplicial modules over the simplicial ring $A$. Assumptions 1.1.0.1, 1.1.0.2, 1.1.0.3 and 1.1.0.4 are again well known to be satisfied. For $\mathcal{C}_0$ one can take the whole $\mathcal{C}$.

Dually, one could also let $\mathcal{C} = csMod_k$ be the category of co-simplicial $k$-modules, and our assumptions would again be satisfied. In this case, $\mathcal{C}_0$ could be for example the full subcategory of co-simplicial modules $E$ such that $\pi_{-i}(E) = H^i(E) = 0$ for any $i > 0$. This subcategory is stable under homotopy colimits as the functor $E \mapsto H^0(E)$ is right Quillen and right adjoint to the inclusion functor $k - Mod \longrightarrow csMod_k$.

(4) Let $Sp^{\Sigma}$ be the category of $\mathbb{U}$-symmetric spectra and its smash product, endowed with the positive stable model structure of [**Shi**]. Then, $Comm(\mathcal{C})$ is the category of commutative symmetric ring spectra, and assumptions 1.1.0.1, 1.1.0.2, 1.1.0.3 and 1.1.0.4 are known to be satisfied (see [**Shi**, Thm. 3.1, Thm. 3.2, Cor. 4.3]). The two canonical choices for $\mathcal{C}_0$ are the whole $\mathcal{C}$ or the full subcategory of connective spectra.

It is important to note that one can also take for $\mathcal{C}$ the category of $Hk$-modules in $Sp^{\Sigma}$ for some commutative ring $k$. This will give a model for the homotopy theory of $E_{\infty}$-$k$-algebras that were not provided by our example 2.

(5) Finally, the above three examples can be sheafified over some Grothendieck site, giving the corresponding relative theories over a base Grothendieck topos.

In few words, let $\mathcal{S}$ be a $\mathbb{U}$-small Grothendieck site, and $\mathcal{C}$ be one the three symmetric monoidal model category $C(k)$, $sMod_k$, $Sp^{\Sigma}$ discussed above. One considers the corresponding categories of presheaves on $\mathcal{S}$, $Pr(\mathcal{S}, C(k))$, $Pr(\mathcal{S}, sMod_k)$, $Pr(\mathcal{S}, Sp^{\Sigma})$. They can be endowed with the projective model structures for which fibrations and equivalences are defined levelwise. This first model structure does not depend on the topology on $\mathcal{S}$, and will be called the *strong model structure*: its (co)fibrations and equivalences will be called *global (co)fibrations* and *global equivalences*.

The next step is to introduce notions of *local equivalences* in the model categories $Pr(\mathcal{S}, C(k))$, $Pr(\mathcal{S}, sMod_k)$, $Pr(\mathcal{S}, Sp^{\Sigma})$. This notion is defined by first defining reasonable *homotopy sheaves*, as done for the notion of local equivalences in the theory of simplicial presheaves, and then define a morphism to be a local equivalence if it induces isomorphisms on all homotopy sheaves (for various choice of base points, see [**Jo1, Ja1**] for more details). The final model structures on $Pr(\mathcal{S}, C(k))$, $Pr(\mathcal{S}, sMod_k)$, $Pr(\mathcal{S}, Sp^{\Sigma})$ are the one for which equivalences are the local equivalences, and cofibrations are the global cofibrations. The proof that this indeed defines model categories is not given here and is very similar to the proof of the existence of the local projective model structure on the category of simplicial presheaves (see for example [**HAGI**]).

Finally, the symmetric monoidal structures on the categories $C(k)$, $sMod_k$ and $Sp^{\Sigma}$ induces natural symmetric monoidal structures on the categories $Pr(\mathcal{S}, C(k))$, $Pr(\mathcal{S}, sMod_k)$, $Pr(\mathcal{S}, Sp^{\Sigma})$. These symmetric monoidal structures make them into symmetric monoidal model categories when $\mathcal{S}$ has finite products. One can also check that the symmetric monoidal model categories $Pr(\mathcal{S}, C(k))$, $Pr(\mathcal{S}, sMod_k)$, $Pr(\mathcal{S}, Sp^{\Sigma})$ constructed that way all satisfy the assumptions 1.1.0.1, 1.1.0.2, 1.1.0.3 and 1.1.0.4.

An important example is the following. Let $\mathcal{O}$ be a sheaf of commutative rings on the site $\mathcal{S}$, and let $H\mathcal{O} \in Pr(\mathcal{S}, Sp^{\Sigma})$ be the presheaf of symmetric spectra it defines. The object $H\mathcal{O}$ is a commutative monoid in $Pr(\mathcal{S}, Sp^{\Sigma})$ and one can therefore consider the model category $H\mathcal{O} - Mod$, of $H\mathcal{O}$-modules. The category $H\mathcal{O} - Mod$ is a symmetric monoidal model category and its homotopy category is equivalent to the unbounded derived category $D(\mathcal{S}, \mathcal{O})$ of $\mathcal{O}$-modules on the site $\mathcal{S}$. This gives a way to define all the standard constructions as derived tensor products, derived internal $Hom's$ etc., in the context of unbounded complexes of $\mathcal{O}$-modules.

Let $f : A \longrightarrow B$ be a morphism of commutative monoids in $\mathcal{C}$. We deduce an adjunction between the categories of modules

$$f^* : A - Mod \longrightarrow B - Mod \qquad A - Mod \longleftarrow B - Mod : f_*,$$

where $f^*(M) := B \otimes_A M$, and $f_*$ is the forgetful functor that sees a $B$-module as an $A$-module through the morphism $f$. Assumption 1.1.0.2 tells us that this adjunction is a Quillen adjunction, and assumption 1.1.0.3 implies it is furthermore a Quillen equivalence when $f$ is an equivalence (this is one of the main reasons for assumption 1.1.0.3).

The morphism $f$ induces a pair of adjoint derived functors

$$\mathbb{L}f^* : \mathrm{Ho}(A - Mod) \longrightarrow \mathrm{Ho}(B - Mod) \qquad \mathrm{Ho}(A - Mod) \longleftarrow \mathrm{Ho}(B - Mod) : \mathbb{R}f_* \simeq f_*,$$

and, as usual, we will also use the notation

$$\mathbb{L}f^*(M) =: B \otimes^{\mathbb{L}}_A M \in \mathrm{Ho}(B - Mod).$$

Finally, let

$$\begin{array}{ccc} A & \xrightarrow{f} & B \\ {\scriptstyle p}\downarrow & & \downarrow{\scriptstyle p'} \\ A' & \xrightarrow[f']{} & B' \end{array}$$

be a homotopy cofiber square in $Comm(\mathcal{C})$. Then, for any $A'$-module $M$ we have the well known base change morphism

$$\mathbb{L}f^* p_*(M) \longrightarrow (p')_* \mathbb{L}(f')^*(M).$$

PROPOSITION 1.1.0.8. *Let us keep the notations as above. Then, the morphism*

$$\mathbb{L}f^* p_*(M) \longrightarrow (p')_* \mathbb{L}(f')^*(M)$$

*is an isomorphism in* $\mathrm{Ho}(B - Mod)$ *for any* $A'$*-module* $M$.

PROOF. As the homotopy categories of modules are invariant under equivalences of commutative monoids, one can suppose that the diagram

$$\begin{array}{ccc} A & \xrightarrow{f} & B \\ {\scriptstyle p}\downarrow & & \downarrow{\scriptstyle p'} \\ A' & \xrightarrow[f']{} & B' \end{array}$$

is cocartesian in $Comm(\mathcal{C})$, and consists of cofibrations in $Comm(\mathcal{C})$. Then, using 1.1.0.3 and 1.1.0.4 (2) one sees that the natural morphisms

$$M \otimes^{\mathbb{L}}_A B \longrightarrow M \otimes_A B \qquad M \otimes^{\mathbb{L}}_{A'} B' \longrightarrow M \otimes_{A'} B',$$

are isomorphism in $\mathrm{Ho}(B-Mod)$. Therefore, the proposition follows from the natural isomorphism of $B$-modules

$$M \otimes_{A'} B' \simeq M \otimes_{A'} (A' \otimes_A B) \simeq M \otimes_A B.$$

□

REMARK 1.1.0.9. The above base change formula will be extremely important in the sequel, and most often used implicitly. It should be noticed that it implies that the homotopy coproduct in the model category of commutative monoids is given by the derived tensor product. This last property is only satisfied because we have used commutative monoids, and is clearly wrong for simply associative monoids. This is one major reason why our setting cannot be used, at least without some modifications, to develop truly non-commutative geometries. Even partially commutative structures, like $E_n$-monoids for $n > 1$ would not satisfy the base change formula, and one really needs $E_\infty$-monoids at least. This fact also prevents us to generalize our setting by replacing the category of commutative monoids by more general categories, like some category of algebras over more general operads.

We now consider $A \in Comm(\mathcal{C})$ and the natural inclusion $A-Mod_0 \longrightarrow A-Mod$. We consider the restricted Yoneda embedding

$$\mathbb{R}\underline{h}_0^- : \mathrm{Ho}(A - Mod^{op}) \longrightarrow \mathrm{Ho}((A - Mod_0^{op})^\wedge),$$

sending an $A$-module $M$ to the functor

$$Map(M, -) : A - Mod_0^{op} \longrightarrow SSet_{\mathbb{V}}.$$

We recall here from [**HAGI**, §4.1] that for a model category $M$, and a full subcategory stable by equivalences $M_0 \subset M$, $M_0^\wedge$ is the left Bousfield localization of $SPr(M_0)$ along equivalences in $M_0$.

DEFINITION 1.1.0.10. *We will say that* $A \in Comm(\mathcal{C})$ *is* good with respect to $\mathcal{C}_0$ *(or simply* $\mathcal{C}_0$-good*) if the functor*

$$\mathrm{Ho}(A - Mod^{op}) \longrightarrow \mathrm{Ho}((A - Mod_0^{op})^\wedge)$$

*is fully faithful.*

In usual category theory, a full subcategory $D \subset C$ is called *dense* if the restricted Yoneda functor $C \longrightarrow Pr(D) := \underline{Hom}(D^{op}, Ens)$ is fully faithful (in [**SGA4-I**] this notion is equivalent to the fact that $D$ *generates* $C$ *through strict epimorphisms*). This implies for example that any object of $C$ is the colimit of objects of $D$, but is a slightly stronger condition because any object $x \in C$ is in fact isomorphic to the colimit of the canonical diagram $D/x \longrightarrow C$ (see e.g. [**SGA4-I**, ExpI-Prop. 7.2]).

Our notion of being good (Def. 1.1.0.10) essentially means that $A - Mod_0^{op}$ is *homotopically dense* in $A - Mod^{op}$. This of course implies that any object in $A - Mod$ is equivalent to a homotopy limit of objects in $A - Mod_0$, and is equivalent to the fact that any cofibrant object $M \in A - Mod$ is equivalent to the homotopy limit of the natural diagram $(M/A - Mod_0)^c \longrightarrow A - Mod$, where $(M/A - Mod_0)^c$ denotes the category of cofibrations under $M$. Dually, one could say that $A$ being good with respect to $\mathcal{C}_0$ means that $A - Mod_0$ *cogenerates* $A - Mod$ *through strict monomorphisms* in a homotopical sense.

We finish this first chapter by the following definition, gathering our assumptions 1.1.0.1, 1.1.0.2, 1.1.0.3, 1.1.0.4 and 1.1.0.6 all together.

DEFINITION 1.1.0.11. *A* Homotopical Algebraic context *(or simply* HA context*) is a triplet $(\mathcal{C}, \mathcal{C}_0, \mathcal{A})$, consisting of a symmetric monoidal model category $\mathcal{C}$, two full sub-categories stable by equivalences*

$$\mathcal{C}_0 \subset \mathcal{C} \qquad \mathcal{A} \subset Comm(\mathcal{C}),$$

*such that any $A \in \mathcal{A}$ is $\mathcal{C}_0$-good, and assumptions 1.1.0.1, 1.1.0.2, 1.1.0.3, 1.1.0.4, 1.1.0.6 are satisfied.*

CHAPTER 1.2

# Preliminaries on linear and commutative algebra in an HA context

All along this chapter we fix once for all a HA context $(\mathcal{C}, \mathcal{C}_0, \mathcal{A})$, in the sense of Def. 1.1.0.11. The purpose of this chapter is to show that the assumptions of the last chapter imply that many general notions of linear and commutative algebra generalize in some reasonable sense in our base category $\mathcal{C}$.

## 1.2.1. Derivations and the cotangent complex

This section is nothing else than a rewriting of the first pages of [**Ba**], which stay valid in our general context.

Let $A \in Comm(\mathcal{C})$ be a commutative monoid in $\mathcal{C}$, and $M$ be an $A$-module. We define a new commutative monoid $A \oplus M$ in the following way. The underlying object of $A \oplus M$ is the coproduct $A \coprod M$. The multiplicative structure is defined by the morphism

$$(A \coprod M) \otimes (A \coprod M) \simeq A \otimes A \coprod A \otimes M \coprod A \otimes M \coprod M \otimes M \longrightarrow A \coprod M$$

given by the three morphisms

$$\mu \coprod * : A \otimes A \longrightarrow A \coprod M$$
$$* \coprod \rho : A \otimes M \longrightarrow A \coprod M$$
$$* : M \otimes M \longrightarrow M,$$

where $\mu : A \otimes A \longrightarrow A$ is the multiplicative structure of $A$, and $\rho : A \otimes M \longrightarrow M$ is the module structure of $M$. The monoid $A \oplus M$ is commutative and unital, and defines an object in $Comm(\mathcal{C})$. It comes furthermore with a natural morphism of commutative monoids $id \coprod * : A \oplus M \longrightarrow A$, which has a natural section $id \coprod * : A \longrightarrow A \oplus M$.

Now, if $A \longrightarrow B$ is a morphism in $Comm(\mathcal{C})$, and $M$ is a $B$-module, the morphism $B \oplus M \longrightarrow B$ can be seen as a morphism of commutative $A$-algebras. In other words, $B \oplus M$ can be seen as an object of the double comma model category $A - Comm(\mathcal{C})/B$.

DEFINITION 1.2.1.1. *Let $A \longrightarrow B$ be a morphism of commutative monoids, and $M$ be a $B$-module. The simplicial set of* derived $A$-derivations from $B$ to $M$, *is the object*

$$\mathbb{D}er_A(B, M) := Map_{A-Comm(\mathcal{C})/B}(B, B \oplus M) \in \mathrm{Ho}(SSet_{\mathbb{U}}).$$

Clearly, $M \mapsto \mathbb{D}er_A(B, M)$ defines a functor from the homotopy category of $B$-module $\mathrm{Ho}(B - Mod)$ to the homotopy category of simplicial sets $\mathrm{Ho}(SSet)$. More precisely, the functoriality of the construction of mapping spaces implies that one can also construct a genuine functor

$$\mathbb{D}er_A(B, -) : B - Mod \longrightarrow SSet_{\mathbb{U}},$$

lifting the previous functor on the homotopy categories. This last functor will be considered as an object in the model category of pre-stacks $(B-Mod^{op})^{\wedge}$ as defined in [**HAGI**, §4.1]. Recall from [**HAGI**, §4.2] that there exists a Yoneda embedding

$$\mathrm{Ho}(B-Mod)^{op} \longrightarrow \mathrm{Ho}((B-Mod^{op})^{\wedge})$$

sending a $B$-module $M$ to the simplicial presheaf $N \mapsto Map_{B-Mod}(M,N)$, and objects in the essential image will be called *co-representable.*

PROPOSITION 1.2.1.2. *For any morphism $A \longrightarrow B$ in $Comm(\mathcal{C})$, there exists a $B$-module $\mathbb{L}_{B/A}$, and an element $d \in \pi_0(\mathbb{D}er_A(B, \mathbb{L}_{B/A}))$, such that for any $B$-module $M$, the natural morphism obtained by composing with $d$*

$$d^* : Map_{B-Mod}(\mathbb{L}_{B/A}, M) \longrightarrow \mathbb{D}er_A(B,M)$$

*is an isomorphism in* $\mathrm{Ho}(SSet)$.

PROOF. The proof is the same as in [**Ba**], and uses our assumptions 1.1.0.1, 1.1.0.2, 1.1.0.3 and 1.1.0.4. We will reproduce it for the reader convenience.

We first consider the Quillen adjunction

$$-\otimes_A B : A\text{–}Comm(\mathcal{C})/B \longrightarrow B\text{–}Comm(\mathcal{C})/B \qquad A\text{–}Comm(\mathcal{C})/B \longleftarrow B\text{–}Comm(\mathcal{C})/B : F,$$

where $F$ is the forgetful functor. This induces an adjunction on the level of homotopy categories

$$-\otimes^{\mathbb{L}}_A B : \mathrm{Ho}(A-Comm(\mathcal{C})/B) \longrightarrow \mathrm{Ho}(B-Comm(\mathcal{C})/B)$$

$$\mathrm{Ho}(A-Comm(\mathcal{C})/B) \longleftarrow \mathrm{Ho}(B-Comm(\mathcal{C})/B) : F.$$

We consider a second Quillen adjunction

$$K : B\text{–}Comm_{nu}(\mathcal{C}) \longrightarrow B\text{–}Comm(\mathcal{C})/B \qquad B\text{–}Comm_{nu}(\mathcal{C}) \longleftarrow B\text{–}Comm(\mathcal{C})/B : I,$$

where $B-Comm_{nu}(\mathcal{C})$ is the category of non-unital commutative $B$-algebras (i.e. non-unital commutative monoids in $B-Mod$). The functor $I$ takes a diagram of commutative monoids $B \xrightarrow{s} C \xrightarrow{p} B$ to the kernel of $p$ computed in the category of non-unital commutative $B$-algebras. In the other direction, the functor $K$ takes a non-unital commutative $B$-algebra $C$ to the trivial extension of $B$ by $C$ (defined as our $B \oplus M$ but taking into account the multiplication on $C$). Clearly, $I$ is a right Quillen functor, and the adjunction defines an adjunction on the homotopy categories

$$\mathbb{L}K : \mathrm{Ho}(B-Comm_{nu}(\mathcal{C})) \longrightarrow \mathrm{Ho}(B-Comm(\mathcal{C})/B)$$

$$\mathrm{Ho}(B-Comm_{nu}(\mathcal{C})) \longleftarrow \mathrm{Ho}(B-Comm(\mathcal{C})/B) : \mathbb{R}I.$$

LEMMA 1.2.1.3. *The adjunction $(\mathbb{L}K, \mathbb{R}I)$ is an equivalence.*

PROOF. This follows easily from our assumption 1.1.0.1. Indeed, it implies that for any fibration in $\mathcal{C}$, $f : X \longrightarrow Y$, which has a section $s : Y \longrightarrow X$, the natural morphism

$$i \coprod s : F \coprod Y \longrightarrow X,$$

where $i : F \longrightarrow X$ is the fiber of $f$, is an equivalence. It also implies that the homotopy fiber of the natural morphism $id \coprod * : X \coprod Y \longrightarrow X$ is naturally equivalent to $Y$. These two facts imply the lemma. □

Finally, we consider a third adjunction

$$Q : B - Comm_{nu}(\mathcal{C}) \longrightarrow B - Mod \qquad B - Comm_{nu}(\mathcal{C}) \longleftarrow B - Mod : Z,$$

where $Q$ of an object $C \in B - Comm_{nu}(\mathcal{C})$ is the push-out of $B$-modules

$$\begin{array}{ccc} C \otimes_B C & \xrightarrow{\mu} & C \\ \downarrow & & \downarrow \\ \bullet & \longrightarrow & Q(C), \end{array}$$

and $Z$ sends a $B$-module $M$ to the non-unital $B$-algebra $M$ endowed with the zero multiplication. Clearly, $(Q, Z)$ is a Quillen adjunction and gives rise to an adjunction on the homotopy categories

$$\mathbb{L}Q : \mathrm{Ho}(B - Comm_{nu}(\mathcal{C})) \longrightarrow \mathrm{Ho}(B - Mod)$$

$$\mathrm{Ho}(B - Comm_{nu}(\mathcal{C})) \longleftarrow \mathrm{Ho}(B - Mod) : Z.$$

We can now conclude the proof of the proposition by chaining up the various adjunction to get a string of isomorphisms in $\mathrm{Ho}(SSet)$

$$\mathbb{D}er_A(B, M) \simeq Map_{A-Comm(\mathcal{C})/B}(B, F(B \oplus M)) \simeq Map_{B-Comm(\mathcal{C})/B}(B \otimes^{\mathbb{L}}_A B, B \oplus M)$$

$$\simeq Map_{B-Comm_{nu}(\mathcal{C})}(\mathbb{R}I(B \otimes^{\mathbb{L}}_A B), \mathbb{R}I(B \oplus M)) \simeq Map_{B-Comm_{nu}(\mathcal{C})}(\mathbb{R}I(B \otimes^{\mathbb{L}}_A B), Z(M))$$

$$\simeq Map_{B-Mod}(\mathbb{L}Q\mathbb{R}I(B \otimes^{\mathbb{L}}_A B), M).$$

Therefore, $\mathbb{L}_{B/A} := \mathbb{L}Q\mathbb{R}I(B \otimes^{\mathbb{L}}_A B)$ and the image of $id \in Map_{B-Mod}(\mathbb{L}_{B/A}, \mathbb{L}_{B/A})$ gives what we were looking for. $\square$

REMARK 1.2.1.4. Proposition 1.2.1.2 implies that the two functors

$$M \mapsto Map_{B-Mod}(\mathbb{L}_{B/A}, M) \qquad M \mapsto \mathbb{D}er_A(B, M)$$

are isomorphic as objects in $\mathrm{Ho}((B - Mod^{op})^{\wedge})$. In other words, Prop. 1.2.1.2 implies that the functor $\mathbb{D}er_A(B, -)$ is *co-representable* in the sense of [**HAGI**].

DEFINITION 1.2.1.5. *Let $A \longrightarrow B$ be a morphism in $Comm(\mathcal{C})$.*

(1) *The $B$-module $\mathbb{L}_{B/A} \in \mathrm{Ho}(B - Mod)$ is called the* cotangent complex of $B$ over $A$.
(2) *When $A = \mathbf{1}$, we will use the following notation*

$$\mathbb{L}_B := \mathbb{L}_{B/\mathbf{1}},$$

*and $\mathbb{L}_B$ will be called the* cotangent complex of $B$.

Using the definition and proposition 1.2.1.2, it is easy to check the following facts.

PROPOSITION 1.2.1.6. (1) *Let $A \longrightarrow B \longrightarrow C$ be two morphisms in $Comm(\mathcal{C})$. Then, there is a homotopy cofiber sequence in $C - Mod$*

$$\mathbb{L}_{B/A} \otimes^{\mathbb{L}}_B C \longrightarrow \mathbb{L}_{C/A} \longrightarrow \mathbb{L}_{C/B}.$$

(2) *Let*

$$\begin{array}{ccc} A & \longrightarrow & B \\ \downarrow & & \downarrow \\ A' & \longrightarrow & B' \end{array}$$

*be a homotopy cofiber square in $Comm(\mathcal{C})$. Then, the natural morphism*

$$\mathbb{L}_{B/A} \otimes^{\mathbb{L}}_B B' \longrightarrow \mathbb{L}_{B'/A'}$$

*is an isomorphism in* $\mathrm{Ho}(B' - Mod)$. *Furthermore, the natural morphism*

$$\mathbb{L}_{B/A} \otimes^{\mathbb{L}}_B B' \coprod \mathbb{L}_{A'/A} \otimes^{\mathbb{L}}_{A'} B' \longrightarrow \mathbb{L}_{B'/A}$$

*is an isomorphism in* $\mathrm{Ho}(B' - Mod)$.

(3) *Let*

$$\begin{array}{ccc} A & \longrightarrow & B \\ \downarrow & & \downarrow \\ A' & \longrightarrow & B' \end{array}$$

*be a homotopy cofiber square in* $Comm(\mathcal{C})$. *Then the following square is homotopy cocartesian in* $B' - Mod$

$$\begin{array}{ccc} \mathbb{L}_A \otimes^{\mathbb{L}}_A B' & \longrightarrow & \mathbb{L}_B \otimes^{\mathbb{L}}_B B' \\ \downarrow & & \downarrow \\ \mathbb{L}_{A'} \otimes^{\mathbb{L}}_{A'} B' & \longrightarrow & \mathbb{L}_{B'}. \end{array}$$

(4) *For any commutative monoid* $A$ *and any* $A$*-module* $M$, *one has a natural isomorphism in* $\mathrm{Ho}(A - Mod)$

$$\mathbb{L}_{A\oplus M} \otimes^{\mathbb{L}}_{A\oplus M} A \simeq \mathbb{L}_A \coprod \mathbb{L}QZ(M),$$

*where*

$$Q : A - Comm_{nu}(\mathcal{C}) \longrightarrow A - Mod \qquad A - Comm_{nu}(\mathcal{C}) \longleftarrow A - Mod : Z$$

*is the Quillen adjunction used during the proof of 1.2.1.2.*

Proof. (1) to (3) are simple exercises, using the definitions and that for any morphism of commutative monoids $A \longrightarrow B$, and any $B$-module $M$, the following square is homotopy cartesian in $Comm(\mathcal{C})$ (because of our assumption 1.1.0.1)

$$\begin{array}{ccc} A \oplus M & \longrightarrow & B \oplus M \\ \downarrow & & \downarrow \\ A & \longrightarrow & B. \end{array}$$

(4) We note that for any commutative monoid $A$, and any $A$-modules $M$ and $N$, one has a natural homotopy fiber sequence

$$Map_{A-Comm(\mathcal{C})/A}(A \oplus M, A \oplus N) \longrightarrow Map_{Comm(\mathcal{C})/A}(A \oplus M, A \oplus N) \rightarrow$$
$$\longrightarrow Map_{Comm(\mathcal{C})/A}(A, A \oplus N),$$

or equivalently using lemma 1.2.1.3

$$Map_{A-Mod}(\mathbb{L}QZ(M), N) \longrightarrow \mathbb{D}er(A \oplus M, N) \longrightarrow \mathbb{D}er(A, N).$$

This implies the existence of a natural homotopy cofiber sequence of $A$-modules

$$\mathbb{L}_A \longrightarrow \mathbb{L}_{A\oplus M} \otimes^{\mathbb{L}}_{A\oplus M} A \longrightarrow \mathbb{L}QZ(M).$$

Clearly this sequence splits in $\mathrm{Ho}(A - Mod)$ and gives rise to a natural isomorphism

$$\mathbb{L}_{A\oplus M} \otimes^{\mathbb{L}}_{A\oplus M} A \simeq \mathbb{L}_A \coprod \mathbb{L}QZ(M).$$

□

The importance of derivations come from the fact that they give rise to infinitesimal extensions in the following way. Let $A \longrightarrow B$ be a morphism of commutative monoids in $\mathcal{C}$, and $M$ be a $B$-module. Let $d : \mathbb{L}_{B/A} \longrightarrow M$ be a morphism in $\mathrm{Ho}(B - Mod)$, corresponding to a derivation $d \in \pi_0(\mathbb{D}er_A(B, M))$. This derivation can be seen as a section $d : B \longrightarrow B \oplus M$ of the morphism of commutative $A$-algebras $B \oplus M \longrightarrow B$. We consider the following homotopy cartesian diagram in the category of commutative $A$-algebras

$$\begin{array}{ccc} C & \longrightarrow & B \\ \downarrow & & \downarrow{\scriptstyle d} \\ B & \xrightarrow[s]{} & B \oplus M \end{array}$$

where $s : B \longrightarrow B \oplus M$ is the natural section corresponding to the zero morphism $\mathbb{L}_{B/A} \longrightarrow M$. Then, $C \longrightarrow B$ is a morphism of commutative $A$-algebras such that its fiber is a non-unital commutative $A$-algebra isomorphic in $\mathrm{Ho}(A - Comm_{nu}(\mathcal{C}))$ to the loop $A$-module $\Omega M := * \times^h_M *$ with the zero multiplication. In other words, $C$ is a *square zero extension of* $B$ *by* $\Omega M$. It will be denoted by $B \oplus_d \Omega M$. The most important case is of course when $A = B$, and we make the following definition.

DEFINITION 1.2.1.7. *Let* $A$ *be a commutative monoid,* $M$ *and* $A$*-module and* $d \in \pi_0\mathbb{D}er(A, M)$ *be a derivation given by a morphism in* $d : A \longrightarrow A \oplus M$ *in* $\mathrm{Ho}(Comm(\mathcal{C})/A)$*. The* square zero extension associated to $d$*, denoted by* $A \oplus_d \Omega M$*, is defined as the homotopy pullback diagram of commutative monoids*

$$\begin{array}{ccc} A \oplus_d \Omega M & \longrightarrow & A \\ \downarrow & & \downarrow{\scriptstyle d} \\ A & \xrightarrow[s]{} & A \oplus M, \end{array}$$

*where* $s$ *is the natural morphism corresponding to the zero derivation. The top horizontal morphism* $A \oplus_d \Omega M \longrightarrow A$ *will be called the natural projection.*

### 1.2.2. Hochschild homology

For a commutative monoid $A \in Comm(\mathcal{C})$, we set

$$THH(A) := S^1 \otimes^{\mathbb{L}} A \in \mathrm{Ho}(Comm(\mathcal{C})),$$

where $S^1 \otimes^{\mathbb{L}}$ denotes the derived external product of object in $Comm(\mathcal{C})$ by the simplicial circle $S^1 := \Delta^1/\partial\Delta^1$. Presenting the circle $S^1$ has the homotopy push-out

$$* \coprod^{\mathbb{L}}_{* \coprod *} *$$

one gets that

$$THH(A) \simeq A \otimes^{\mathbb{L}}_{A \otimes^{\mathbb{L}} A} A.$$

The natural point $* \longrightarrow S^1$ induces a natural morphism in $\mathrm{Ho}(Comm(\mathcal{C}))$

$$A \longrightarrow THH(A)$$

making $THH(A)$ as a commutative $A$-algebra, and as a natural object in $\mathrm{Ho}(A - Comm(\mathcal{C}))$.

DEFINITION 1.2.2.1. *Let* $A$ *be a commutative monoid in* $\mathcal{C}$*. The* topological Hochschild homology of $A$ *(or simply* Hochschild homology*) is the commutative* $A$*-algebra* $THH(A) := S^1 \otimes^{\mathbb{L}} A$.

*More generally, if $A \longrightarrow B$ is a morphism of commutative monoids in $\mathcal{C}$, the* relative topological Hochschild homology of $B$ over $A$ *(or simply* relative Hochschild homology*) is the commutative $A$-algebra*

$$THH(B/A) := THH(B) \otimes^{\mathbb{L}}_{THH(A)} A.$$

By definition, we have for any commutative monoid $B$,

$$Map_{Comm(\mathcal{C})}(THH(A), B) \simeq Map_{SSet}(S^1, Map_{Comm(\mathcal{C})}(A, B)).$$

This implies that if $f : A \longrightarrow B$ is a morphism of commutative monoids in $\mathcal{C}$, then we have

$$Map_{A-Comm(\mathcal{C})}(THH(A), B) \simeq \Omega_f Map_{Comm(\mathcal{C})}(A, B),$$

where $\Omega_f Map_{Comm(\mathcal{C})}(A, B)$ is the loop space of $Map_{Comm(\mathcal{C})}(A, B)$ at the point $f$. More generally, if $B$ and $C$ are commutative $A$-algebras, then

$$Map_{A-Comm(\mathcal{C})}(THH(B/A), C) \simeq Map_{SSet}(S^1, Map_{A-Comm(\mathcal{C})}(B, C)),$$

and for a morphism $f : B \longrightarrow C$ of commutative $A$-algebras

$$Map_{B-Comm(\mathcal{C})}(THH(B/A), C) \simeq \Omega_f Map_{A-Comm(\mathcal{C})}(B, C).$$

PROPOSITION 1.2.2.2. (1) *Let $A \longrightarrow B \longrightarrow C$ be two morphisms in $Comm(\mathcal{C})$. Then, the natural morphism*

$$THH(C/A) \otimes^{\mathbb{L}}_{THH(B/A)} B \longrightarrow THH(C/B)$$

*is an isomorphism in* $\mathrm{Ho}(B - Comm(\mathcal{C}))$.

(2) *Let*

$$\begin{array}{ccc} A & \longrightarrow & B \\ \downarrow & & \downarrow \\ A' & \longrightarrow & B' \end{array}$$

*be a homotopy cofiber square in $Comm(\mathcal{C})$. Then, the natural morphism*

$$THH(B/A) \otimes^{\mathbb{L}}_{A} THH(A'/A) \longrightarrow THH(B'/A)$$

*is an isomorphism in* $\mathrm{Ho}(A - Comm(\mathcal{C}))$.

PROOF. Exercise. □

### 1.2.3. Finiteness conditions

We present here two different finiteness conditions for objects in model categories. The first one is valid in any model category, and is a homotopy analog of the notion of finitely presented object in a category. The second one is only valid for symmetric monoidal categories, and is a homotopy generalization of the notion of rigid objects in monoidal categories.

DEFINITION 1.2.3.1. *A morphism $x \longrightarrow y$ in a proper model category $M$ is* finitely presented *(we also say that $y$* is finitely presented over *$x$) if for any filtered diagram of objects under $x$, $\{z_i\}_{i \in I} \in x/M$, the natural morphism*

$$Hocolim_{i \in I} Map_{x/M}(y, z_i) \longrightarrow Map_{x/M}(y, Hocolim_{i \in I} z_i)$$

*is an isomorphism in* $\mathrm{Ho}(SSet)$.

REMARK 1.2.3.2. If the model category $M$ is not proper the definition 1.2.3.1 has to be modified by replacing $Map_{x/M}$ with $Map_{Qx/M}$, where $Qx$ is a cofibrant model for $x$. By our assumption 1.1.0.1 all the model categories we will use are proper.

PROPOSITION 1.2.3.3. *Let $M$ be a proper model category.*

(1) *Finitely presented morphisms in $M$ are stable by equivalences. In other words, if one has a commutative diagram in $M$*

$$\begin{array}{ccc} x & \xrightarrow{f} & y \\ {\scriptstyle p}\downarrow & & \downarrow{\scriptstyle q} \\ x' & \xrightarrow[f']{} & y' \end{array}$$

*such that $p$ and $q$ are equivalences, then $f$ is finitely presented if and only if $f'$ is finitely presented.*

(2) *Finitely presented morphisms in $M$ are stable by compositions and retracts.*

(3) *Finitely presented morphisms in $M$ are stable by homotopy push-outs. In other words, if one has a homotopy push-out diagram in $M$*

$$\begin{array}{ccc} x & \xrightarrow{f} & y \\ {\scriptstyle p}\downarrow & & \downarrow{\scriptstyle q} \\ x' & \xrightarrow[f']{} & y' \end{array}$$

*then $f'$ is finitely presented if $f$ is so.*

PROOF. (1) is clear as $Map_{x/M}(a, b)$ only depends on the isomorphism class of $a$ and $b$ as objects in the homotopy category $\mathrm{Ho}(x/M)$.

(2) Let $x \longrightarrow y \longrightarrow z$ be two finitely presented morphisms in $M$, and let $\{z_i\}_{i\in I} \in x/M$ be a filtered diagrams of objects. Then, one has for any object $t \in x/M$ a fibration sequence of simplicial sets

$$Map_{y/M}(z,t) \longrightarrow Map_{x/M}(z,t) \longrightarrow Map_{x/M}(y,t).$$

As fibration sequences are stable by filtered homotopy colimits, one gets a morphism of fibration sequences

$$\begin{array}{ccccc} Hocolim_{i\in I} Map_{y/M}(z, z_i) & \longrightarrow & Hocolim_{i\in I} Map_{x/M}(z, z_i) & \longrightarrow & Hocolim_{i\in I} Map_{x/M}(y, z_i) \\ \downarrow & & \downarrow & & \downarrow \\ Map_{y/M}(z, Hocolim_{i\in I} z_i) & \longrightarrow & Map_{x/M}(z, Hocolim_{i\in I} z_i) & \longrightarrow & Map_{x/M}(y, Hocolim_{i\in I} z_i), \end{array}$$

and the five lemma tells us that the vertical arrow in the middle is an isomorphism in $\mathrm{Ho}(SSet)$. This implies that $z$ is finitely presented over $x$.

The assertion concerning retracts is clear since, if $x \longrightarrow y$ is a retract of $x' \longrightarrow y'$, for any $z \in x/M$ the simplicial set $Map_{x/M}(y, z)$ is a retract of $Map_{x'/M}(y', z)$.

(3) This is clear since we have for any object $t \in x'/M$, a natural equivalence $Map_{x'/M}(y', t) \simeq Map_{x/M}(y, t)$. □

Let us now fix $I$, a set of generating cofibrations in $M$.

DEFINITION 1.2.3.4. (1) *An object $X$ is a* strict finite $I$-cell object, *if there exists a finite sequence*

$$X_0 = \emptyset \longrightarrow X_1 \longrightarrow \cdots \longrightarrow X_n = X,$$

*and for any $0 \leq i < n$ a push-out square*

$$\begin{array}{ccc} X_i & \longrightarrow & X_{i+1} \\ \uparrow & & \uparrow \\ A & \xrightarrow[u_i]{} & B, \end{array}$$

*with $u_i \in I$.*

(2) *An object $X$ is a* finite $I$-cell object *(or simply a* finite cell object *when $I$ is clear) if it is equivalent to a strict finite $I$-cell object.*

(3) *The model category $M$ is* compactly generated *if it satisfies the following conditions.*
   (a) *The model category $M$ is cellular (in the sense of* [**Hi**, §12]*).*
   (b) *There exists a set of generating cofibrations $I$, and generating trivial cofibrations $J$ whose domains and codomains are cofibrant, $\omega$-compact (in the sense of* [**Hi**, §10.8]*) and $\omega$-small with respect to the whole category $M$.*
   (c) *Filtered colimits commute with finite limits in $M$.*

The following proposition identifies finitely presented objects when $M$ is compactly generated.

PROPOSITION 1.2.3.5. *Let $M$ be a compactly generated model category, and $I$ be a set of generating cofibrations whose domains and codomains are cofibrant, $\omega$-compact and $\omega$-small with respect to the whole category $M$.*

(1) *A filtered colimit of fibrations (resp. trivial fibrations) is a fibration (resp. a trivial fibration).*

(2) *For any filtered diagram $X_i$ in $M$, the natural morphism*

$$Hocolim_i X_i \longrightarrow Colim_i X_i$$

*is an isomorphism in* $\mathrm{Ho}(M)$.

(3) *Any object $X$ in $M$ is equivalent to a filtered colimit of strict finite $I$-cell objects.*

(4) *An object $X$ in $M$ is finitely presented if and only if it is equivalent to a retract, in* $\mathrm{Ho}(M)$*, of a strict finite $I$-cell object.*

PROOF. (1) By assumption the domain and codomain of morphisms of $I$ are $\omega$-small, so $M$ is finitely generated in the sense of [**Ho1**, §7]. Property (1) is then proved in [**Ho1**, §7].

(2) For a filtered category $A$, the colimit functor

$$Colim : M^A \longrightarrow M$$

is a left Quillen functor for the levelwise projective model structure on $M$. By (1) we know that $Colim$ preserves trivial fibrations, and thus that it also preserves equivalences. We therefore have isomorphisms of functors $Hocolim \simeq \mathbb{L}Colim \simeq Colim$.

(3) The small object argument (e.g. [**Ho1**, Thm. 2.1.14]) gives that any object $X$ is equivalent to a $I$-cell complex $Q(X)$. By $\omega$-compactness of the domains and codomains of $I$, $Q(X)$ is the filtered colimit of its finite sub-$I$-cell complexes. This implies that $X$ is equivalent to a filtered colimit of strict finite $I$-cell objects.

(4) Let $A$ be a filtered category, and $Y \in M^A$ be a $A$-diagram. Let $c(Y) \longrightarrow R_*(Y)$ be a Reedy fibrant replacement of the constant simplicial object $c(Y)$ with

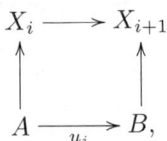

values $Y$ (in the model category of simplicial objects in $M^A$, see [**Ho1**, §5.2]). By (2), the induced morphism

$$c(Colim_{a\in A}Y_a) \longrightarrow Colim_{a\in A}R_*(Y_a)$$

is an equivalence of simplicial objects in $M^A$. Moreover, (1) and the exactness of filtered colimits implies that $Colim_{a\in A}R_*(Y_a)$ is a Reedy fibrant object in the model categroy of simplicial objects in $M$ (as filtered colimits commute with matching objects for the Reedy category $\Delta^{op}$, see [**Ho1**, §5.2]). This implies that for any cofibrant and $\omega$-small object $K$ in $M$, we have

$$Hocolim_{a\in A}Map(K,Y_a) \simeq Colim_{a\in A}Map(K,Y_a) \simeq Colim_{a\in A}Hom(K,R_*(Y_a)) \simeq$$
$$Hom(K,Colim_{a\in A}R_*(Y_a)) \simeq Map(K,Colim_{a\in A}Y_a).$$

This implies that the domains and codomains of $I$ are homotopically finitely presented.

As filtered colimits of simplicial sets preserve homotopy pull-backs, we deduce that any finite cell objects is also homotopically finitely presented, as they are constructed from domains and codomains of $I$ by iterated homotopy push-outs (we use here that domains and codomains of $I$ are cofibrant). This implies that any retract of a finite cell object is homotopically finitely presented. Conversely, let $X$ be a homotopically finitely presented object in $Ho(M)$, and by (3) let us write it as $Colim_i X_i$, where $X_i$ is a filtered diagram of finite cell objects. Then, $[X,X] \simeq Colim_i[X,X_i]$, which implies that this identity of $X$ factors through some $X_i$, or in other words that $X$ is a retract in $Ho(M)$ of some $X_i$. □

Now, let $M$ be a symmetric monoidal model category in the sense of [**Ho1**, §4]. We remind that this implies in particular that the monoidal structure on $M$ is closed, and therefore possesses $Hom$'s objects $\underline{\mathbf{Hom}}_M(x,y) \in M$ satisfying the usual adjunction rule

$$Hom(x,\underline{\mathbf{Hom}}_M(y,z))) \simeq Hom(x\otimes y,z).$$

The internal structure can be derived, and gives on one side a symmetric monoidal structure $-\otimes^{\mathbb{L}}-$ on $\mathrm{Ho}(M)$, as well as $Hom$'s objects $\mathbb{R}\underline{\mathbf{Hom}}_M(x,y) \in \mathrm{Ho}(M)$ satisfying the derived version of the previous adjunction

$$[x,\mathbb{R}\underline{\mathbf{Hom}}_M(y,z))] \simeq [x\otimes^{\mathbb{L}} y,z].$$

In particular, if $\mathbf{1}$ is the unit of the monoidal structure of $M$, then

$$[\mathbf{1},\mathbb{R}\underline{\mathbf{Hom}}_M(x,y)] \simeq [x,y],$$

and more generally

$$Map_M(\mathbf{1},\mathbb{R}\underline{\mathbf{Hom}}_M(x,y)) \simeq Map_M(x,y).$$

Moreover, the adjunction between $-\otimes^{\mathbb{L}}$ and $\mathbb{R}\underline{\mathbf{Hom}}_M$ extends naturally to an adjunction isomorphism

$$\mathbb{R}\underline{\mathbf{Hom}}_M(x,\mathbb{R}\underline{\mathbf{Hom}}_M(y,z))) \simeq \mathbb{R}\underline{\mathbf{Hom}}_M(x\otimes^{\mathbb{L}} y,z).$$

The derived dual of an object $x\in M$ will be denoted by

$$x^{\vee} := \mathbb{R}\underline{\mathbf{Hom}}_M(x,\mathbf{1}).$$

DEFINITION 1.2.3.6. *Let $M$ be a symmetric monoidal model category. An object $x\in M$ is called* perfect *if the natural morphism*

$$x\otimes^{\mathbb{L}} x^{\vee} \longrightarrow \mathbb{R}\underline{\mathbf{Hom}}_M(x,x)$$

*is an isomorphism in* $\mathrm{Ho}(M)$.

PROPOSITION 1.2.3.7. *Let $M$ be a symmetric monoidal model category.*

(1) *If $x$ and $y$ are perfect objects in $M$, then so is $x \otimes^{\mathbb{L}} y$.*
(2) *If $x$ is a perfect object in $M$, then for any objects $y$ and $z$, the natural morphism*

$$\mathbb{R}\underline{\mathbf{Hom}}_M(y, x \otimes^{\mathbb{L}} z) \longrightarrow \mathbb{R}\underline{\mathbf{Hom}}_M(y \otimes^{\mathbb{L}} x^{\vee}, z)$$

*is an isomorphism in* $\mathrm{Ho}(M)$.
(3) *If $x$ in perfect in $M$, and $y \in M$, then the natural morphism*

$$\mathbb{R}\underline{\mathbf{Hom}}_M(x, y) \longrightarrow x^{\vee} \otimes^{\mathbb{L}} y$$

*is an isomorphism in* $\mathrm{Ho}(M)$.
(4) *If* $\mathbf{1}$ *is finitely presented in $M$, then so is any perfect object.*
(5) *If $M$ is furthermore a stable model category then perfect objects are stable by homotopy push-outs and homotopy pullbacks. In other words, if $x \longrightarrow y \longrightarrow z$ is a homotopy fiber sequence in $M$, and if two of the objects $x$, $y$, and $z$ are perfect then so is the third.*
(6) *Perfect objects are stable by retracts in* $\mathrm{Ho}(M)$.

PROOF. (1), (2) and (3) are standard, as perfect objects are precisely the strongly dualizable objects of the closed monoidal category $\mathrm{Ho}(M)$ (see for example [**May2**]).

(4) Let $x$ be a perfect object in $M$, and $\{z_i\}_{i\in I}$ be a filtered diagram of objects in $M$. Let $x^{\vee} := \mathbb{R}\underline{Hom}(x, \mathbf{1})$ the dual of $x$ in $\mathrm{Ho}(M)$. Then, we have

$$Map_M(x, Hocolim_i z_i) \simeq Map_M(\mathbf{1}, x^{\vee} \otimes^{\mathbb{L}} Hocolim_i z_i) \simeq Map_M(\mathbf{1}, Hocolim_i x^{\vee} \otimes^{\mathbb{L}} z_i)$$

$$\simeq Hocolim_i Map_M(\mathbf{1}, x^{\vee} \otimes^{\mathbb{L}} z_i) \simeq Hocolim_i Map_M(x, z_i).$$

(5) Let $x \longrightarrow y \longrightarrow z$ be a homotopy fiber sequence in $M$. It is enough to prove that if $y$ and $z$ are perfect then so is $x$. For this, let $x^{\vee}$, $y^{\vee}$ and $z^{\vee}$ the duals of $x$, $y$ and $z$.

One has a morphism of homotopy fiber sequences

$$\begin{array}{ccccc} x \otimes^{\mathbb{L}} z^{\vee} & \longrightarrow & x \otimes^{\mathbb{L}} y^{\vee} & \longrightarrow & x \otimes^{\mathbb{L}} x^{\vee} \\ \downarrow & & \downarrow & & \downarrow \\ \mathbb{R}\underline{\mathbf{Hom}}_M(z, x) & \longrightarrow & \mathbb{R}\underline{\mathbf{Hom}}_M(y, x) & \longrightarrow & \mathbb{R}\underline{\mathbf{Hom}}_M(x, x). \end{array}$$

The five lemma and point (3) implies that the last vertical morphism is isomorphism, and that $x$ is perfect.

(6) If $x$ is a retract of $y$, then the natural morphism

$$x \otimes^{\mathbb{L}} x^{\vee} \longrightarrow \mathbb{R}\underline{\mathbf{Hom}}_M(x, x)$$

is a retract of

$$y \otimes^{\mathbb{L}} y^{\vee} \longrightarrow \mathbb{R}\underline{\mathbf{Hom}}_M(y, y).$$

□

The following corollary gives a condition under which perfect and finitely presented objects are the same.

COROLLARY 1.2.3.8. *Suppose that $M$ is a stable and compactly generated symmetric monoidal model category with* $\mathbf{1}$ *being $\omega$-compact and cofibrant. We assume that the set $I$ of morphisms of the form*

$$S^n \otimes \mathbf{1} \longrightarrow \Delta^{n+1} \otimes \mathbf{1}$$

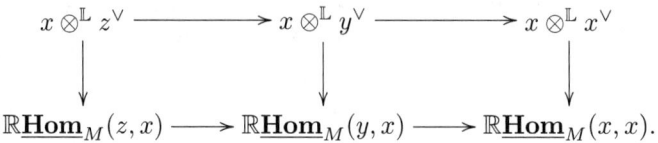

*is a set of generating cofibration for $M$. Then an object $x$ in $M$ is perfect if and only if it is finitely presented, and if and only if it is a retract of a finite $I$-cell object.*

PROOF. This essentially follows from Prop. 1.2.3.5 and Prop. 1.2.3.7, the only statement which remains to be proved is that retract of finite $I$-cell objects are perfect. But this follows from the fact that $\mathbf{1}$ is always perfect, and from Prop. 1.2.3.7 (5) and (6). $\square$

REMARK 1.2.3.9. The notion of finitely presented morphisms and perfect objects depend on the model structure and not only on the underlying category $M$. They specialize to the corresponding usual categorical notions when $M$ is endowed with the trivial model structure.

### 1.2.4. Some properties of modules

In this Section we give some general notions of flatness and projectiveness of modules over a commutative monoid in $\mathcal{C}$.

DEFINITION 1.2.4.1. *Let $A \in Comm(\mathcal{C})$ be a commutative monoid, and $M$ be an $A$-module.*

(1) *The $A$-module $M$ is* flat *if the functor*
$$- \otimes_A^{\mathbb{L}} M : \mathrm{Ho}(A - Mod) \longrightarrow \mathrm{Ho}(A - Mod)$$
*preserves homotopy pullbacks.*
(2) *The $A$-module $M$ is* projective *if it is a retract in $\mathrm{Ho}(A - Mod)$ of $\coprod_E^{\mathbb{L}} A$ for some $\mathbb{U}$-small set $E$.*

PROPOSITION 1.2.4.2. *Let $A \longrightarrow B$ be a morphism of commutative monoids in $\mathcal{C}$.*

(1) *The free $A$-module $A^n$ of rank $n$ is flat. Moreover, if infinite direct sums in $\mathrm{Ho}(A - Mod)$ commute with homotopy pull-backs, then for any $\mathbb{U}$-small set $E$, the free $A$-module $\coprod_E^{\mathbb{L}} A$ is flat.*
(2) *Flat modules in $\mathrm{Ho}(A - Mod)$ are stable by derived tensor products, finite coproducts and retracts.*
(3) *Projective modules in $\mathrm{Ho}(A - Mod)$ are stable by derived tensor products, finite coproducts and retracts.*
(4) *If $M$ is a flat (resp. projective) $A$-module, then $M \otimes_A^{\mathbb{L}} B$ is a flat (resp. projective) $B$-module.*
(5) *A perfect $A$-module is flat.*
(6) *Let us suppose that $\mathbf{1}$ is a finitely presented object in $\mathcal{C}$. Then, a projective $A$-module is finitely presented if and only if it is a retract of $\coprod_E^{\mathbb{L}} A$ for some finite set $E$.*
(7) *Let us assume that $\mathbf{1}$ is a finitely presented object in $\mathcal{C}$. Then, a projective $A$-module is perfect if and only it is finitely presented.*

PROOF. (1) to (4) are easy and follow from the definitions.

(5) Let $M$ be a perfect $A$-module, and $M^{\vee} := \mathbb{R}\underline{Hom}_A(M, A)$ be its dual. Then, for any homotopy cartesian square of $A$-modules
$$\begin{array}{ccc} P & \longrightarrow & Q \\ \downarrow & & \downarrow \\ P' & \longrightarrow & Q', \end{array}$$

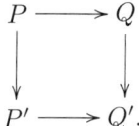

the diagram

$$\begin{array}{ccc} P \otimes^{\mathbb{L}}_A M & \longrightarrow & Q \otimes^{\mathbb{L}}_A M \\ \downarrow & & \downarrow \\ P' \otimes^{\mathbb{L}}_A M & \longrightarrow & Q' \otimes^{\mathbb{L}}_A M \end{array}$$

is equivalent to

$$\begin{array}{ccc} \mathbb{R}\underline{Hom}_A(M^\vee, P) & \longrightarrow & \mathbb{R}\underline{Hom}_A(M^\vee, Q) \\ \downarrow & & \downarrow \\ \mathbb{R}\underline{Hom}_A(M^\vee, P') & \longrightarrow & \mathbb{R}\underline{Hom}_A(M^\vee, Q'), \end{array}$$

which is again homotopy cartesian by the general properties of derived internal $Hom$'s. This shows that $- \otimes^{\mathbb{L}} M$ preserves homotopy pullbacks, and hence that $M$ is flat.

(6) Clearly, if $E$ a finite set, then $\coprod^{\mathbb{L}}_E A$ is finitely presented object, as for any $A$-module $M$ one has

$$Map_{A-Mod}(\coprod^{\mathbb{L}}_E A, M) \simeq Map_{\mathcal{C}}(\mathbf{1}, M)^E.$$

Therefore, a retract of $\coprod^{\mathbb{L}}_E A$ is also finitely presented.

Conversely, let $M$ be a projective $A$-module which is also finitely presented. Let $i : M \longrightarrow \coprod^{\mathbb{L}}_E A$ be a morphism which admits a retraction. As $\coprod^{\mathbb{L}}_E A$ is the colimit of $\coprod^{\mathbb{L}}_{E_0} A$, for $E_0$ running over the finite subsets of $E$, the morphism $i$ factors as

$$M \longrightarrow \coprod^{\mathbb{L}}_{E_0} A \longrightarrow \coprod^{\mathbb{L}}_E A$$

for some finite subset $E_0 \subset E$. This shows that $M$ is in fact a retract of $\coprod^{\mathbb{L}}_E A$.

(7) Using Prop. 1.2.3.7 (4) one sees that if $M$ is perfect then it is finitely presented. Conversely, let $M$ be a finitely presented projective $A$-module. By (6) we know that $M$ is a retract of $\coprod^{\mathbb{L}}_E A$ for some finite set $E$. But, as $E$ is finite, $\coprod^{\mathbb{L}}_E A$ is perfect, and therefore so is $M$ as a retract of a perfect module. □

### 1.2.5. Formal coverings

The following notion will be highly used, and is a categorical version of faithful morphisms of affine schemes.

DEFINITION 1.2.5.1. *A family of morphisms of commutative monoids* $\{f_i : A \to B_i\}_{i\in I}$ *is a* formal covering *if the family of functors*

$$\{\mathbb{L}f_i^* : \mathrm{Ho}(A - Mod) \longrightarrow \mathrm{Ho}(B_i - Mod)\}_{i \in I}$$

*is conservative (i.e. a morphism* $u$ *in* $\mathrm{Ho}(A - Mod)$ *is an isomorphism if and only if all the* $\mathbb{L}f_i^*(u)$ *are isomorphisms).*

The formal covering families are stable by equivalences, homotopy push-outs and compositions and therefore do form a model topology in the sense of [**HAGI**, Def. 4.3.1] (or Def. 1.3.1.1).

PROPOSITION 1.2.5.2. *Formal covering families form a model topology (Def. 1.3.1.1) on the model category* $Comm(\mathcal{C})$.

PROOF. Stability by equivalences and composition is clear. The stability by homotopy push-outs is an easy consequence of the transfer formula Prop. 1.1.0.8. □

### 1.2.6. Some properties of morphisms

In this part we review several classes of morphisms of commutative monoids in $\mathcal{C}$, generalizing the usual notions of Zariski open immersions, unramified, étale, smooth and flat morphisms of affine schemes. It is interesting to notice that these notions make sense in our general context, but specialize to very different notions in various specific cases (see the examples at the beginning of each section §2.1, §2.2, §2.3 and §2.4).

DEFINITION 1.2.6.1. *Let $f : A \longrightarrow B$ be a morphism of commutative monoids in $\mathcal{C}$.*

(1) *The morphism $f$ is an* epimorphism *if for any commutative $A$-algebra $C$ the simplicial set $Map_{A-Comm(\mathcal{C})}(B, C)$ is either empty or contractible.*
(2) *The morphism $f$ is* flat *if the induced functor*
$$\mathbb{L}f^* : \mathrm{Ho}(A - Mod) \longrightarrow \mathrm{Ho}(B - Mod)$$
*commutes with finite homotopy limits.*
(3) *The morphism $f$ is a* formal Zariski open immersion *if it is flat and if the functor*
$$f_* : \mathrm{Ho}(B - Mod) \longrightarrow \mathrm{Ho}(A - Mod)$$
*is fully faithful.*
(4) *The morphism $f$ is* formally unramified *if $\mathbb{L}_{B/A} \simeq 0$ in* $\mathrm{Ho}(B - Mod)$.
(5) *The morphism $f$ is* formally étale *if the natural morphism*
$$\mathbb{L}_A \otimes^{\mathbb{L}}_A B \longrightarrow \mathbb{L}_B$$
*is an isomorphism in* $\mathrm{Ho}(B - Mod)$.
(6) *The morphism $f$ is* formally thh-étale *if the natural morphism*
$$B \longrightarrow THH(B/A)$$
*is an isomorphism in* $\mathrm{Ho}(Comm(\mathcal{C}))$.

REMARK 1.2.6.2. One remark concerning our notion of epimorphism of commutative monoids is in order. First of all, in a category $\mathcal{C}$ (without any model structure) having fiber products, a morphism $x \longrightarrow y$ is a monomorphism if and only the diagonal morphism $x \longrightarrow x \times_y x$ is an isomorphism. A natural generalization of this fact gives a notion of monomorphism in any model category $M$, as a morphism $x \longrightarrow y$ whose diagonal $x \longrightarrow x \times^h_y x$ is an isomorphism in $\mathrm{Ho}(M)$. Equivalently, the morphism $x \longrightarrow y$ is a monomorphism if and and only if for any $z \in M$ the induced morphism of simplicial sets $Map_M(z, x) \longrightarrow Map_M(z, y)$ is a monomorphism in the model category $SSet$. Furthermore, it is easy to check that a morphism $f : K \longrightarrow L$ is a monomorphism in the model category $SSet$ if and only if for any $s \in L$ the homotopy fiber of $f$ at $s$ is either empty or contractible. Therefore we see that a morphism $A \longrightarrow B$ in $Comm(\mathcal{C})$ is an epimorphism in the sense of Def. 1.2.6.1 (1) if and only if it is a monomorphism when considered as a morphism in the model category $Comm(\mathcal{C})^{op}$, or equivalently if the induced morphism $B \otimes^{\mathbb{L}}_A B \longrightarrow B$ is an isomorphism in $\mathrm{Ho}(Comm(\mathcal{C}))$. This justifies our terminology, and moreover shows that when the model structure on the category $M$ is trivial, an epimorphism in the sense of our definition is nothing else than an epimorphism in $M$ in the usual categorical sense.

PROPOSITION 1.2.6.3. *Epimorphisms, flat morphisms, formal Zariski open immersions, formally unramified morphisms, formally étale morphisms, and formally thh-étale morphisms, are all stable by compositions, equivalences and homotopy push-outs.*

PROOF. This is a simple exercise using the definitions and Propositions 1.1.0.8, 1.2.1.6, 1.2.2.2, and 1.2.3.7. □

The relations between all these notions are given by the following proposition.

PROPOSITION 1.2.6.4.
(1) *A morphism $f : A \longrightarrow B$ is an epimorphism if and only if the functor*
$$f_* : \mathrm{Ho}(B - Mod) \longrightarrow \mathrm{Ho}(A - Mod)$$
*is fully faithful.*
(2) *A formal Zariski open immersion is an epimorphism. A flat epimorphism is a formal Zariski open immersion.*
(3) *A morphism $f : A \longrightarrow B$ of commutative monoids is formally thh-étale if and only if for any commutative $A$-algebra $C$ the simplicial set*
$$Map_{A-Comm(\mathcal{C})}(B, C)$$
*is discrete (i.e. equivalent to a set).*
(4) *A formally étale morphism is formally unramified.*
(5) *An epimorphism is formally unramified and formally thh-étale.*
(6) *A morphism $f : A \longrightarrow B$ in $Comm(\mathcal{C})$ is formally unramified if and only the morphism*
$$B \otimes^{\mathbb{L}}_A B \longrightarrow B$$
*is formally étale.*

PROOF. (1) Let $f : A \longrightarrow B$ be a morphism such that the right Quillen functor $f_* : B-Mod \longrightarrow A-Mod$ induces a fully faithful functor on the homotopy categories. Therefore, the adjunction morphism
$$M \otimes^{\mathbb{L}}_A B \longrightarrow M$$
is an isomorphism for any $M \in \mathrm{Ho}(B - Mod)$. In particular, the functor
$$f_* : \mathrm{Ho}(B - Comm(\mathcal{C})) \longrightarrow \mathrm{Ho}(A - Comm(\mathcal{C}))$$
is also fully faithful. Let $C$ be a commutative $A$-algebra, and let us suppose that $Map_{A-Comm(\mathcal{C})}(B, C)$ is not empty. This implies that $C$ is isomorphic in $\mathrm{Ho}(A - Comm(\mathcal{C}))$ to some $f_*(C')$ for $C' \in \mathrm{Ho}(B - Comm(\mathcal{C}))$. Therefore, we have
$$Map_{A-Comm(\mathcal{C})}(B, C) \simeq Map_{A-Comm(\mathcal{C})}(f_*(B), f_*(C')) \simeq Map_{B-Comm(\mathcal{C})}(B, C') \simeq *,$$
showing that $f$ is a epimorphism.

Conversely, let $f : A \longrightarrow B$ be epimorphism. For any commutative $A$-algebra $C$, we have
$$Map_{A-Comm(\mathcal{C})}(B, C) \simeq Map_{A-Comm(\mathcal{C})}(B, C) \times Map_{A-Comm(\mathcal{C})}(B, C),$$
showing that the natural morphism $B \otimes^{\mathbb{L}}_A B \longrightarrow B$ is an isomorphism in $\mathrm{Ho}(A - Comm(\mathcal{C}))$. This implies that for any $B$-module $M$, we have
$$M \otimes^{\mathbb{L}}_A B \simeq M \otimes^{\mathbb{L}}_B (B \otimes^{\mathbb{L}}_A B) \simeq M,$$
or in other words, that the adjunction morphism $M \longrightarrow f_*\mathbb{L}f^*(M) \simeq M \otimes^{\mathbb{L}}_A B$ is an isomorphism in $\mathrm{Ho}(B - Mod)$. This means that $f_* : \mathrm{Ho}(B - Mod) \longrightarrow \mathrm{Ho}(A - Mod)$ is fully faithful.

(2) This is clear by (1) and the definitions.

(3) For any morphism of commutative $A$-algebras, $f : B \longrightarrow C$ we have

$$Map_{B-Comm(\mathcal{C})}(THH(B/A), C)) \simeq \Omega_f Map_{A-Comm(\mathcal{C})}(B, C).$$

Therefore, $B \longrightarrow THH(B/A)$ is an equivalence if and only if for any such $f : B \longrightarrow C$, the simplicial set $\Omega_f Map_{A-Comm(\mathcal{C})}(B, C)$ is contractible. Equivalently, $f$ is formally thh-étale if and only if $Map_{A-Comm(\mathcal{C})}(B, C)$ is discrete.

(4) Let $f : A \longrightarrow B$ be a formally étale morphism of commutative monoids in $\mathcal{C}$. By Prop. 1.2.1.6 (1), there is a homotopy cofiber sequence of $B$-modules

$$\mathbb{L}_A \otimes^{\mathbb{L}}_A B \longrightarrow \mathbb{L}_B \longrightarrow \mathbb{L}_{B/A},$$

showing that if the first morphism is an isomorphism then $\mathbb{L}_{B/A} \simeq *$ and therefore that $f$ is formally unramified.

(5) Let $f : A \longrightarrow B$ an epimorphism. By definition and (3) we know that $f$ is formally *thh*-étale. Let us prove that $f$ is also formally unramified. Let $M$ be a $B$-module. As we have seen before, for any commutative $B$-algebra $C$, the adjunction morphism $C \longrightarrow C \otimes^{\mathbb{L}}_A B$ is an isomorphism. In particular, the functor

$$f_* : \mathrm{Ho}(B - Comm(\mathcal{C})/B) \longrightarrow \mathrm{Ho}(A - Comm(\mathcal{C})/B)$$

is fully faithful. Therefore we have

$$\mathbb{D}er_A(B, M) \simeq Map_{A-Comm(\mathcal{C})/B}(B, B \oplus M) \simeq Map_{B-Comm(\mathcal{C})/B}(B, B \oplus M) \simeq *,$$

showing that $\mathbb{D}er_A(B, M) \simeq *$ for any $B$-module $M$, or equivalently that $\mathbb{L}_{B/A} \simeq *$.

(6) Finally, Prop. 1.2.1.6 (2) shows that the morphism $B \otimes^{\mathbb{L}}_A B \longrightarrow B$ is formally étale if and only if the natural morphism

$$\mathbb{L}_{B/A} \coprod \mathbb{L}_{B/A} \longrightarrow \mathbb{L}_{B/A}$$

is an isomorphism in $\mathrm{Ho}(B - Mod)$. But this is equivalent to $\mathbb{L}_{B/A} \simeq *$. □

In order to state the next results we recall that as $\mathcal{C}$ is a pointed model category one can define a suspension functor (see [**Ho1**, §7])

$$\begin{array}{cccc} S: & \mathrm{Ho}(\mathcal{C}) & \longrightarrow & \mathrm{Ho}(\mathcal{C}) \\ & X & \mapsto & S(X) := * \coprod^{\mathbb{L}}_X *. \end{array}$$

This functor possesses a right adjoint, the loop functor

$$\begin{array}{cccc} \Omega: & \mathrm{Ho}(\mathcal{C}) & \longrightarrow & \mathrm{Ho}(\mathcal{C}) \\ & X & \mapsto & \Omega(X) := * \times^h_X *. \end{array}$$

PROPOSITION 1.2.6.5. *Assume that the base model category $\mathcal{C}$ is such that the suspension functor $S : \mathrm{Ho}(\mathcal{C}) \longrightarrow \mathrm{Ho}(\mathcal{C})$ is fully faithful.*

1. *A morphism of commutative monoids in $\mathcal{C}$ is formally étale if and only if it is formally unramified.*
2. *A formally thh-étale morphism of commutative monoids in $\mathcal{C}$ is a formally étale morphism.*
3. *An epimorphism of commutative monoids in $\mathcal{C}$ is formally étale.*

PROOF. (1) By the last proposition we only need to prove that a formally unramified morphism is also formally étale. Let $f : A \longrightarrow B$ be such a morphism. By Prop. 1.2.1.6 (1) there is a homotopy cofiber sequence of $B$-modules

$$\mathbb{L}_A \otimes^{\mathbb{L}}_A B \longrightarrow \mathbb{L}_B \longrightarrow *.$$

This implies that for any $B$-module $M$, the homotopy fiber of the morphism

$$\mathbb{D}er(B, M) \longrightarrow \mathbb{D}er(A, M)$$

is contractible, and in particular that this morphism induces isomorphisms on all higher homotopy groups. It remains to show that this morphism induces also an isomorphism on $\pi_0$. For this, we can use the hypothesis on $\mathcal{C}$ which implies that the suspension functor on $\mathrm{Ho}(B - Mod)$ is fully faithful. Therefore, we have

$$\pi_0(\mathbb{D}er(B, M)) \simeq \pi_0(\mathbb{D}er(B, \Omega S(M))) \simeq \pi_1(\mathbb{D}er(B, SM)) \simeq$$
$$\simeq \pi_1(\mathbb{D}er(A, SM)) \simeq \pi_0(\mathbb{D}er(A, M)).$$

(2) Let $A \longrightarrow B$ be a formally *thh*-étale morphism in $Comm(\mathcal{C})$.
As $Map_{A-Comm(\mathcal{C})}(B, C)$ is discrete for any commutative $A$-algebra $C$, the simplicial set $\mathbb{D}er_A(B, M)$ is discrete for any $B$-module $M$. Using the hypothesis on $\mathcal{C}$ we get that for any $B$-module $M$

$$\pi_0(\mathbb{D}er_A(B, M)) \simeq \pi_0(\mathbb{D}er_A(B, \Omega S(M))) \simeq \pi_1(\mathbb{D}er_A(B, SM)) \simeq 0,$$

showing that $\mathbb{D}er_A(B, M) \simeq *$, and therefore that $\mathbb{L}_{B/A} \simeq *$. This implies that $f$ is formally unramified, and therefore is formally étale by the first part of the proposition.

(3) This follows from (2) and Prop. 1.2.6.4 (5). □

The hypothesis of Proposition 1.2.6.5 saying that the suspension is fully faithful will appear in many places in the sequel. It is essentially equivalent to saying that the homotopy theory of $\mathcal{C}$ can be embedded in a stable homotopy theory in such a way that homotopy colimits are preserved (it could be in fact called *left semi-stable*). It is a very natural condition as many statements will then simplify, as shown for example by our infinitesimal theory in §3.

COROLLARY 1.2.6.6. *Assume furthermore that $\mathcal{C}$ is a stable model category, and let $f : A \longrightarrow B$ be a morphism of commutative monoids in $\mathcal{C}$. The following are equivalent.*

(1) *The morphism $f$ is a formal Zariski open immersion.*
(2) *The morphism $f$ is an epimorphism.*
(3) *The morphism $f$ is a formally étale epimorphism.*

PROOF. Indeed, in the stable case all base change functors commute with limits, so Prop. 1.2.6.4 (1) and (2) shows that formal Zariski open immersions are exactly the epimorphisms. Furthermore, by Prop. 1.2.6.4 (5) and Prop. 1.2.6.5 all epimorphisms are formally étale. □

DEFINITION 1.2.6.7. *A morphism of commutative monoids in $\mathcal{C}$ is a* Zariski open immersion *(resp.* unramified*, resp.* étale*, resp.* *thh*-étale*) if it is finitely presented (as a morphism in the model category $Comm(\mathcal{C})$) and is a formal Zariski open immersion (resp. formally unramified, resp. formally étale, resp. formally thh-étale).*

Clearly, using what we have seen before, Zariksi open immersions, unramified morphisms, étale morphisms, and *thh*-étale morphisms are all stable by equivalences, compositions and push-outs.

### 1.2.7. Smoothness

We define two general notions of smoothness, both different generalizations of the usual notion, and both useful in certain contexts. A third, and still different, notion of smoothness will be given in the next section.

DEFINITION 1.2.7.1. *Let $f : A \longrightarrow B$ be a morphism of commutative algebras.*

(1) *The morphism $f$ is* formally perfect *(or simply* fp*) if the $B$-module $\mathbb{L}_{B/A}$ is perfect (in the sense of Def. 1.2.3.6).*
(2) *The morphism $f$ is* formally smooth *if the $B$-module $\mathbb{L}_{B/A}$ is projective (in the sense of Def. 1.2.4.1) and if the morphism*
$$\mathbb{L}_A \otimes^{\mathbb{L}}_A B \longrightarrow \mathbb{L}_B$$
*has a retraction in* $\mathrm{Ho}(B - Mod)$.

DEFINITION 1.2.7.2. *Let $f : B \longrightarrow C$ be a morphism of commutative algebras. The morphism $f$ is* perfect, *or simply* p, *(resp.* smooth*) if it is finitely presented (as a morphism in the model category $Comm(\mathcal{C})$) and is fp (resp. formally smooth).*

Of course, (formally) étale morphisms are (formally) smooth morphisms as well as (formally) perfect morphisms.

PROPOSITION 1.2.7.3. *The fp, perfect, formally smooth and smooth morphisms are all stable by compositions, homotopy push outs and equivalences.*

PROOF. Exercise. □

### 1.2.8. Infinitesimal lifting properties

While the first two notions of smoothness only depend on the underlying symmetric monoidal model category $\mathcal{C}$, the third one, to be defined below, will also depend on the HA context we are working in.
Recall from Definition 1.1.0.11 that an *HA context* is a triplet $(\mathcal{C}, \mathcal{C}_0, \mathcal{A})$, consisting of a symmetric monoidal model category $\mathcal{C}$, two full sub-categories stable by equivalences
$$\mathcal{C}_0 \subset \mathcal{C} \qquad \mathcal{A} \subset Comm(\mathcal{C}),$$
such that:

- $\mathbf{1} \in \mathcal{C}_0$, $\mathcal{C}_0$ is closed under by $\mathbb{U}$-small homotopy colimits, and $X \otimes^{\mathbb{L}} Y \in \mathrm{Ho}(\mathcal{C}_0)$ if $X$ and $Y$ in $\mathrm{Ho}(\mathcal{C}_0)$.
- any $A \in \mathcal{A}$ is $\mathcal{C}_0$-good (i.e. the functor
$$\mathrm{Ho}(A - Mod) \longrightarrow \mathrm{Ho}((A - Mod_0^{op})^{\wedge})$$
is fully faithful);
- assumptions 1.1.0.1, 1.1.0.2, 1.1.0.3, 1.1.0.4, 1.1.0.6 are satisfied.

Recall also that $\mathcal{C}_1$ is the full subcategory of $\mathcal{C}$ consisting of all objects equivalent to suspensions of objects in $\mathcal{C}_0$, $Comm(\mathcal{C})_0$ the full subcategory of $Comm(\mathcal{C})$ consisting of commutative monoids whose underlying $\mathcal{C}$-object is in $\mathcal{C}_0$, and, for $A \in Comm(\mathcal{C})$, $A - Mod_0$ (resp. $A - Mod_1$, resp. $A - Comm(\mathcal{C})_0$) is the full subcategory of $A - Mod$ consisting of $A$-modules whose underlying $\mathcal{C}$-object is in $\mathcal{C}_0$ (resp. of $A - Mod$ consisting of $A$-modules whose underlying $\mathcal{C}$-object is in $\mathcal{C}_1$, resp. of $A - Comm(\mathcal{C})$ consisting of commutative $A$-algebras whose underlying $\mathcal{C}$-object is in $\mathcal{C}_0$).

DEFINITION 1.2.8.1. *Let $f : A \longrightarrow B$ be a morphism in $Comm(\mathcal{C})$.*

(1) *The morphism $f$ is called* formally infinitesimally smooth relative to the HA context $(\mathcal{C}, \mathcal{C}_0, \mathcal{A})$ *(or simply* formally i-smooth *when the HA context is clear) if for any $R \in \mathcal{A}$, any morphism $A \longrightarrow R$ of commutative monoids, any $M \in R - Mod_1$, and any $d \in \pi_0(\mathbb{D}er_A(R, M))$, the natural morphism*

$$\pi_0\left(Map_{A-Comm(\mathcal{C})}(B, R \oplus_d \Omega M)\right) \longrightarrow \pi_0\left(Map_{A-Comm(\mathcal{C})}(B, R)\right)$$

*is surjective.*

(2) *The morphism $f$ is called* i-smooth *if it is formally i-smooth and finitely presented.*

The following proposition is immediate from the definition.

PROPOSITION 1.2.8.2. *Formally i-smooth and i-smooth morphisms are stable by equivalences, composition and homotopy push-outs.*

The next result provides a criterion for formally i-smooth morphisms in terms of their cotangent complexes.

PROPOSITION 1.2.8.3. *A morphism $f : A \longrightarrow B$ is formally i-smooth if and only if for any morphism $B \longrightarrow R$ with $R \in \mathcal{A}$, and any $R$-module $M \in R - Mod_1$, the natural morphism*

$$[\mathbb{L}_{R/A}, M] \simeq \pi_0(\mathbb{D}er_A(R, M))) \longrightarrow \pi_0(\mathbb{D}er_A(B, M)) \simeq [\mathbb{L}_{B/A}, M]_{B-Mod}$$

*is zero.*

PROOF. Let us first assume that the condition of the proposition is satisfied. Let us consider $R \in \mathcal{A}$, a morphism $A \longrightarrow R$, an $R$-module $M \in R - Mod_1$ and $d \in \pi_0(\mathbb{D}er_A(R, M))$. The homotopy fiber of the natural morphism

$$Map_{A-Comm(\mathcal{C})}(B, R \oplus_d \Omega M) \longrightarrow Map_{A-Comm(\mathcal{C})}(B, R)$$

taken at some morphism $B \longrightarrow R$ in $\mathrm{Ho}(Comm(\mathcal{C}))$, can be identified with the path space $Path_{0,d'}\mathbb{D}er_A(B, M)$ from $0$ to $d'$ in $\mathbb{D}er_A(B, M)$, where $d'$ is the image of $d$ under the natural morphism

$$\pi_0(\mathbb{D}er_A(R, M)) \longrightarrow \pi_0(\mathbb{D}er_A(B, M)).$$

By assumption, $0$ and $d'$ belong to the same connected component of $\mathbb{D}er_A(B, M)$, and thus we see that $Path_{0,d'}\mathbb{D}er_A(B, M)$ is non-empty. We have thus shown that

$$Map_{A-Comm(\mathcal{C})}(B, R \oplus_d \Omega M) \longrightarrow Map_{A-Comm(\mathcal{C})}(B, R)$$

has non-empty homotopy fibers and therefore that

$$\pi_0\left(Map_{A-Comm(\mathcal{C})}(B, R \oplus_d \Omega M)\right) \longrightarrow \pi_0\left(Map_{A-Comm(\mathcal{C})}(B, R)\right)$$

is surjective. The morphism $f$ is therefore formally i-smooth.

Conversely, let $R \in \mathcal{A}$, $A \longrightarrow R$ a morphism, $M \in R - Mod_1$ and $d \in \pi_0(\mathbb{D}er_A(R, M))$. Let $A \to R$ be a morphism of commutative monoids. We consider the diagram

$$\begin{array}{ccccc} A & \longrightarrow & R[\Omega_d M] & \longrightarrow & R \\ \downarrow & & \downarrow & & \downarrow \\ B & \longrightarrow & R & \underset{s}{\longrightarrow} & R \oplus M. \end{array}$$

The homotopy fiber of

$$Map_{A-Comm(\mathcal{C})}(B, R \oplus_d \Omega M) \longrightarrow Map_{A-Comm(\mathcal{C})}(B, R)$$

is non-empty because $f$ is formally i-smooth. By definition this means that the image of $d$ by the morphism

$$\pi_0(\mathbb{D}er_A(R,M)) \longrightarrow \pi_0(\mathbb{D}er_A(B,M))$$

is zero. As this is true for any $d$, this finishes the proof of the proposition. □

COROLLARY 1.2.8.4. (1) *Any formally unramified morphism is formally i-smooth.*

(2) *Assume that the suspension functor $S$ is fully faithful, that $\mathcal{C}_1 = \mathcal{C}$ (so that in particular $\mathcal{C}$ is stable), and $\mathcal{A} = Comm(\mathcal{C})$. Then formally i-smooth morphisms are precisely the formally étale morphisms.*

PROOF. It follows immediately from 1.2.8.3. □

COROLLARY 1.2.8.5. *We assume that for any $M \in \mathcal{C}_1$ one has $[\mathbf{1}, M] = 0$. Then any formally smooth morphism is formally i-smooth.*

PROOF. By Prop. 1.2.8.3 and definition of formal smoothness, it is enough to show that for any commutative monoid $A \in \mathcal{C}$, any $A$-module $M \in A-Mod_1$ and any projective $A$-module $P$ we have $[P, M] = 0$. By assumption this is true for $P = A$, and thus also true for free $A$-modules and their retracts. □

### 1.2.9. Standard localizations and Zariski open immersions

For any object $A \in \mathcal{C}$, we can define its *underlying space* as

$$|A| := Map_{\mathcal{C}}(\mathbf{1}, A) \in \mathrm{Ho}(SSet_{\mathbb{U}}).$$

The model category $\mathcal{C}$ being pointed, the simplicial set $|A|$ has a natural base point $* \in |A|$, and one can therefore define the homotopy groups of $A$

$$\pi_i(A) := \pi_i(|A|, *).$$

When $A$ is the underlying object of a commutative monoid $A \in Comm(\mathcal{C})$ one has by adjunction

$$\pi_0(A) = \pi_0(Map_{\mathcal{C}}(\mathbf{1}, A)) \simeq [\mathbf{1}, A]_{\mathcal{C}} \simeq [A, A]_{A-Mod}.$$

Since $A$ is the unit of the monoidal structure on the additive category $\mathrm{Ho}(A - Mod)$, the composition of endomorphisms endows $\pi_0(A)$ with a multiplicative structure making it into a commutative ring. More generally, the category $\mathrm{Ho}(A - Mod)$ has a natural structure of a graded category (i.e. has a natural enrichment into the symmetric monoidal category of $\mathbb{N}$-graded abelian groups), defined by

$$[M, N]_* := \oplus_i [S^i(N), M] \simeq \oplus_i [N, \Omega^i(M)],$$

where $S^i$ is the $i$-fold iterated suspension functor, and $\Omega^i$ the $i$-fold iterated loop functor. Therefore, the graded endomorphism ring of the unit $A$ has a natural structure of a graded commutative ring. As this endomorphism ring is naturally isomorphic to

$$\pi_*(A) := \oplus_i \pi_i(A),$$

we obtain this way a natural structure of a graded commutative ring on $\pi_*(A)$. This clearly defines a functor

$$\pi_* : \mathrm{Ho}(Comm(\mathcal{C})) \longrightarrow GComm,$$

from $\mathrm{Ho}(Comm(\mathcal{C}))$ to the category of $\mathbb{N}$-graded commutative rings.

In the same way, for a commutative monoid $A \in Comm(\mathcal{C})$, and a $A$-module $M$, one can define

$$\pi_*(M) := \pi_*(|M|) = \pi_*(Map_{A-Mod}(A, M)),$$

which has a natural structure of a graded $\pi_*(A)$-module. This defines a functor

$$\pi_* : \mathrm{Ho}(A - Mod) \longrightarrow \pi_*(A) - GMod,$$

from $\mathrm{Ho}(A - Mod)$ to the category of $\mathbb{N}$-graded $\pi_*(A)$-modules.

PROPOSITION 1.2.9.1. *Let $A \in Comm(\mathcal{C})$ be a commutative monoid in $\mathcal{C}$, and $a \in \pi_0(A)$. There exists an epimorphism $A \longrightarrow A[a^{-1}]$, such that for any commutative $A$-algebra $C$, the simplicial set $Map_{A-Comm(\mathcal{C})}(A[a^{-1}], C)$ is non-empty (and thus contractible) if and only if the image of $a$ in $\pi_0(C)$ by the morphism $\pi_0(A) \to \pi_0(C)$ is an invertible element.*

PROOF. We represent the element $a$ as a morphism in $\mathrm{Ho}(A - Mod)$ $a : A \longrightarrow A$. Taking the image of $a$ by the left derived functor of the free commutative $A$-algebra functor (which is left Quillen)

$$F_A : A - Mod \longrightarrow A - Comm(\mathcal{C})$$

we find a morphism in $\mathrm{Ho}(A - Comm(\mathcal{C}))$

$$a : \mathbb{L}F_A(A) \longrightarrow \mathbb{L}F_A(A).$$

As the model category $A - Comm(\mathcal{C})$ is a $\mathbb{U}$-combinatorial model category, one can apply the localization techniques of in order to invert any $\mathbb{U}$-small set of morphisms in $A - Comm(\mathcal{C})$ (see [**Sm, Du2**]). We let $L_a A - Comm(\mathcal{C})$ be the left Bousfield localization of $A - Comm(\mathcal{C})$ along the set of morphisms (with one element)

$$S_a := \{a : \mathbb{L}F_A(A) \longrightarrow \mathbb{L}F_A(A)\}.$$

We define $A[a^{-1}] \in \mathrm{Ho}(A - Comm(\mathcal{C}))$ as a local model of $A$ in $L_a A - Comm(\mathcal{C})$, the left Bousfield localization of $A - Comm(\mathcal{C})$ along $S_a$.

First of all, the $S_a$-local objects are the commutative $A$-algebras $B$ such that the induced morphism

$$a^* : Map_{A-Mod}(A, B) \longrightarrow Map_{A-Mod}(A, B)$$

is an equivalence. Equivalently, the multiplication by $a \in \pi_0(A)$

$$\times a : \pi_*(B) \longrightarrow \pi_*(B)$$

is an isomorphism. This shows that the $S_a$-local objects are the commutative $A$-algebras $A \longrightarrow B$ such that the image of $a$ by $\pi_0(A) \longrightarrow \pi_0(B)$ is invertible.

Suppose now that $C \in A - Comm(\mathcal{C})$ is such that $Map_{A-Comm(\mathcal{C})}(A[a^{-1}], C)$ is not empty. The morphism $\pi_0(A) \longrightarrow \pi_0(C)$ then factors through $\pi_0(A[a^{-1}])$, and thus the image of $a$ is invertible in $\pi_0(C)$. Therefore, $C$ is a $S_a$-local object, and thus

$$Map_{A-Comm(\mathcal{C})}(A[a^{-1}], C) \simeq Map_{A-Comm(\mathcal{C})}(A, C) \simeq *.$$

This implies that $A \longrightarrow A[a^{-1}]$ is an epimorphism. It only remain to prove that if $C \in A - Comm(\mathcal{C})$ is such that the image of $a$ is invertible in $\pi_0(C)$, then $Map_{A-Comm(\mathcal{C})}(A[a^{-1}], C)$ is non-empty. But such a $C$ is a $S_a$-local object, and therefore

$$* \simeq Map_{A-Comm(\mathcal{C})}(A, C) \simeq Map_{A-Comm(\mathcal{C})}(A[a^{-1}], C).$$

□

DEFINITION 1.2.9.2. *Let $A \in Comm(\mathcal{C})$ and $a \in \pi_0(A)$. The commutative $A$-algebra $A[a^{-1}]$ is called* the standard localization of $A$ with respect to $a$.

A useful property of standard localizations is given by the following corollary of the proof of Prop. 1.2.9.1. In order to state it, we will use the following notations. For any $A \in Comm(\mathcal{C})$ and $a \in \pi_0(A)$, we represent $a$ as a morphism in $\mathrm{Ho}(A - Mod)$, $A \longrightarrow A$. Tensoring this morphism with $M$ gives a morphism in $\mathrm{Ho}(A - Mod)$, denoted by

$$\times a : M \longrightarrow M.$$

COROLLARY 1.2.9.3. *Let $A \in Comm(\mathcal{C})$, $a \in \pi_0(A)$ and let $f : A \longrightarrow A[a^{-1}]$ be as in Prop. 1.2.9.1.*

(1) *The functor*

$$\mathrm{Ho}(A[a^{-1}] - Comm(\mathcal{C})) \longrightarrow \mathrm{Ho}(A - Comm(\mathcal{C}))$$

*is fully faithful, and its essential image consists of all commutative $B$-algebras such that the image of $a$ is invertible in $\pi_0(B)$.*

(2) *The functor*

$$f_* : \mathrm{Ho}(A[a^{-1}] - Mod) \longrightarrow \mathrm{Ho}(A - Mod)$$

*is fully faithful and its essential image consists of all $A$-modules $M$ such that the multiplication by $a$*

$$\times a : M \longrightarrow M$$

*is an isomorphism in* $\mathrm{Ho}(A - Mod)$.

PROOF. (1) The fact that the functor $f_*$ is fully faithful is immediate as $f$ is an epimorphism. The fact that the functor $f_*$ takes its values in the required subcategory is clear by functoriality of the construction $\pi_*$.

Let $B$ be a commutative $A$-algebra such that the image of $a$ is invertible in $\pi_0(B)$. Then, we know that $B$ is a $S_a$-local object. Therefore, one has

$$Map_{A-Comm(\mathcal{C})}(A[a^{-1}], B) \simeq Map_{A-Comm(\mathcal{C})}(A, B) \simeq *,$$

showing that $B$ is in the image of $f_*$.

(2) The fact that $f_*$ is fully faithful follows from Prop. 1.2.6.4 (1). Let $M \in A[a^{-1}] - Mod$ and let us prove that the morphism $\times a : M \longrightarrow M$ is an isomorphism in $\mathrm{Ho}(A - Mod)$. Using that $A \longrightarrow A[a^{-1}]$ is an epimorphism, one finds $M \simeq M \otimes^{\mathbb{L}}_A A[a^{-1}]$, which reduces the problem to the case where $M = A[a^{-1}]$. But then, the morphism $\times a : A[a^{-1}] \longrightarrow A[a^{-1}]$, as a morphism in $\mathrm{Ho}(A[a^{-1}] - Mod)$ lives in $[A[a^{-1}], A[a^{-1}]] \simeq \pi_0(A[a^{-1}])$, and correspond to the image of $a$ by the morphism $\pi_0(A) \longrightarrow \pi_0(A[a^{-1}])$, which is then invertible. In other words, $\times a : A[a^{-1}] \longrightarrow A[a^{-1}]$ is an isomorphism.

Conversely, let $M$ be an $A$-module such that the morphism $\times a : M \longrightarrow M$ is an isomorphism in $\mathrm{Ho}(A - Mod)$. We need to show that the adjunction morphism

$$M \longrightarrow M \otimes^{\mathbb{L}}_A A[a^{-1}]$$

is an isomorphism. For this, we use that the morphism $A \longrightarrow A[a^{-1}]$ can be constructed using a small object argument with respect to the horns over the morphism $a : \mathbb{L}F_A(A) \longrightarrow \mathbb{L}F_A(A)$ (see [**Hi**, 4.2]). Therefore, a transfinite induction argument shows that it is enough to prove that the morphism induced by tensoring

$$a \otimes Id : \mathbb{L}F_A(A) \otimes^{\mathbb{L}}_A M \simeq \mathbb{L}F_A(M) \longrightarrow \mathbb{L}F_A(A) \otimes^{\mathbb{L}}_A M \simeq \mathbb{L}F_A(M)$$

is an isomorphism in $\mathrm{Ho}(A - Mod)$. But this morphism is the image by the functor $\mathbb{L}F_A$ of the morphism $\times a : M \longrightarrow M$, and is therefore an isomorphism. $\square$

PROPOSITION 1.2.9.4. *Let $A \in Comm(\mathcal{C})$, $a \in \pi_0(A)$ and $A \longrightarrow A[a^{-1}]$ the standard localization with respect to $a$. Assume that the model category $\mathcal{C}$ is finitely generated (in the sense of* [**Ho1**]*).*

1. *The morphism $A \longrightarrow A[a^{-1}]$ is a formal Zariski open immersion.*
2. *If $1$ is a finitely presented object in $\mathcal{C}$, then $A \longrightarrow A[a^{-1}]$ is a Zariski open immersion.*

PROOF. (1) It only remains to show that the morphism $A \longrightarrow A[a^{-1}]$ is flat. In other words, we need to prove that the functor $M \mapsto M \otimes^{\mathbb{L}}_A A[a^{-1}]$, preserves homotopy fiber sequences of $A$-modules.

The model category $A - Mod$ is $\mathbb{U}$-combinatorial and finitely generated. In particular, there exists a $\mathbb{U}$-small set $G$ of $\omega$-small cofibrant objects in $A - Mod$, such that a morphism $N \longrightarrow P$ is an equivalence in $A - Mod$ if and only if for any $X \in G$ the induced morphism $Map_{A-Mod}(X, N) \longrightarrow Map_{A-Mod}(X, P)$ is an isomorphism in $\mathrm{Ho}(SSet)$. Furthermore, filtered homotopy colimits preserve homotopy fiber sequences. For any $A$-module $M$, we let $M_a$ be the transfinite homotopy colimit

$$M \xrightarrow{\times a} M \xrightarrow{\times a} \cdots$$

where the morphism $\times a$ is composed with itself $\omega$-times.

The functor $M \mapsto M_a$ commutes with homotopy fiber sequences. Therefore, it only remain to show that $M_a$ is naturally isomorphic in $\mathrm{Ho}(A-Mod)$ to $M \otimes^{\mathbb{L}}_A A[a^{-1}]$. For this, it is enough to check that the natural morphism $M \longrightarrow M_a$ induces an isomorphism in $\mathrm{Ho}(A - Mod)$

$$M \otimes^{\mathbb{L}}_A A[a^{-1}] \simeq (M_a) \otimes^{\mathbb{L}}_A A[a^{-1}],$$

and that the natural morphism

$$(M_a) \otimes^{\mathbb{L}}_A A[a^{-1}] \longrightarrow M_a$$

is an isomorphism in $\mathrm{Ho}(A - Mod)$. The first assumption follows easily from the fact that $- \otimes^{\mathbb{L}}_A A[a^{-1}]$ commutes with homotopy colimits, and the fact that $\times a : A[a^{-1}] \longrightarrow A[a^{-1}]$ is an isomorphism in $\mathrm{Ho}(A - Mod)$. For the second assumption we use Cor. 1.2.9.3 (2), which tells us that it is enough to check that the morphism

$$\times a : M_a \longrightarrow M_a$$

is an isomorphism in $\mathrm{Ho}(A - Mod)$. For this, we need to show that for any $X \in G$ the induced morphism

$$Map_{A-Mod}(X, M_a) \longrightarrow Map_{A-Mod}(X, M_a)$$

is an isomorphism in $\mathrm{Ho}(SSet)$. But, as the objects in $G$ are cofibrant and $\omega$-small, $Map_{A-Mod}(X, -)$ commutes with $\omega$-filtered homotopy colimits, and the morphism

$$Map_{A-Mod}(X, M_a) \longrightarrow Map_{A-Mod}(X, M_a)$$

is then obviously an isomorphism in $\mathrm{Ho}(SSet)$ by the construction of $M_a$.

(2) When $1$ is finitely presented, one has for any filtered diagram of commutative $A$-algebras $B_i$ an isomorphism

$$Colim_i \pi_*(B_i) \simeq \pi_*(Hocolim_i B_i).$$

Using that $Map_{A-Comm(\mathcal{C})}(A[a^{-1}], B)$ is either empty or contractible, depending whether of not $a$ goes to a unit in $\pi_*(B)$, we easily deduce that

$$Hocolim_i Map_{A-Comm(\mathcal{C})}(A[a^{-1}], B_i) \simeq Map_{A-Comm(\mathcal{C})}(A[a^{-1}], Hocolim_i B_i).$$

□

We can also show that the natural morphism $A \longrightarrow A[a^{-1}]$ is a formally étale morphism in the sense of Def. 1.2.6.1.

PROPOSITION 1.2.9.5. *Let $A \in Comm(\mathcal{C})$ and $a \in \pi_0(A)$. Then, the natural morphism $A \longrightarrow A[a^{-1}]$ is formally étale.*

PROOF. Let $M$ be any $A[a^{-1}]$-module. We need to show that the natural morphism

$$Map_{A-Comm(\mathcal{C})/A[a^{-1}]}(A[a^{-1}], A[a^{-1}]\oplus M) \longrightarrow Map_{A-Comm(\mathcal{C})/A[a^{-1}]}(A, A[a^{-1}]\oplus M)$$

is an isomorphism in Ho($SSet$). Using the universal property of $A \longrightarrow A[a^{-1}]$ given by Prop. 1.2.9.1 we see that it is enough to prove that for any $B \in Comm(\mathcal{C})$, and any $B$-module $M$ the natural projection $\pi_0(B \oplus M) \longrightarrow \pi_0(B)$ reflects invertible elements (i.e. an element in $\pi_0(B \oplus M)$ is invertible if and only its image in $\pi_0(B)$ is so). But, clearly, $\pi_0(B \oplus M)$ can be identified with the trivial square zero extension of the commutative ring $\pi_0(B)$ by $\pi_0(M)$, which implies the required result. □

COROLLARY 1.2.9.6. *Assume that the model category $\mathcal{C}$ is finitely presented, and that the unit **1** is finitely presented in $\mathcal{C}$. Then for any $A \in Comm(\mathcal{C})$ and $a \in \pi_0(A)$, the morphism $A \longrightarrow A[a^{-1}]$ is an étale, flat epimorphism.*

PROOF. Put 1.2.9.4 and 1.2.9.5 together. □

### 1.2.10. Zariski open immersions and perfect modules

Let $A$ be a commutative monoid in $\mathcal{C}$ and $K$ be a perfect $A$-module in the sense of Def. 1.2.3.6. We are going to define a Zariski open immersion $A \longrightarrow A_K$, which has to be thought as the complement of the support of the $A$-module $K$.

PROPOSITION 1.2.10.1. *Assume that $\mathcal{C}$ is stable model category. Then there exists a formal Zariski open immersion $A \longrightarrow A_K$, such that for any commutative $A$-algebra $C$, the simplicial set*

$$Map_{A-Comm(\mathcal{C})}(A_K, C)$$

*is non-empty (and thus contractible) if and only if $K \otimes^{\mathbb{L}}_A C \simeq *$ in* Ho($C - Mod$). *If the unit **1** is furthermore finitely presented, then $A \longrightarrow A_K$ is finitely presented and thus is a Zariski open immersion.*

PROOF. The commutative $A$-algebra is constructed using a left Bousfield localization of the model category $A - Comm(\mathcal{C})$, as done in the proof of Prop. 1.2.9.1.

We let $I$ be a generating $\mathbb{U}$-small set of cofibrations in $A - Comm(\mathcal{C})$, and $K^\vee := \mathbb{R}\underline{Hom}_{A-Mod}(K, A)$ be the dual of $K$ in Ho($A - Mod$). For any morphism $X \longrightarrow Y$ in $I$, we consider the morphism of free commutative $A$-algebras

$$\mathbb{L}F_A(K^\vee \otimes^{\mathbb{L}}_A X) \longrightarrow \mathbb{L}F_A(K^\vee \otimes^{\mathbb{L}}_A Y),$$

where $F_A : A - Mod \longrightarrow A - Comm(\mathcal{C})$ is the left Quillen functor sending an $A$-module to the free commutative $A$-algebra it generates. When $X \longrightarrow Y$ varies in $I$ this gives a $\mathbb{U}$-small set of morphisms denoted by $S_K$ in $A - Comm(\mathcal{C})$. We consider $L_K A - Comm(\mathcal{C})$, the left Bousfield localization of $A - Comm(\mathcal{C})$ along the set $S_K$. By definition, $A \longrightarrow A_K$ is an $S_K$-local model of $A$ in the localized model category $L_K A - Comm(\mathcal{C})$.

LEMMA 1.2.10.2. *The $S_K$-local objects in $L_K A - Comm(\mathcal{C})$ are precisely the commutative $A$-algebras $B$ such that $K \otimes^{\mathbb{L}}_A B \simeq *$ in* Ho($A - Mod$).

PROOF. First of all, one has an adjunction isomorphism in $\mathrm{Ho}(SSet)$

$$Map_{A-Comm(\mathcal{C})}(\mathbb{L}F_A(K^\vee \otimes^{\mathbb{L}}_A X), B) \simeq Map_{A-Mod}(X, K \otimes^{\mathbb{L}}_A B).$$

This implies that an object $B \in A - Comm(\mathcal{C})$ is $S_K$-local if and only if for all morphism $X \longrightarrow Y$ in $I$ the induced morphism

$$Map_{A-Mod}(Y, K \otimes^{\mathbb{L}}_A B) \longrightarrow Map_{A-Mod}(X, K \otimes^{\mathbb{L}}_A B)$$

is an isomorphism in $\mathrm{Ho}(SSet)$. As $I$ is a set of generating cofibrations in $A - Mod$ this implies that $B$ is $S_K$-local if and only if $K \otimes^{\mathbb{L}}_A C \simeq *$ in $\mathrm{Ho}(A - Mod)$. □

We now finish the proof of proposition 1.2.10.1. First of all, Lem. 1.2.10.2 implies that for any commutative $A$-algebra $B$, if the mapping space $Map_{A-Comm(\mathcal{C})}(A_K, B)$ is non-empty then

$$B \otimes^{\mathbb{L}}_A K \simeq B \otimes^{\mathbb{L}}_{A_K} (A_K \otimes^{\mathbb{L}}_A K) \simeq *,$$

and thus $B$ is an $S_K$-local object. This shows that if $Map_{A-Comm(\mathcal{C})}(A_K, B)$ is non-empty then one has

$$Map_{A-Comm(\mathcal{C})}(A_K, B) \simeq Map_{A-Comm(\mathcal{C})}(A, B) \simeq *.$$

In other words $A \longrightarrow A_K$ is a formal Zariski open immersion by Cor. 1.2.6.6 and the stability assumption on $\mathcal{C}$. It only remains to prove that $A \longrightarrow A_K$ is also finitely presented when $\mathbf{1}$ is a finitely presented object in $\mathcal{C}$.

For this, let $\{C_i\}_{i\in I}$ be a filtered diagram of commutative $A$-algebras and $C$ be its homotopy colimit. By the property of $A \longrightarrow A_K$, we need to show that if $K \otimes^{\mathbb{L}}_A C \simeq *$ then there is an $i \in I$ such that $K \otimes^{\mathbb{L}}_A C_i \simeq *$. For this, we consider the two elements $Id$ and $*$ in $[K \otimes^{\mathbb{L}}_A C, K \otimes^{\mathbb{L}}_A C]_{C-Mod}$. As $K$ is perfect and $\mathbf{1}$ is a finitely presented object, $K$ is a finitely presented $A$-module by Prop. 1.2.3.7 (4). Therefore, one has

$$* \simeq [K \otimes^{\mathbb{L}}_A C, K \otimes^{\mathbb{L}}_A C] \simeq Colim_{i\in I}[K, K \otimes^{\mathbb{L}}_A C_i]_{A-Mod}.$$

As the two elements $Id$ and $*$ becomes equal in the colimit, there is an $i$ such that they are equal as elements in

$$[K, K \otimes^{\mathbb{L}}_A C_i]_{A-Mod} \simeq [K \otimes^{\mathbb{L}}_A C_i, K \otimes^{\mathbb{L}}_A C_i]_{C_i-Mod},$$

showing that $K \otimes^{\mathbb{L}}_A C_i \simeq *$ in $\mathrm{Ho}(C_i - Mod)$. □

COROLLARY 1.2.10.3. *Assume that $\mathcal{C}$ is a stable model category, and let $f : A \longrightarrow A_K$ be as in Prop. 1.2.10.1.*

(1) *The functor*

$$\mathrm{Ho}(A_K - Comm(\mathcal{C})) \longrightarrow \mathrm{Ho}(A - Comm(\mathcal{C}))$$

*is fully faithful, and its essential image consists of all commutative $B$-algebras such that $B \otimes^{\mathbb{L}}_A K \simeq *$.*

(2) *The functor*

$$f_* : \mathrm{Ho}(A_K - Mod) \longrightarrow \mathrm{Ho}(A - Mod)$$

*is fully faithful and its essential image consists of all $A$-modules $M$ such that $M \otimes^{\mathbb{L}}_A K \simeq *$.*

PROOF. As the morphism $A \longrightarrow A_K$ is an epimorphism, we know by Prop. 1.2.6.4 that the functors

$$\mathrm{Ho}(A_K - Comm(\mathcal{C})) \longrightarrow \mathrm{Ho}(A - Comm(\mathcal{C})) \qquad \mathrm{Ho}(A_K - Mod) \longrightarrow \mathrm{Ho}(A - Mod)$$

are both fully faithful. Furthermore, a commutative $A$-algebra $B$ is in the essential image of the first one if and only if $Map_{A-Comm(\mathcal{C})}(A_K, B) \simeq *$, and therefore if and only if $B \otimes^{\mathbb{L}}_A K \simeq *$. For the second functor, its clear that if $M$ is a $A_K$-module, then

$$M \otimes^{\mathbb{L}}_A K \simeq M \otimes^{\mathbb{L}}_{A_K} A_K \otimes^{\mathbb{L}}_A K \simeq *.$$

Conversely, let $M$ be an $A$-module such that $M \otimes^{\mathbb{L}}_A K \simeq *$.

LEMMA 1.2.10.4. *For any $A$-module $M$, $K \otimes^{\mathbb{L}}_A M \simeq *$ if and only if $K^{\vee} \otimes^{\mathbb{L}}_A M \simeq *$.*

PROOF. Indeed, as $K$ is perfect, $K^{\vee}$ is a retract of $K^{\vee} \otimes^{\mathbb{L}}_A K \otimes^{\mathbb{L}}_A K^{\vee}$ in $\mathrm{Ho}(A - Mod)$. This implies that $K^{\vee} \otimes^{\mathbb{L}}_A M$ is a retract of $K^{\vee} \otimes^{\mathbb{L}}_A K \otimes^{\mathbb{L}}_A K^{\vee} \otimes^{\mathbb{L}}_A M$, showing that

$$\left(K \otimes^{\mathbb{L}}_A M \simeq *\right) \Rightarrow \left(K^{\vee} \otimes^{\mathbb{L}}_A M \simeq *\right).$$

By symmetry this proves the lemma. □

We need to prove that the adjunction morphism

$$M \longrightarrow M \otimes^{\mathbb{L}}_A A_K$$

is an isomorphism in $\mathrm{Ho}(A - Mod)$. For this, we use the fact that the morphism $A \longrightarrow A_K$ can be constructed using a small object argument on the set of horns on the set $S_K$ (see [**Hi**, 4.2]). By a transfinite induction we are therefore reduced to show that for any morphism $X \longrightarrow Y$ in $\mathcal{C}$, the natural morphism

$$\mathbb{L}F_A(K^{\vee} \otimes^{\mathbb{L}}_A X) \otimes^{\mathbb{L}}_A M \longrightarrow \mathbb{L}F_A(K^{\vee} \otimes^{\mathbb{L}}_A Y) \otimes^{\mathbb{L}}_A M$$

is an isomorphism in $\mathrm{Ho}(\mathcal{C})$. But, using that $M \otimes^{\mathbb{L}}_A K \simeq *$ and lemma 1.2.10.4, this is clear by the explicit description of the functor $F_A$. □

### 1.2.11. Stable modules

We recall that as $\mathcal{C}$ is a pointed model category one can define a suspension functor (see [**Ho1**, §7])

$$\begin{array}{cccc} S: & \mathrm{Ho}(\mathcal{C}) & \longrightarrow & \mathrm{Ho}(\mathcal{C}) \\ & X & \mapsto & S(X) := * \coprod^{\mathbb{L}}_X *. \end{array}$$

This functor possesses a right adjoint, the loop functor

$$\begin{array}{cccc} \Omega: & \mathrm{Ho}(\mathcal{C}) & \longrightarrow & \mathrm{Ho}(\mathcal{C}) \\ & X & \mapsto & \Omega(X) := * \times^h_X *. \end{array}$$

We fix, once for all an object $S^1_{\mathcal{C}} \in \mathcal{C}$, which is a cofibrant model for $S(\mathbf{1}) \in \mathrm{Ho}(\mathcal{C})$. For any commutative monoid $A \in Comm(\mathcal{C})$, we let

$$S^1_A := S^1_{\mathcal{C}} \otimes A \in A - Mod$$

be the free $A$-module on $S^1_{\mathcal{C}}$. It is a cofibrant object in $A - Mod$, which is a model for the suspension $S(A)$ (note that $S^1_A$ is cofibrant in $A - Mod$, but not in $\mathcal{C}$ unless $A$ is itself cofibrant in $\mathcal{C}$). The functor

$$S^1_A \otimes_A - : A - Mod \longrightarrow A - Mod$$

has a right adjoint

$$\underline{Hom}_A(S^1_A, -) : A - Mod \longrightarrow A - Mod.$$

Furthermore, assumption 1.1.0.2 implies that $S^1_A \otimes_A -$ is a left Quillen functor. We can therefore apply the general construction of [**Ho2**] in order to produce a model category $Sp^{S^1_A}(A - Mod)$, of spectra in $A - Mod$ with respect to the left Quillen endofunctor $S^1_A \otimes_A -$.

DEFINITION 1.2.11.1. *Let $A \in Comm(\mathcal{C})$ be a commutative monoid in $\mathcal{C}$. The model category of* stable $A$-modules *is the model category $Sp^{S^1_A}(A - Mod)$, of spectra in $A - Mod$ with respect to the left Quillen endo-functor*

$$S^1_A \otimes_A - : A - Mod \longrightarrow A - Mod.$$

*It will simply be denoted by $Sp(A - Mod)$, and its objects will be called* stable $A$-modules.

Recall that objects in the category $Sp(A - Mod)$ are families of objects $M_n \in A - Mod$ for $n \geq 0$, together with morphisms $\sigma_n : S^1_A \otimes_A M_n \longrightarrow M_{n+1}$. Morphisms in $Sp(A - Mod)$ are simply families of morphisms $f_n : M_n \to N_n$ commuting with the morphisms $\sigma_n$. One starts by endowing $Sp(A - Mod)$ with the levelwise model structure, for which equivalences (resp. fibrations) are the morphisms $f : M_* \longrightarrow N_*$ such that each morphism $f_n : M_n \longrightarrow N_n$ is an equivalence in $A - Mod$ (resp. a fibration). The definitive model structure, called the stable model structure, is the left Bousfield localization of $Sp(A - Mod)$ whose local objects are the stable modules $M_* \in Sp(A - Mod)$ such that each induced morphism

$$M_n \longrightarrow \mathbb{R}\underline{Hom}_A(S^1_A, M_{n+1})$$

is an isomorphism in $\mathrm{Ho}(A - Mod)$. These local objects will be called *$\Omega$-stable $A$-modules*. We refer to [**Ho2**] for details concerning the existence and the properties of this model structure.

There exists an adjunction

$$S_A : A - Mod \longrightarrow Sp(A - Mod) \qquad A - Mod \longleftarrow Sp(A - Mod) : (-)_0,$$

where the right adjoint sends a stable $A$-module $M_*$ to $M_0$. The left adjoint is defined by $S_A(M)_n := (S^1_A)^{\otimes_A n} \otimes_A M$, with the natural transition morphisms. This adjunction is a Quillen adjunction, and can be derived into an adjunction on the level of homotopy categories

$$\mathbb{L}S_A \simeq S_A : \mathrm{Ho}(A - Mod) \longrightarrow \mathrm{Ho}(Sp(A - Mod))$$

$$\mathrm{Ho}(A - Mod) \longleftarrow \mathrm{Ho}(Sp(A - Mod)) : \mathbb{R}(-)_0.$$

Note that by 1.1.0.2, $S_A$ preserves equivalences, so $\mathbb{L}S_A \simeq S_A$. On the contrary, the functor $(-)_0$ does not preserve equivalences and must be derived on the right. In particular, the functor $S_A : \mathrm{Ho}(A - Mod) \longrightarrow \mathrm{Ho}(Sp(A - Mod))$ is not fully faithful in general.

LEMMA 1.2.11.2. (1) *Assume that the suspension functor*

$$S : \mathrm{Ho}(\mathcal{C}) \longrightarrow \mathrm{Ho}(\mathcal{C})$$

*is fully faithful. Then, for any commutative monoid $A \in Comm(\mathcal{C})$, the functor*

$$S_A : \mathrm{Ho}(A - Mod) \longrightarrow \mathrm{Ho}(Sp(A - Mod))$$

*is fully faithful.*

(2) *If furthermore, $\mathcal{C}$ is a stable model category then $S_A$ is a Quillen equivalence.*

PROOF. (1) As the adjunction morphism $M \longrightarrow S_A(M)_0$ is always an isomorphism in $A - Mod$, it is enough to show that for any $M \in A - Mod$ the stable $A$-module $S_A(M)$ is a $\Omega$-stable $A$-module. For this, it is enough to show that for any $M \in A - Mod$, the adjunction morphism

$$M \longrightarrow \mathbb{R}\underline{Hom}_A(S^1_A, S^1_A \otimes_A M)$$

is an isomorphism in $\mathrm{Ho}(\mathcal{C})$. But, one has natural isomorphisms in $\mathrm{Ho}(\mathcal{C})$

$$S^1_A \otimes_A M \simeq S^1_{\mathcal{C}} \otimes M \simeq S(M)$$

$$\mathbb{R}\underline{Hom}_A(S^1_A, S^1_A \otimes_A M) \simeq \mathbb{R}\underline{Hom}_1(S^1, S(M)) \simeq \Omega(S(M)).$$

This shows that the above morphism is in fact isomorphic in $\mathrm{Ho}(\mathcal{C})$ to the adjunction morphism

$$M \longrightarrow \Omega(S(M))$$

which is an isomorphism by hypothesis on $\mathcal{C}$.

(2) As $\mathcal{C}$ is a stable model category, the functor $S^1_{\mathcal{C}} \otimes - : \mathcal{C} \longrightarrow \mathcal{C}$ is a Quillen equivalence. This also implies that for any $A \in Comm(\mathcal{C})$, the functor $S^1_A : A-Mod \longrightarrow A-Mod$ is a Quillen equivalence. We know by [**Ho2**] that $S_A : A-Mod \longrightarrow Sp(A-Mod)$ is a Quillen equivalence. □

The Quillen adjunction

$$S_A : A-Mod \longrightarrow Sp(A-Mod) \qquad A-Mod \longleftarrow Sp(A-Mod) : (-)_0,$$

is furthermore functorial in $A$. Indeed, for $A \longrightarrow B$ a morphism in $Comm(\mathcal{C})$, one defines a functor

$$- \otimes_A B : Sp(A-Mod) \longrightarrow Sp(B-Mod)$$

defined by

$$(M_* \otimes_A B)_n := M_n \otimes_A B.$$

The transitions morphisms are given by

$$S^1_B \otimes_B (M_n \otimes_A B) \simeq (S^1_A \otimes_A M_n) \otimes_A B \longrightarrow M_{n+1} \otimes_A B.$$

Clearly, the square of left Quillen functors

$$\begin{array}{ccc} A-Mod & \xrightarrow{S_A} & Sp(A-Mod) \\ {\scriptstyle -\otimes_A B}\downarrow & & \downarrow{\scriptstyle -\otimes_A B} \\ B-Mod & \xrightarrow[S_B]{} & Sp(B-Mod) \end{array}$$

commutes up to a natural isomorphism. So does the square of right Quillen functors

$$\begin{array}{ccc} B-Mod & \longrightarrow & Sp(B-Mod) \\ \downarrow & & \downarrow \\ A-Mod & \longrightarrow & Sp(A-Mod). \end{array}$$

Finally, using techniques of symmetric spectra, as done in [**Ho2**], it is possible to show that the homotopy category of stable $A$-modules inherits from $\mathrm{Ho}(A-Mod)$ a symmetric monoidal structure, still denoted by $- \otimes^{\mathbb{L}}_A -$. This makes the homotopy category $\mathrm{Ho}(Sp(A-Mod))$ into a closed symmetric monoidal category. In particular, for two stable $A$-modules $M_*$ and $N_*$ one can define a stable $A$-modules of morphisms

$$\mathbb{R}\underline{Hom}^{Sp}_A(M_*, N_*) \in \mathrm{Ho}(Sp(A-Mod)).$$

We now consider the category $(A-Mod_0^{op})^\wedge$ of pre-stacks over $A-Mod_0^{op}$, as defined in [**HAGI**]. Recall it is the category of $\mathbb{V}$-simplicial presheaves on $A-Mod_0^{op}$, and that its model structure is obtained from the projective levelwise model structure by a left Bousfield localization inverting the equivalences in $A-Mod$. The homotopy category $\mathrm{Ho}((A - Mod_0^{op})^{\wedge})$ can be naturally identified with the full subcategory of $\mathrm{Ho}(SPr(A - Mod_0^{op}))$ consisting of functors

$$F : A - Mod_0 \longrightarrow SSet_{\mathbb{V}},$$

sending equivalences of $A$-modules to equivalences of simplicial sets.

We define a functor

$$\begin{array}{cccc} \underline{h}_s^- : & Sp(A - Mod)^{op} & \longrightarrow & (A - Mod_0^{op})^{\wedge} \\ & M_* & \mapsto & \underline{h}_s^{M_*} \end{array}$$

by

$$\begin{array}{cccc} \underline{h}_s^{M_*} : & A - Mod_0 & \longrightarrow & SSet_{\mathbb{V}} \\ & N & \mapsto & Hom(M_*, \Gamma_*(S_A(N))), \end{array}$$

where $\Gamma_*$ is a simplicial resolution functor on the model category $Sp(A - Mod)$.

Finally we will need some terminology. A stable $A$-module $M_* \in \mathrm{Ho}(Sp(A - Mod))$ is called *0-connective*, if it is isomorphic to some $S_A(M)$ for an $A$-module $M \in \mathrm{Ho}(A - Mod_0)$. By induction, for an integer $n > 0$, a stable $A$-module $M_* \in \mathrm{Ho}(Sp(A - Mod))$ is called *$(-n)$-connective*, if it is isomorphic to $\Omega(M'_*)$ for some $-(n-1)$-connective stable $A$-module $M'_*$ (here $\Omega$ is the loop functor on $\mathrm{Ho}(Sp(A - Mod)))$. Note that if the suspension functor is fully faithful, connective stable modules are exactly connective objects with respect to the natural $t$-structure on $A$-modules.

PROPOSITION 1.2.11.3. *For any $A \in Comm(\mathcal{C})$, the functor $\underline{h}_s^-$ has a total right derived functor*

$$\mathbb{R}\underline{h}_s^- : \mathrm{Ho}(Sp(A - Mod))^{op} \longrightarrow \mathrm{Ho}((A - Mod_0^{op})^{\wedge}),$$

*which commutes with homotopy limits* [1]. *If the suspension functor*

$$S : \mathrm{Ho}(\mathcal{C}) \longrightarrow \mathrm{Ho}(\mathcal{C})$$

*is fully faithful, and if $A \in \mathcal{A}$, then for any integer $n \geq 0$, the functor $\mathbb{R}\underline{h}_s^-$ is fully faithful when restricted to the full subcategory of $(-n)$-connective objects.*

PROOF. As $S_A : A - Mod_0 \longrightarrow Sp(A - Mod)$ preserves equivalences, one checks easily that $\underline{h}_s^{M_*}$ is a fibrant object in $(A - Mod_0^{op})^{\wedge}$ when $M_*$ is cofibrant in $Sp(A - Mod)$. This easily implies that $M_* \mapsto \underline{h}_s^{QM_*}$ is a right derived functor for $\underline{h}_s^-$, and the standard properties of mapping spaces imply that it commutes with homotopy limits.

We now assume that the suspension functor

$$S : \mathrm{Ho}(\mathcal{C}) \longrightarrow \mathrm{Ho}(\mathcal{C})$$

is fully faithful, and that $A \in \mathcal{A}$. Let $n \geq 0$ be an integer.

Let $S^n : A - Mod \longrightarrow A - Mod$ be a left Quillen functor which is a model for the suspension functor iterated $n$ times (e.g. $S^n(N) := S_A^n \otimes_A N$, where $S_A^n := (S_A^1)^{\otimes_A n}$). There is a pullback functor

$$(S^n)^* : \mathrm{Ho}((A - Mod_0^{op})^{\wedge}) \longrightarrow \mathrm{Ho}((A - Mod_0^{op})^{\wedge})$$

defined by $(S^n)^*(F)(N) := F(S^n(N))$ for any $N \in A - Mod_0$ (note that $S^n$ stablizes the subcategory $A - Mod_0$ because of our assumption 1.1.0.6). For an $(-n)$-connective object $M_*$, we claim there exists a natural isomorphism in $\mathrm{Ho}((A - Mod_0^{op})^{\wedge})$ between

[1] This makes sense as the functor $\mathbb{R}\underline{h}_s^-$ is naturally defined on the level of the Dwyer-Kan simplicial localizations with respect to equivalences

$$LSp(A - Mod)^{op} \longrightarrow L((A - Mod_0^{op})^{\wedge}).$$

$\mathbb{R}\underline{h}_s^{M_*}$ and $(S^n)^*(\mathbb{R}\underline{h}^M)$, where $\mathbb{R}\underline{h}^M$ is the value at $M$ of the restricted Yoneda embedding

$$\mathbb{R}\underline{h}_0^M : \mathrm{Ho}(A - Mod) \longrightarrow \mathrm{Ho}((A - Mod_0^{op})^\wedge).$$

Indeed, let us write $M$ as $\Omega^n(S_A(M))$ for some object $M \in \mathrm{Ho}(A - Mod)$, where $\Omega^n$ is the loop functor of $Sp(A - Mod)$, iterated $n$ times. Then, using our lemma Lem. 1.2.11.2, for any $N \in \mathrm{Ho}(A - Mod)$, we have natural isomorphisms in $\mathrm{Ho}(SSet)$

$$Map_{Sp(A-Mod)}(M_*, S_A(N)) \simeq Map_{Sp(A-Mod)}(S_A(M), S^n(S_A(N))) \simeq$$

$$\simeq Map_{A-Mod}(M, S^n(N)),$$

where $S^n$ denotes the suspension functor iterated $n$-times. Using that $A$ is $\mathcal{C}_0$-good, this shows that

$$\mathbb{R}\underline{h}_s^{M_*} \simeq (S^n)^*(\mathbb{R}\underline{h}_0^M).$$

Moreover, for any $(-n)$-connective objects $M_*$ and $N_*$ in $\mathrm{Ho}(Sp(A - Mod))$ one has natural isomorphisms in $\mathrm{Ho}(SSet)$

$$Map_{Sp(A-Mod)}(M_*, N_*) \simeq Map_{A-Mod}(M, N)$$

where $M_* \simeq \Omega^n(S_A(M))$ and $N_* \simeq \Omega^n(S_A(N))$. We are therefore reduced to show that for any $A$-modules $M$ and $N$ in $\mathrm{Ho}(A-Mod)$, the natural morphism in $\mathrm{Ho}(SSet)$

$$Map_{A-Mod}(M, N) \longrightarrow Map_{(A-Mod_0^{op})^\wedge}((S^n)^*(\mathbb{R}\underline{h}_0^N), (S^n)^*(\mathbb{R}\underline{h}_0^M))$$

is an isomorphism. To see this, we define a morphism in the opposite direction in the following way. Taking the $n$-th loop functor on each side gives a morphism

$$Map_{(A-Mod_0^{op})^\wedge}((S^n)^*(\mathbb{R}\underline{h}_0^N), (S^n)^*(\mathbb{R}\underline{h}_0^M)) \rightarrow$$

$$\longrightarrow Map_{(A-Mod_0^{op})^\wedge}(\Omega^n(S^n)^*(\mathbb{R}\underline{h}_0^N), \Omega^n(S^n)^*(\mathbb{R}\underline{h}_0^M).$$

Moreover, there are isomorphisms

$$\Omega^n Map_{A-Mod}(M, S^n(P)) \simeq Map_{A-Mod}(M, \Omega^n(S^n(P))) \simeq Map_{A-Mod}(M, P) \simeq \mathbb{R}\underline{h}_0^M(P),$$

showing that there exists a natural isomorphism in $\mathrm{Ho}((A - Mod_0^{op})^\wedge)$ between $\Omega^n(S^n)^*(\mathbb{R}\underline{h}_0^M)$ and $\mathbb{R}\underline{h}_0^M$. One therefore gets a morphism

$$Map_{(A-Mod_0^{op})^\wedge}((S^n)^*(\mathbb{R}\underline{h}_0^N), (S^n)^*(\mathbb{R}\underline{h}_0^M)) \rightarrow$$

$$\longrightarrow Map_{(A-Mod_0^{op})^\wedge}(\Omega^n(S^n)^*(\mathbb{R}\underline{h}_0^N), \Omega^n(S^n)^*(\mathbb{R}\underline{h}_0^M) \simeq$$

$$\simeq Map_{(A-Mod_0^{op})^\wedge}(\mathbb{R}\underline{h}_0^N, \mathbb{R}\underline{h}_0^M).$$

Using that $A$ is $\mathcal{C}_0$-good we get the required morphism

$$Map_{(A-Mod_0^{op})^\wedge}((S^n)^*(\mathbb{R}\underline{h}_0^N), (S^n)^*(\mathbb{R}\underline{h}_0^M)) \longrightarrow Map_{A-Mod}(M, N),$$

and it is easy to check it is an inverse in $\mathrm{Ho}(SSet)$ to the natural morphism

$$Map_{A-Mod}(M, N) \longrightarrow Map_{(A-Mod_0^{op})^\wedge}((S^n)^*(\mathbb{R}\underline{h}_0^N), (S^n)^*(\mathbb{R}\underline{h}_0^M)).$$

□

### 1.2.12. Descent for modules and stable modules

In this last section we present some definitions concerning descent for modules and stable modules in our general context. In a few words, a co-augmented co-simplicial object $A \longrightarrow B_*$ in $Comm(\mathcal{C})$, is said to *have the descent property for modules* (resp. *for stable modules*) if the homotopy theory of $A$-modules (resp. of stable $A$-modules) is equivalent to the homotopy theory of certain co-simplicial $B_*$-modules (resp. stable $B_*$-modules). From the geometric, dual point of view, the object $A \longrightarrow B_*$ should be thought as an augmented simplicial space $Y_* \longrightarrow X$, and having the descent property essentially means that the theory of sheaves on $X$ is equivalent to the theory of certain sheaves on $Y_*$ (see [**SGA4-II**, Exp. $V^{bis}$] for more details on this point of view).

The purpose of this section is only to introduce the basic set-up for descent that will be used later to state that the theory of quasi-coherent modules is local with respect to the topology, or in other words that hypercovers have the descent property for modules. This is a very important property allowing local-to-global arguments.

Let $A_*$ be a co-simplicial object in the category $Comm(\mathcal{C})$ of commutative monoids. Therefore, $A_*$ is given by a functor

$$\begin{array}{ccc} \Delta & \longrightarrow & Comm(\mathcal{C}) \\ [n] & \mapsto & A_n. \end{array}$$

A co-simplicial $A_*$-module $M_*$ is by definition the following datum.

- A $A_n$-module $M_n \in A_n - Mod$ for any $n \in \Delta$.
- For any morphism $u : [n] \to [m]$ in $\Delta$, a morphism of $A_n$-modules $\alpha_u : M_n \longrightarrow M_m$, such that $\alpha_v \circ \alpha_u = \alpha_{v \circ u}$ for any $[n] \xrightarrow{u} [m] \xrightarrow{v} [p]$ in $\Delta$.

In the same way, a morphism of co-simplicial $A_*$-modules $f : M_* \longrightarrow N_*$ is the data of morphisms $f_n : M_n \longrightarrow N_n$ for any $n$, commuting with the $\alpha$'s

$$\alpha_u^N \circ f_n = f_m \circ \alpha_u^M$$

for any $u : [n] \to [m]$.

The co-simplicial $A_*$-modules and morphisms of $A_*$-modules form a category, denoted by $csA_* - Mod$. It is furthermore a $\mathbb{U}$-combinatorial model category for which the equivalences (resp. fibrations) are the morphisms $f : M_* \longrightarrow N_*$ such that each $f_n : M_n \longrightarrow N_n$ is an equivalence (resp. a fibration) in $A_n - Mod$.

Let $A$ be a commutative monoid, and $B_*$ be a co-simplicial commutative $A$-algebra. We can also consider $B_*$ as a co-simplicial commutative monoid together with a co-augmentation

$$A \longrightarrow B_*,$$

which is a morphism of co-simplicial objects when $A$ is considered as a constant co-simplicial object. As a co-simplicial commutative monoid $A$ possesses a category of co-simplicial $A_*$-modules $csA - Mod$ which is nothing else than the model category of co-simplicial objects in $A - Mod$ with its projective levelwise model structure.

For a co-simplicial $A$-module $M_*$, we define a co-simplicial $B_*$-module $B_* \otimes_A M_*$ by the formula

$$(B_* \otimes_A M_*) := B_n \otimes_A M_n,$$

and for which the transitions morphisms are given by the one of $B_*$ and of $M_*$. This construction defines a functor

$$B_* \otimes_A - : csA - Mod \longrightarrow csB_* - Mod,$$

which has a right adjoint

$$csB_* - Mod \longrightarrow csA - Mod.$$

This right adjoint, sends a $B_*$-module $M_*$ to its underlying co-simplicial $A$-module. Clearly, this defines a Quillen adjunction. There exists another adjunction

$$ct : \mathrm{Ho}(A-Mod) \longrightarrow \mathrm{Ho}(csA-Mod) \qquad \mathrm{Ho}(A-Mod) \longleftarrow \mathrm{Ho}(csA-Mod) : Holim,$$

where $Holim$ is defined as the total right derived functor of the functor lim (see [**Hi**, 8.5], [**DS**, 10.13]).

Composing these two adjunctions gives a new adjunction

$$B_*\otimes^{\mathbb{L}}_A - : \mathrm{Ho}(A-Mod) \longrightarrow \mathrm{Ho}(csB_*-Mod) \qquad \mathrm{Ho}(A-Mod) \longleftarrow \mathrm{Ho}(csB_*-Mod) : \int.$$

DEFINITION 1.2.12.1. (1) *Let $B_*$ be a co-simplicial commutative monoid and $M_*$ be a co-simplicial $B_*$-module. We say that $M_*$* is homotopy cartesian *if for any $u : [n] \to [m]$ in $\Delta$ the morphism, induced by $\alpha_u$,*

$$M_n \otimes^{\mathbb{L}}_{B_n} B_m \longrightarrow M_m$$

*is an isomorphism in* $\mathrm{Ho}(B_m - Mod)$.

(2) *Let $A$ be a commutative monoid and $B_*$ be a co-simplicial commutative $A$-algebra. We say that the co-augmentation morphism $A \longrightarrow B_*$* satisfies the descent condition, *if in the adjunction*

$$B_*\otimes^{\mathbb{L}}_A - : \mathrm{Ho}(A-Mod) \longrightarrow \mathrm{Ho}(csB_*-Mod) \qquad \mathrm{Ho}(A-Mod) \longleftarrow \mathrm{Ho}(csB_*-Mod) : \int$$

*the functor $B_* \otimes^{\mathbb{L}}_A -$ is fully faithful and induces an equivalence between* $\mathrm{Ho}(A - Mod)$ *and the full subcategory of* $\mathrm{Ho}(csB_* - Mod)$ *consisting of homotopy cartesian objects.*

REMARK 1.2.12.2. If a morphism $A \longrightarrow B_*$ satisfies the descent condition, then so does any co-simplicial object equivalent to it (as a morphism in the model category of co-simplicial commutative monoids).

LEMMA 1.2.12.3. *Let $A \longrightarrow B_*$ be a co-augmented co-simplicial commutative monoid in $Comm(\mathcal{C})$ which satisfies the descent condition. Then, the natural morphism*

$$A \longrightarrow Holim_n B_n$$

*is an isomorphism in* $\mathrm{Ho}(Comm(\mathcal{C}))$.

PROOF. This is clear as $\int B_* = Holim_n B_n$. □

We now pass to descent for stable modules. For this, let again $A_*$ be a co-simplicial object in the category $Comm(\mathcal{C})$ of commutative monoids. A co-simplicial stable $A_*$-module $M_*$ is by definition the following datum.

- A stable $A_n$-module $M_n \in Sp(A_n - Mod)$ for any $n \in \Delta$.
- For any morphism $u : [n] \to [m]$ in $\Delta$, a morphism of stable $A_n$-modules $\alpha_u : M_n \longrightarrow M_m$, such that $\alpha_v \circ \alpha_u = \alpha_{v \circ u}$ for any $[n] \xrightarrow{u} [m] \xrightarrow{v} [p]$ in $\Delta$.

In the same way, a morphism of co-simplicial stable $A_*$-modules $f : M_* \longrightarrow N_*$ is the data of morphisms $f_n : M_n \longrightarrow N_n$ in $Sp(A_n - Mod)$ for any $n$, commuting with the $\alpha$'s

$$\alpha^N_u \circ f_n = f_m \circ \alpha^M_u$$

for any $u : [n] \to [m]$.

The co-simplicial stable $A_*$-modules and morphisms of $A_*$-modules form a category, denoted by $Sp(csA_*-Mod)$. It is furthermore a $\mathbb{U}$-combinatorial model category for which the equivalences (resp. fibrations) are the morphisms $f : M_* \longrightarrow N_*$ such that each $f_n : M_n \longrightarrow N_n$ is an equivalence (resp. a fibration) in $Sp(A_n - Mod)$. We note that $Sp(csA_* - Mod)$ is also naturally equivalent as a category to the category of $S^1_A$-spectra in $csA_*Mod$, hence the notation $Sp(csA_*-Mod)$ is not ambiguous.

Let $A$ be a commutative monoid, and $B_*$ be a co-simplicial commutative $A$-algebra. We can also consider $B_*$ as a co-simplicial commutative monoid together with a co-augmentation

$$A \longrightarrow B_*,$$

which is a morphism of co-simplicial objects when $A$ is considered as a constant co-simplicial object. As a co-simplicial commutative monoid $A$ possesses a category of co-simplicial stable $A_*$-modules $Sp(csA - Mod)$ which is nothing else than the model category of co-simplicial objects in $Sp(A - Mod)$ with its projective levelwise model structure.

For a co-simplicial stable $A$-module $M_*$, we define a co-simplicial stable $B_*$-module $B_* \otimes_A M_*$ by the formula

$$(B_* \otimes_A M_*) := B_n \otimes_A M_n,$$

and for which the transitions morphisms are given by the one of $B_*$ and of $M_*$. This construction defines a functor

$$B_* \otimes_A - : Sp(csA - Mod) \longrightarrow Sp(csB_* - Mod),$$

which has a right adjoint

$$Sp(csB_* - Mod) \longrightarrow Sp(csA - Mod).$$

This right adjoint, sends a co-simplicial stable $B_*$-module $M_*$ to its underlying co-simplicial stable $A$-module. Clearly, this is a Quillen adjunction. One also has an adjunction

$$ct : \mathrm{Ho}(Sp(A - Mod)) \longrightarrow \mathrm{Ho}(Sp(csA - Mod))$$
$$\mathrm{Ho}(Sp(A - Mod)) \longleftarrow \mathrm{Ho}(Sp(csA - Mod)) : Holim,$$

where $Holim$ is defined as the total right derived functor of the functor lim (see [**Hi**, 8.5], [**DS**, 10.13]).

Composing these two adjunctions gives a new adjunction

$$B_* \otimes^{\mathbb{L}}_A - : \mathrm{Ho}(Sp(A - Mod)) \longrightarrow \mathrm{Ho}(Sp(csB_* - Mod))$$
$$\mathrm{Ho}(Sp(A - Mod)) \longleftarrow \mathrm{Ho}(Sp(csB_* - Mod)) : \int .$$

DEFINITION 1.2.12.4. (1) *Let $B_*$ be a co-simplicial commutative monoid and $M_*$ be a co-simplicial stable $B_*$-module. We say that $M_*$* is homotopy cartesian *if for any $u : [n] \to [m]$ in $\Delta$ the morphism, induced by $\alpha_u$,*

$$M_n \otimes^{\mathbb{L}}_{B_n} B_m \longrightarrow M_m$$

*is an isomorphism in* $\mathrm{Ho}(Sp(B_m - Mod))$.

(2) *Let $A$ be a commutative monoid and $B_*$ be a co-simplicial commutative $A$-algebra. We say that the co-augmentation morphism $A \longrightarrow B_*$* satisfies the stable descent condition, *if in the adjunction*

$$B_* \otimes^{\mathbb{L}}_A - : \mathrm{Ho}(Sp(A - Mod)) \longrightarrow \mathrm{Ho}(Sp(csB_* - Mod))$$
$$\mathrm{Ho}(Sp(A - Mod)) \longleftarrow \mathrm{Ho}(Sp(csB_* - Mod)) : \int$$

*the functor $B_* \otimes^{\mathbb{L}}_A -$ is fully faithful and induces an equivalence between* $\mathrm{Ho}(Sp(A-Mod))$ *and the full subcategory of* $\mathrm{Ho}(Sp(csB_*-Mod))$ *consisting of homotopy cartesian objects.*

The following proposition insures that the unstable descent condition implies the stable one under certain conditions.

PROPOSITION 1.2.12.5. *Let $A$ be a commutative monoid and $B_*$ be a co-simplicial commutative $A$-algebra, such that $A \longrightarrow B_*$ satisfies the descent condition. Assume that the two following conditions are satisfied.*

(1) *The suspension functor $S : \mathrm{Ho}(\mathcal{C}) \longrightarrow \mathrm{Ho}(\mathcal{C})$ is fully faithful.*
(2) *For any $n$, the morphism $A \longrightarrow B_n$ is flat in the sense of Def. 1.2.4.1.*

*Then $A \longrightarrow B_*$ satisfies the stable descent property.*

PROOF. The second condition insures that the diagram

$$\begin{array}{ccc} \mathrm{Ho}(Sp(A-Mod)) & \longrightarrow & \mathrm{Ho}(Sp(csB_*-Mod) \\ \downarrow & & \downarrow \\ \mathrm{Ho}(A-Mod) & \longrightarrow & \mathrm{Ho}(csB_*-Mod) \end{array}$$

commutes up to a natural isomorphism (here the vertical functors are the right adjoint to the suspensions inclusion functors, and send a spectrum to its 0-th level). So, for any $M \in Sp(A-Mod)$, the adjunction morphism

$$M \longrightarrow \int (B_* \otimes^{\mathbb{L}} M)$$

is easily seen to induces an isomorphism on the 0-th level objects in $\mathrm{Ho}(A-Mod)$. As this is true for any $M$, and in particular for the suspensions of $M$, we find that this adjunction morphism induces isomorphisms on each $n$-th level objects in $\mathrm{Ho}(A-Mod)$. Therefore it is an isomorphism. In the same way, one proves that for any cartesian object $M_* \in \mathrm{Ho}(Sp(csB_*-Mod))$ the adjunction morphism

$$B_* \otimes_A \int (M_*) \longrightarrow M_*$$

is an isomorphism. □

## 1.2.13. Comparison with the usual notions

In this last section we present what the general notions introduced before give, when $\mathcal{C}$ is the model category of $k$-modules with the trivial model structure. The other non trivial examples will be given at the beginning of the various chapters §2.2, §2.3 and §2.4 where the case of simplicial modules, complexes and symmetric spectra will be studied.

Let $k$ be a commutative ring in $\mathbb{U}$, and we let $\mathcal{C}$ be the category of $k$-modules belonging to the universe $\mathbb{U}$, endowed with its trivial model structure for which equivalences are isomorphisms and all morphisms are fibrations and cofibrations. The category $\mathcal{C}$ is then a symmetric monoidal model category for the tensor product $- \otimes_k -$. Furthermore, all of our assumption 1.1.0.1, 1.1.0.3, 1.1.0.2 and 1.1.0.4 are satisfied. The category $Comm(\mathcal{C})$ is of course the category of commutative $k$-algebras in $\mathbb{U}$, endowed with its trivial model structure. In the same way, for any commutative $k$-algebra $A$, the category $A-Mod$ is the usual category of $A$-modules in $\mathbb{U}$ together with its trivial model structure. We set $\mathcal{C}_0 = \mathcal{C}$ and $\mathcal{A} = Comm(\mathcal{C})$. The notions we have presented before restrict essentially to the usual notions of algebraic geometry with some remarkable caveats.

- For any morphism of commutative $k$-algebras $A \longrightarrow B$ and any $B$-module $M$, the simplicial set $\mathbb{D}er_A(B, M)$ is discrete and naturally isomorphic to the set $Der_A(B, M)$ of derivations from $B$ to $M$ over $A$. Equivalently, the $B$-module $\mathbb{L}_{B/A}$ (defined in Def. 1.2.1.5) is simply the usual $B$-module of Kähler differentials $\Omega^1_{B/A}$. Note that this $\mathbb{L}_{B/A}$ is *not* the Quillen-Illusie cotangent complex of $A \to B$.
- For any morphism of commutative $k$-algebras $A \longrightarrow B$ the natural morphism $B \longrightarrow THH(B/A)$ is always an isomorphism. Indeed

$$THH(A) \simeq A \otimes_{A \otimes_k A} A \simeq A.$$

- A morphism of commutative $k$-algebras $A \longrightarrow B$ is finitely presented in the sense of Def. 1.2.3.1 if and only if it $B$ is a finitely presented $A$-algebra in the usual sense. In the same way, finitely presented objects in $A - Mod$ in the sense of Def. 1.2.3.1 are the finitely presented $A$-modules. Also, perfect objects in $A - Mod$ are the projective $A$-modules of finite type.
- A morphism $A \longrightarrow B$ of commutative $k$-algebras is a formal covering if and only if it is a faithful morphism of rings.
- Let $f : A \longrightarrow B$ be a morphism of commutative $k$-algebras, and $Spec\, B \longrightarrow Spec\, A$ the associated morphism of schemes. We have the following comparison board.

| **In the sense of Def. 1.2.6.1, 1.2.6.7, 1.2.7.1** | **As a morphism of affine schemes** |
| --- | --- |
| *epimorphism* | *monomorphism* |
| *flat* | *flat* |
| *formal Zariski open immersion* | *flat monomorphism* |
| *Zariski open immersion* | *open immersion* |
| *(formally) unramified* | *(formally) unramified* |
| *formally thh − etale* | *always satisfied* |
| *formally i − smooth* | *always satisfied* |
| *thh − etale* | *finitely presented* |

The reader will immediately notice the absence of (formally) smooth and (formally) étale maps in the previous table. This is essentially due to the fact that there are no easy *general* characterizations of this maps in terms of the module of Kähler differentials alone (which in this trivial model structure context is our cotangent complex). On the contrary such characterizations do exist in terms of the "correct" cotangent complex which is the Quillen-Illusie one. But this correct cotangent complex of a morphism of usual commutative $k$-algebras $A \to B$ will appear as the cotangent complex according to our definition 1.2.1.5 only if we consider this morphism in the category of simplicial $k$-modules, i.e. if we replace the category $\mathcal{C}$ of $k$-modules with the category of simplicial $k$-modules. In other words our definitions of (formally) smooth and (formally) étale maps reduce to the usual ones (e.g. [**EGAIV**]) between commutative $k$-algebras only if we consider them in the context of *derived algebraic geometry* (see §2.2). This is consistent with the general philosophy that some aspects of usual algebraic geometry, especially those related to infinitesimal lifting properties and deformation theory, are conceptually more transparent in (and actually already a part of) derived algebraic geometry. See also Remark 2.1.2.2 for another instance of this point of view.

- For commutative monoid $A$, one has $\pi_*(A) = A$, and for any $a \in A$ the commutative $A$-algebra $A \longrightarrow A[a^{-1}]$ is the usual localization of $A$ inverting $a$.
- One has that $A_K \simeq 0$ for any projective $A$-module $K$ of constant finite rank $n > 0$ (for example if the scheme $Spec\, A$ is connected and $K$ is non zero).

CHAPTER 1.3

# Geometric stacks: Basic theory

In this chapter, after a brief reminder of [**HAGI**], we present the key definition of this work, the notion of *n-geometric stack*. The definition we present here is a generalization of the original notion of *geometric n-stack* introduced by C. Simpson in [**S3**]. As already remarked in [**S3**], the notion of $n$-geometric stack only depends on a topology on the opposite category of commutative monoids $Comm(\mathcal{C})^{op}$, and on a class $\mathbf{P}$ of morphisms. Roughly speaking, geomeric stacks are the stacks obtained by taking quotient of representable stacks by some equivalence relations in $P$. By choosing different classes $\mathbf{P}$ one gets different notions of geometric stacks. For example, in the classical situation where $\mathcal{C} = \mathbb{Z} - Mod$, and the topology is chosen to be the étale topology, Deligne-Mumford algebraic stacks correspond to the case where $\mathbf{P}$ is the class of étale morphisms, whereas Artin algebraic stacks correspond to the case where $\mathbf{P}$ is the class of smooth morphisms. We think it is important to leave the choice of the class $\mathbf{P}$ open in the general definition, so that it can be specialized differently depending of the kind of objects one is willing to consider.

From the second section on, we will fix a HA context $(\mathcal{C}, \mathcal{C}_0, \mathcal{A})$, in the sense of Def. 1.1.0.11.

## 1.3.1. Reminders on model topoi

To make the paper essentially self-contained, we briefly summarize in this subsection the basic notions and results of the theory of stacks in homotopical contexts as exposed in [**HAGI**]. We will limit ourselves to recall only the topics that will be needed in the sequel; the reader is addressed to [**HAGI**] for further details and for proofs.

Let $M$ be a $\mathbb{U}$-small model category, and $W(M)$ its class of weak equivalences. We let $SPr(M) := SSets^{M^{op}}$ be the category of simplicial presheaves on $M$ with its projective model structure, i.e. with equivalences and fibrations defined objectwise.

The model category of *prestacks* $M^{\wedge}$ on $M$ is the model category obtained as the left Bousfield localization of $SPr(M)$ at $\{h_u \mid u \in W(M)\}$, where $h : M \longrightarrow Pr(M) \hookrightarrow SPr(M)$ is the (constant) Yoneda embedding. The homotopy category $\mathrm{Ho}(M^{\wedge})$ can be identified with the full subcategory of $\mathrm{Ho}(SPr(M))$ consisting of those simplicial presheaves $F$ on $M$ that preserve weak equivalences ([**HAGI**, Def. 4.1.4]); any such simplicial presheaf (i.e. any object in $\mathrm{Ho}(M^{\wedge})$) will be called a *prestack on M*. $M^{\wedge}$ is a $\mathbb{U}$-cellular and $\mathbb{U}$-combinatorial ([**HAGI**, App. A]) simplicial model category and its derived simplicial $Hom$'s will be denoted simply by $\mathbb{R}\underline{Hom}$ (denoted as $\mathbb{R}_w\underline{Hom}$ in [**HAGI**, §4.1]).

If $\Gamma_* : M \longrightarrow M^{\Delta}$ is a cofibrant resolution functor for $M$ ([**Hi**, 16.1]), and we define

$$\underline{h} : M \to M^{\wedge} \ : \ x \longmapsto (\underline{h}_x : y \mapsto Hom_M(\Gamma_*(y), x)),$$

we have that $\underline{h}$ preserves fibrant objects and weak equivalences between fibrant objects ([**HAGI**, Lem. 4.2.1]). Therefore we can right-derive $\underline{h}$ to get a functor $\mathbb{R}\underline{h} := \underline{h} \circ R$ : $\mathrm{Ho}(M) \to \mathrm{Ho}(M^\wedge)$, where $R$ is a fibrant replacement functor in $M$; $\mathbb{R}\underline{h}$ is in fact fully faithful ([**HAGI**, Thm. 4.2.3]) and is therefore called the *(model) Yoneda embedding* for the model category $M$ ($\mathbb{R}\underline{h}$, as opposed to $\underline{h}$, does not depend, up to a unique isomorphism, on the choice of the cofibrant resolution functor $\Gamma_*$).

We also recall that the canonical morphism $h_x \to \mathbb{R}\underline{h}_x$ is always an isomorphism in $\mathrm{Ho}(M^\wedge)$ ([**HAGI**, Lem. 4.2.2]), and that with the notations introduced above for the derived simplicial *Hom*'s in $M^\wedge$, the model Yoneda lemma ([**HAGI**, Cor. 4.2.4]) is expressed by the isomorphisms in $\mathrm{Ho}(SSets)$

$$\mathbb{R}\underline{Hom}(\mathbb{R}\underline{h}_x, F) \simeq \mathbb{R}\underline{Hom}(h_x, F) \simeq F(x)$$

for any fibrant object $F$ in $M^\wedge$.

A convenient homotopical replacement of the notion of a Grothendieck topology in the case of model categories, is the following ([**HAGI**, Def. 4.3.1])

DEFINITION 1.3.1.1. *A* model (pre-)topology $\tau$ *on a* $\mathbb{U}$*-small model category* $M$*, is the datum for any object* $x \in M$*, of a set* $Cov_\tau(x)$ *of subsets of objects in* $\mathrm{Ho}(M)/x$*, called* $\tau$-covering families *of* $x$*, satisfying the following three conditions*

1. (Stability) *For all* $x \in M$ *and any isomorphism* $y \to x$ *in* $\mathrm{Ho}(M)$*, the one-element set* $\{y \to x\}$ *is in* $Cov_\tau(x)$*.*
2. (Composition) *If* $\{u_i \to x\}_{i\in I} \in Cov_\tau(x)$*, and for any* $i \in I$*,* $\{v_{ij} \to u_i\}_{j\in J_i} \in Cov_\tau(u_i)$*, the family* $\{v_{ij} \to x\}_{i\in I, j\in J_i}$ *is in* $Cov_\tau(x)$*.*
3. (Homotopy base change) *Assume the two previous conditions hold. For any* $\{u_i \to x\}_{i\in I} \in Cov_\tau(x)$*, and any morphism in* $\mathrm{Ho}(M)$*,* $y \to x$*, the family* $\{u_i \times^h_x y \to y\}_{i\in I}$ *is in* $Cov_\tau(y)$*.*

*A* $\mathbb{U}$*-small model category* $M$ *together with a model pre-topology* $\tau$ *will be called a* $\mathbb{U}$*-small* model site.

By [**HAGI**, Prop. 4.3.5] a model pre-topology $\tau$ on $M$ induces and is essentially the same thing as a Grothendieck topology, still denoted by $\tau$, on the homotopy category $\mathrm{Ho}(M)$.

Given a model site $(M, \tau)$ we have, as in [**HAGI**, Thm. 4.6.1], a model category $M^{\sim,\tau}$ ($\mathbb{U}$-combinatorial and left proper) of *stacks* on the model site, which is defined as the left Bousfield localization of the model category $M^\wedge$ of prestacks on $M$ along a class $H_\tau$ of *homotopy* $\tau$*-hypercovers* ([**HAGI**, 4.4, 4.5]). To any prestack $F$ we can associate a sheaf $\pi_0$ of connected components on the site $(\mathrm{Ho}(M), \tau)$ defined as the associated sheaf to the presheaf $x \longmapsto \pi_0(F(x))$. In a similar way ([**HAGI**, Def. 4.5.3]), for any $i > 0$, any fibrant object $x \in M$, and any $s \in F(x)_0$, we can define a sheaf of homotopy groups $\pi_i(F, s)$ on the induced comma site $(\mathrm{Ho}(M/x), \tau)$. The weak equivalences in $M^{\sim,\tau}$ turn out to be exactly the $\pi_*$-sheaves isomorphisms ([**HAGI**, Thm. 4.6.1]), i.e. those maps $u : F \to G$ in $M^\wedge$ inducing an isomorphism of sheaves $\pi_0(F) \simeq \pi_0(G)$ on $(\mathrm{Ho}(M), \tau)$, and isomorphisms $\pi_i(F, s) \simeq \pi_i(G, u(s))$ of sheaves on $(\mathrm{Ho}(M/x), \tau)$ for any $i \geq 0$, for any choice of fibrant $x \in M$ and any base point $s \in F(x)_0$.

The left Bousfield localization construction defining $M^{\sim,\tau}$ yields a pair of adjoint Quillen functors

$$\mathrm{Id} : M^\wedge \longrightarrow M^{\sim,\tau} \qquad M^\wedge \longleftarrow M^{\sim,\tau} : \mathrm{Id}$$

which induces an adjunction pair at the level of homotopy categories

$$a := \mathbb{L}\mathrm{Id} : \mathrm{Ho}(M^{\wedge}) \longrightarrow \mathrm{Ho}(M^{\sim,\tau}) \qquad \mathrm{Ho}(M^{\wedge}) \longleftarrow \mathrm{Ho}(M^{\sim,\tau}) : j := \mathbb{R}\mathrm{Id}$$

where $j$ is fully faithful.

DEFINITION 1.3.1.2. (1) *A* stack *on the model site* $(M, \tau)$ *is an object* $F \in SPr(M)$ *whose image in* $\mathrm{Ho}(M^{\wedge})$ *is in the essential image of the functor* $j$.

(2) *If* $F$ *and* $G$ *are stacks on the model site* $(M, \tau)$, *a* morphism of stacks *is a morphism* $F \to G$ *in in* $\mathrm{Ho}(SPr(M))$, *or equivalently in* $\mathrm{Ho}(M^{\wedge})$, *or equivalently in* $\mathrm{Ho}(M^{\sim,\tau})$.

(3) *A morphism of stacks* $f : F \longrightarrow G$ *is a* covering *(or a* cover *or an* epimorphism*) if the induced morphism of sheaves*

$$\pi_0(f) : \pi_0(F) \longrightarrow \pi_0(G)$$

*is an epimorphism in the category of sheaves.*

Recall that a simplicial presheaf $F : M^{\mathrm{op}} \to SSet_{\mathbb{V}}$ is a stack if and only if it preserves weak equivalences and satisfy a *τ-hyperdescent* condition (descent, i.e. sheaf-like, condition with respect to the class $H_\tau$ of homotopy hypercovers): see [**HAGI**, Def. 4.6.5 and Cor. 4.6.3]. We will always consider $\mathrm{Ho}(M^{\sim,\tau})$ embedded in $\mathrm{Ho}(M^{\wedge})$ embedded in $\mathrm{Ho}(SPr(M))$, omitting in particular to mention explicitly the functor $j$ above. With this conventions, the functor $a$ above becomes an endofunctor of $\mathrm{Ho}(M^{\wedge})$, called the *associated stack* functor for the model site $(M, \tau)$. The associated stack functor preserves finite homotopy limits and all homotopy colimits ([**HAGI**, Prop. 4.6.7]).

DEFINITION 1.3.1.3. *A model pre-topology* $\tau$ *on* $M$ *is* sub-canonical *if for any* $x \in M$ *the pre-stack* $\mathbb{R}\underline{h}_x$ *is a stack.*

$M^{\sim,\tau}$ is a left proper (but not right proper) simplicial model category and its derived simplicial *Hom*'s will be denoted by $\mathbb{R}_\tau \underline{Hom}$ (denoted by $\mathbb{R}_{w,\tau}\underline{Hom}$ in [**HAGI**, Def. 4.6.6]); for $F$ and $G$ prestacks on $M$, there is always a morphism in $\mathrm{Ho}(SSet)$

$$\mathbb{R}_\tau \underline{Hom}(F, G) \to \mathbb{R}\underline{Hom}(F, G)$$

which is an isomorphism when $G$ is a stack ([**HAGI**, Prop.4.6.7]).
Moreover $M^{\sim,\tau}$ is a *t*-complete *model topos* ([**HAGI**, Def. 3.8.2]) therefore possesses important exactness properties. For the readers' convenience we collect below (from [**HAGI**]) the definition of (t-complete) model topoi and the main theorem characterizing them (Giraud-like theorem).

For a $\mathbb{U}$-combinatorial model category $N$, and a $\mathbb{U}$-small set $S$ of morphisms in $N$, we denote by $L_S N$ the left Bousfield localization of $N$ along $S$. It is a model category, having $N$ as underlying category, with the same cofibrations as $N$ and whose equivalences are the $S$-local equivalences ([**Hi**, Ch. 3]).

DEFINITION 1.3.1.4. (1) *An object* $x$ *in a model category* $N$ *is* truncated *if there exists an integer* $n \geq 0$, *such that for any* $y \in N$ *the mapping space* $\mathrm{Map}_N(y, x)$ *(*[**Hi**, §17.4]*) is a n-truncated simplicial set (i.e.* $\pi_i(\mathrm{Map}_N(x, y), u) = 0$ *for all* $i > n$ *and for all base point* $u$*).*

(2) *A model category* $N$ *is* t-complete *if a morphism* $u : y \to y'$ *in* $\mathrm{Ho}(N)$ *is an isomorphism if and only if the induced map* $u^* : [y', x] \to [y, x]$ *is a bijection for any truncated object* $x \in N$.

Recall that for any $\mathbb{U}$-small $S$-category $T$ (i.e. a category $T$ enriched over simplicial sets, [**HAGI**, Def. 2.1.1]), we can define a $\mathbb{U}$-combinatorial model category

$SPr(T)$ of *simplicial* functors $T^{op} \longrightarrow SSet_{\mathbb{U}}$, in which equivalences and fibrations are defined levelwise ([**HAGI**, Def. 2.3.2]).

DEFINITION 1.3.1.5. ([**HAGI**, §3.8]) *A* $\mathbb{U}$-model topos *is a* $\mathbb{U}$*-combinatorial model category* $N$ *such that there exists a* $\mathbb{U}$*-small* $S$*-category* $T$ *and a* $\mathbb{U}$*-small set of morphisms* $S$ *in* $SPr(T)$ *satisfying the following two conditions.*

(1) *The model category* $N$ *is Quillen equivalent to* $L_S SPr(T)$.
(2) *The identity functor*

$$\mathrm{Id} : SPr(T) \longrightarrow L_S SPr(T)$$

*preserves homotopy pullbacks.*

*A* $t$-complete model topos *is a* $\mathbb{U}$*-model topos* $N$ *which is* $t$*-complete as a model category.*

We need to recall a few special morphisms in the standard simplicial category $\Delta$. For any $n > 0$, and $0 \leq i < n$ we let

$$\begin{array}{rccc} \sigma_i : & [1] & \longrightarrow & [n] \\ & 0 & \mapsto & i \\ & 1 & \mapsto & i+1. \end{array}$$

DEFINITION 1.3.1.6. *Let* $N$ *be a model category. A* Segal groupoid object in $N$ *is a simplicial object*

$$X_* : \Delta^{op} \longrightarrow N$$

*satisfying the following two conditions.*

(1) *For any* $n > 0$, *the natural morphism*

$$\coprod_{0 \leq i < n} \sigma_i : X_n \longrightarrow \underbrace{X_1 \times^h_{X_0} X_1 \times^h_{X_0} \cdots \times^h_{X_0} X_1}_{n \text{ times}}$$

*is an isomorphism in* $\mathrm{Ho}(N)$.
(2) *The morphism*

$$d_0 \times d_1 : X_2 \longrightarrow X_1 \times^h_{d_0, X_0, d_0} X_1$$

*is an isomorphism in* $\mathrm{Ho}(N)$.

*The homotopy category of Segal groupoid objects in* $N$ *is the full subcategory of* $\mathrm{Ho}(N^{\Delta^{op}})$ *consisting of Segal groupoid objects. It is denoted by* $\mathrm{Ho}(SeGpd(N))$.

The main theorem characterizing model topoi is the following analog of Giraud's theorem.

THEOREM 1.3.1.7. ([**HAGI**, Thm. 4.9.2]) *A model category* $N$ *is a model topos if and only if it satisfies the following conditions.*

(1) *The model category* $N$ *is* $\mathbb{U}$*-combinatorial.*
(2) *For any* $\mathbb{U}$*-small family of objects* $\{x_i\}_{i \in I}$ *in* $N$, *and any* $i \neq j$ *in* $I$ *the following square*

$$\begin{array}{ccc} \emptyset & \longrightarrow & x_i \\ \downarrow & & \downarrow \\ x_j & \longrightarrow & \coprod^{\mathbb{L}}_{k \in I} x_k \end{array}$$

*is homotopy cartesian.*

(3) *For any* $\mathbb{U}$*-small category* $I$*, any morphism* $y \to z$ *and any* $I$*-diagram* $x : I \longrightarrow N/z$*, the natural morphism*

$$Hocolim_{i\in I}(x_i \times^h_z y) \longrightarrow (Hocolim_{i\in I}x_i) \times^h_z y$$

*is an isomorphism in* $\mathrm{Ho}(N)$.

(4) *For any Segal groupoid object (in the sense of Def. 1.3.1.6)*

$$X_* : \Delta^{op} \longrightarrow N,$$

*the natural morphism*

$$X_1 \longrightarrow X_0 \times^h_{|X_*|} X_0$$

*is an isomorphism in* $\mathrm{Ho}(N)$.

An important consequence is the following

COROLLARY 1.3.1.8. *For any* $\mathbb{U}$*-model topos* $N$ *and any fibrant object* $x \in N$*, the category* $\mathrm{Ho}(N/x)$ *is cartesian closed.*

The exactness properties of model topoi will be frequently used all along this work. For instance, we will often use that for any cover of stacks $p : F \longrightarrow G$ (over some model site $(M,\tau)$), the natural morphism

$$|F_*| \longrightarrow G$$

is an isomorphism of stacks, where $F_*$ is the homotopy nerve of $p$ (i.e. the nerve of a fibration equivalent to $p$, computed in the category of simplicial presheaves). This result is also recalled in Lem. 1.3.4.3.

### 1.3.2. Homotopical algebraic geometry context

Let us fix a HA context $(\mathcal{C}, \mathcal{C}_0, \mathcal{A})$.

We denote by $Aff_{\mathcal{C}}$ the opposite of the model category $Comm(\mathcal{C})$: this will be our base model category $M$ to which we will apply the [**HAGI**] constructions recalled in §1.3.1.

An object $X \in Aff_{\mathcal{C}}$ corresponding to a commutative monoid $A \in Comm(\mathcal{C})$ will be symbolically denoted by $X = Spec\,A$. We will consider the model category $Aff_{\mathcal{C}}^{\wedge}$ of pre-stacks on $Aff_{\mathcal{C}}$ as described in §1.3.1 above. By definition, it is the left Bousfield localization of $SPr(Aff_{\mathcal{C}}) := SSet_{\mathbb{V}}^{Aff_{\mathcal{C}}^{op}}$ (the model category of $\mathbb{V}$-simplicial presheaves on $Aff_{\mathcal{C}}$) along the ($\mathbb{V}$-small) set of equivalences of $Aff_{\mathcal{C}}$, and the homotopy category $\mathrm{Ho}(Aff_{\mathcal{C}}^{\wedge})$ will be naturally identified with the full subcategory of $\mathrm{Ho}(SPr(Aff_{\mathcal{C}}))$ consisting of all functors $F : Aff_{\mathcal{C}}^{op} \longrightarrow SSet_{\mathbb{V}}$ preserving weak equivalences. Objects in $\mathrm{Ho}(Aff_{\mathcal{C}}^{\wedge})$ will be called *pre-stacks*, and the derived simplicial $Hom$'s of the simplicial model category $Aff_{\mathcal{C}}^{\wedge}$ will be denoted by $\mathbb{R}\underline{Hom}$.

We will fix once for all a model pre-topology $\tau$ on $Aff_{\mathcal{C}}$ (Def. 1.3.1.1), which induces a Grothendieck topology on $\mathrm{Ho}(Aff_{\mathcal{C}})$, still denoted by the same symbol. As recalled in §1.3.1.1, one can then consider a model category $Aff_{\mathcal{C}}^{\sim,\tau}$, of stacks on the model site $(Aff_{\mathcal{C}}, \tau)$. A morphism $F \longrightarrow G$ of pre-stacks is an equivalence in $Aff_{\mathcal{C}}^{\sim,\tau}$ if it induces isomorphisms on all homotopy sheaves (for any choice of $X \in Aff_{\mathcal{C}}$ and any $s \in F(X)$).

To ease the notation we will write $\mathrm{St}(\mathcal{C},\tau)$ for the homotopy category $\mathrm{Ho}(Aff_{\mathcal{C}}^{\sim,\tau})$ of stacks.

The Bousfield localization construction yields an adjunction

$$a : \mathrm{Ho}(Aff_{\mathcal{C}}^{\wedge}) \longrightarrow \mathrm{St}(\mathcal{C}, \tau) \qquad \mathrm{Ho}(Aff_{\mathcal{C}}^{\wedge}) \longleftarrow \mathrm{St}(\mathcal{C}, \tau) : j$$

where $j$ is fully faithful.

DEFINITION 1.3.2.1. (1) *A* stack *is an object* $F \in SPr(Aff_{\mathcal{C}})$ *whose image in* $\mathrm{Ho}(Aff_{\mathcal{C}}^{\wedge})$ *is in the essential image of the functor* $j$ *above.*

(2) *A* morphism of stacks *is a morphism between stacks in* $\mathrm{Ho}(SPr(Aff_{\mathcal{C}}))$*, or equivalently in* $\mathrm{Ho}(Aff_{\mathcal{C}}^{\wedge})$*, or equivalently in* $\mathrm{St}(\mathcal{C}, \tau)$*.*

(3) *A morphism of stacks* $f : F \longrightarrow G$ *is a* covering *(or a* cover*) if the induced morphism of sheaves*

$$\pi_0(f) : \pi_0(F) \longrightarrow \pi_0(G)$$

*is an epimorphism in the category of sheaves.*

We will always omit mentioning the functor $j$ and consider the category $\mathrm{St}(\mathcal{C}, \tau)$ as embedded in $\mathrm{Ho}(Aff_{\mathcal{C}}^{\wedge})$, and therefore embedded in $\mathrm{Ho}(SPr(Aff_{\mathcal{C}}))$. With these conventions, the endofunctor $a$ of $\mathrm{Ho}(Aff_{\mathcal{C}}^{\wedge})$ becomes the *associated stack functor*, which commutes with finite homotopy limits and arbitrary homotopy colimits.

A functor $F : Aff_{\mathcal{C}}^{op} \longrightarrow SSet_{\mathbb{V}}$ is a stack (§1.3.1) if and only if it preserves equivalences and possesses the descent property with respect to homotopy $\tau$-hypercovers. The derived simplicial $Hom$'s in the model category $Aff_{\mathcal{C}}^{\sim,\tau}$ of stacks will be denoted by $\mathbb{R}_{\tau}\underline{Hom}$. The natural morphism

$$\mathbb{R}\underline{Hom}(F, G) \longrightarrow \mathbb{R}_{\tau}\underline{Hom}(F, G)$$

is an isomorphism in $\mathrm{Ho}(SSet)$ when $G$ is a stack.

The model category $Aff_{\mathcal{C}}^{\sim,\tau}$ is a $t$-complete model topos (Def. 1.3.1.5). We warn the reader that neither of the model categories $Aff_{\mathcal{C}}^{\wedge}$ and $Aff_{\mathcal{C}}^{\sim,\tau}$ is right proper, though they are both left proper. Because of this certain care has to be taken when considering homotopy pullbacks and more generally homotopy limit constructions, as well as comma model categories of objects over a base object. Therefore, even when nothing is specified, adequate fibrant replacement may have been chosen before considering certain constructions.

The model Yoneda embedding (§1.3.1)

$$\underline{h} : Aff_{\mathcal{C}} \longrightarrow Aff_{\mathcal{C}}^{\wedge}$$

has a total right derived functor

$$\mathbb{R}\underline{h} : \mathrm{Ho}(Aff_{\mathcal{C}}) \longrightarrow \mathrm{Ho}(Aff_{\mathcal{C}}^{\wedge}),$$

which is fully faithful. We also have a naive Yoneda functor

$$h : Aff_{\mathcal{C}} \longrightarrow Aff_{\mathcal{C}}^{\wedge}$$

sending an object $X \in Aff_{\mathcal{C}}$ to the presheaf of sets it represents (viewed as a simplicial presheaf). With these notations, the Yoneda lemma reads

$$\mathbb{R}\underline{Hom}(\mathbb{R}\underline{h}_X, F) \simeq \mathbb{R}\underline{Hom}(h_X, F) \simeq F(X)$$

for any fibrant object $F \in Aff_{\mathcal{C}}^{\wedge}$. The natural morphism $h_X \longrightarrow \mathbb{R}\underline{h}_X$ is always an isomorphism in $\mathrm{Ho}(Aff_{\mathcal{C}}^{\wedge})$ for any $X \in Aff_{\mathcal{C}}$.

We will also use the notation $\underline{Spec}\, A$ for $\underline{h}_{Spec\, A}$. We warn the reader that $Spec\, A$ lives in $Aff_{\mathcal{C}}$ whereas $\underline{Spec}\, A$ is an object of the model category of stacks $Aff_{\mathcal{C}}^{\sim,\tau}$.

We will assume the topology $\tau$ satisfies some conditions. In order to state them, recall the category $sAff_{\mathcal{C}}$ of simplicial objects in $Aff_{\mathcal{C}}$ is a simplicial model category for the Reedy model structure. Therefore, for any object $X_* \in sAff_{\mathcal{C}}$ and any $\mathbb{U}$-small simplicial set $K$ we can define an object $X_*^{\mathbb{R}\underline{K}} \in \mathrm{Ho}(sAff_{\mathcal{C}})$, by first taking a Reedy fibrant model for $X_*$ and then the exponential by $K$. The zero-th part of $X_*^{\mathbb{R}\underline{K}}$ will be simply denoted by

$$X_*^{\mathbb{R}K} := (X_*^{\mathbb{R}\underline{K}})_0 \in \mathrm{Ho}(Aff_{\mathcal{C}}).$$

We also refer the reader to [**HAGI**, §4.4] for more details and notations.

The following assumption on the pre-topology will be made.

ASSUMPTION 1.3.2.2. (1) *The topology $\tau$ on $\mathrm{Ho}(Aff_{\mathcal{C}})$ is quasi-compact. In other words, for any covering family $\{U_i \longrightarrow X\}_{i\in I}$ in $Aff_{\mathcal{C}}$ there exists a finite subset $I_0 \subset I$ such that the induced family $\{U_i \longrightarrow X\}_{i\in I_0}$ is a covering.*

(2) *For any finite family of objects $\{X_i\}_{i\in I}$ in $Aff_{\mathcal{C}}$ (including the empty family) the family of morphisms*

$$\{X_i \longrightarrow \coprod_{j\in I}^{\mathbb{L}} X_j\}_{i\in I}$$

*form a $\tau$-covering family of $\coprod_{j\in I}^{\mathbb{L}} X_j$.*

(3) *Let $X_* \longrightarrow Y$ be an augmented simplicial object in $Aff_{\mathcal{C}}$, corresponding to a co-augmented co-simplicial object $A \longrightarrow B_*$ in $Comm(\mathcal{C})$. We assume that for any $n$, the one element family of morphisms*

$$X_n \longrightarrow X_*^{\mathbb{R}\partial\Delta^n} \times^h_{Y^{\mathbb{R}\partial\Delta^n}} Y$$

*form a $\tau$-covering family in $Aff_{\mathcal{C}}$. Then the morphism*

$$A \longrightarrow B_*$$

*satisfies the descent condition in the sense of Def. 1.2.12.1.*

The previous assumption has several consequences on the homotopy theory of stacks. They are subsumed in the following lemma.

LEMMA 1.3.2.3. (1) *For any finite family of objects $X_i$ in $Aff_{\mathcal{C}}$ the natural morphism*

$$\coprod_i \mathbb{R}\underline{h}_{X_i} \longrightarrow \mathbb{R}\underline{h}_{\coprod_i^{\mathbb{L}} X_i}$$

*is an equivalence in $Aff_{\mathcal{C}}^{\sim,\tau}$.*

(2) *Let $H$ be the ($\mathbb{V}$-small) set of augmented simplicial objects $X_* \longrightarrow Y$ in $Aff_{\mathcal{C}}$ such that for any $n$ the one element family of morphisms*

$$X_n \longrightarrow X_*^{\mathbb{R}\partial\Delta^n} \times^h_{Y^{\mathbb{R}\partial\Delta^n}} Y$$

*is a $\tau$-covering family in $Aff_{\mathcal{C}}$. Then, the model category $Aff_{\mathcal{C}}^{\sim,\tau}$ is the left Bousfield localization of $Aff_{\mathcal{C}}^{\wedge}$ along the set of morphisms*

$$|\mathbb{R}\underline{h}_{X_*}| \longrightarrow \mathbb{R}\underline{h}_Y \qquad \coprod_i \mathbb{R}\underline{h}_{U_i} \longrightarrow \mathbb{R}\underline{h}_{\coprod_i^{\mathbb{L}} U_i}$$

*where $X_* \to Y$ runs in $H$ and $\{U_i\}$ runs through the set of all finite families of objects in $Aff_{\mathcal{C}}$.*

PROOF. (1) The case where the set of indices $I$ is empty follows from our assumption 1.3.2.2 (2) with $I$ empty, as it states that the empty family covers the initial object on $Aff_{\mathcal{C}}$.

Let us assume that the set of indices is not empty. By induction, it is clearly enough to treat the case where the finite family consists of two objects $X$ and $Y$. Our assumption 1.3.2.2 (2) then implies that the natural morphism

$$p : \mathbb{R}\underline{h}_X \coprod \mathbb{R}\underline{h}_Y \longrightarrow \mathbb{R}\underline{h}_{X \coprod^{\mathbb{L}} Y}$$

is a covering. Therefore, $\mathbb{R}\underline{h}_{X \coprod^{\mathbb{L}} Y}$ is naturally equivalent to the homotopy colimit of the homotopy nerve of the morphism $p$. Using this remark we see that it is enough to prove that

$$\mathbb{R}\underline{h}_X \times^h_{\mathbb{R}\underline{h}_{X \coprod^{\mathbb{L}} Y}} \mathbb{R}\underline{h}_Y \simeq \emptyset,$$

as then the homotopy nerve of $p$ will be a constant simplicial object with values $\mathbb{R}\underline{h}_X \coprod \mathbb{R}\underline{h}_Y$. As the functor $\mathbb{R}\underline{h}$ commutes with homotopy pullbacks, it is therefore enough to check that

$$A \otimes^{\mathbb{L}}_{A \times^h B} B \simeq 0,$$

for $A$ and $B$ two commutative monoids in $\mathcal{C}$ such that $X = Spec\, A$ and $Y = Spec\, B$ (here 0 is the final object in $Comm(\mathcal{C})$). For this we can of course suppose that $A$ and $B$ are fibrant objects in $\mathcal{C}$.

We define a functor

$$F : A \times B - Comm(\mathcal{C}) \longrightarrow A - Comm(\mathcal{C}) \times B - Comm(\mathcal{C})$$

by the formula

$$F(C) := (C \otimes_{A \times B} A, C \otimes_{A \times B} B).$$

The functor $F$ is left Quillen for the product model structures on the right hand side, and its right adjoint is given by $G(C, D) := C \times D$ for any $(C, D) \in A - Comm(\mathcal{C}) \times B - Comm(\mathcal{C})$. For any $C \in A \times B - Comm(\mathcal{C})$, one has

$$C \simeq C \otimes^{\mathbb{L}}_{A \times B} (A \times B) \simeq \left(C \otimes^{\mathbb{L}}_{A \times B} A\right) \times \left(C \otimes^{\mathbb{L}}_{A \times B} B\right),$$

because of our assumptions 1.1.0.1 and 1.1.0.4, which implies that the adjunction morphism

$$C \longrightarrow \mathbb{R}G(\mathbb{L}F(C))$$

is an isomorphism in $\mathrm{Ho}(A \times B - Comm(\mathcal{C}))$. As the functor $G$ reflects equivalences (because of our assumption 1.1.0.1) this implies that $F$ and $G$ form a Quillen equivalence. Therefore, the functor $\mathbb{R}G$ commutes with homotopy push outs, and we have

$$A \otimes^{\mathbb{L}}_{A \times B} B \simeq \mathbb{R}G(A, 0) \otimes^{\mathbb{L}}_{\mathbb{R}G(A,B)} \mathbb{R}G(0, B) \simeq \mathbb{R}G\left( (A, 0) \coprod^{\mathbb{L}}_{(A,B)} (0, B) \right) \simeq \mathbb{R}G(0) \simeq 0.$$

(2) We know by [**HAGI**] that $Aff^{\sim, \tau}_{\mathcal{C}}$ is the left Bousfield localization of $Aff^{\wedge}_{\mathcal{C}}$ along the set of morphisms $|F_*| \longrightarrow h_X$, where $F_* \longrightarrow h_X$ runs in a certain $\mathbb{V}$-small set of $\tau$-hypercovers. Recall that for each hypercover $F_* \longrightarrow h_X$ in this set, each simplicial presheaf $F_n$ is a coproduct of some $h_U$. Using the quasi-compactness assumption 1.3.2.2 (1) one sees immediately that one can furthermore assume that each $F_n$ is a finite coproduct of some $h_U$. Finally, using the part (1) of the present lemma we see that the descent condition of [**HAGI**] can be stated as two distinct conditions, one concerning finite coproducts and the other one concerning representable hypercovers. From this we deduce part (2) of the lemma. $\square$

Lemma 1.3.2.3 (2) can be reformulated as follows.

COROLLARY 1.3.2.4. *A simplicial presheaf*

$$F : Comm(\mathcal{C}) \longrightarrow SSet_{\mathbb{V}}$$

*is a stack if and only if it satisfies the following three conditions.*

- *For any equivalence $A \longrightarrow B$ in $Comm(\mathcal{C})$ the induced morphism $F(A) \longrightarrow F(B)$ is an equivalence of simplicial sets.*
- *For any finite family of commutative monoids $\{A\}_{i\in I}$ in $\mathcal{C}$ (including the empty family), the natural morphism*

$$F(\prod_{i\in I}{}^{h} A_i) \longrightarrow \prod_{i\in I} F(A_i)$$

*is an isomorphism in* $\mathrm{Ho}(SSet)$.
- *For any co-simplicial commutative $A$-algebra $A \longrightarrow B_*$, corresponding to a $\tau$-hypercover*

$$Spec\, B_* \longrightarrow Spec\, A$$

*in $Aff_{\mathcal{C}}$, the induced morphism*

$$F(A) \longrightarrow Holim_{[n]\in\Delta} F(B_n)$$

*is an isomorphism in* $\mathrm{Ho}(SSet)$.

Another important consequence of lemma 1.3.2.3 is the following.

COROLLARY 1.3.2.5. *The model pre-topology $\tau$ on $Aff_{\mathcal{C}}$ is sub-canonical in the sense of Def. 1.3.1.3.*

PROOF. We need to show that for any $Z \in Aff_{\mathcal{C}}$ the object $G := \mathbb{R}\underline{h}_Z$ is a stack, or in other words is a local object in $Aff_{\mathcal{C}}^{\sim,\tau}$. For this, we use our lemma 1.3.2.3 (2). The descent property for finite coproducts is obviously satisfied because of the Yoneda lemma. Let $X_* \longrightarrow Y$ be a simplicial object in $Aff_{\mathcal{C}}$ such that

$$\mathbb{R}\underline{h}_{X_*} \longrightarrow \mathbb{R}\underline{h}_Y$$

is a $\tau$-hypercover. By lemma 1.2.12.3 the natural morphism

$$Hocolim_n X_n \longrightarrow Y$$

is an isomorphism in $\mathrm{Ho}(Aff_{\mathcal{C}})$. Therefore, the Yoneda lemma implies that one has

$$\mathbb{R}\underline{Hom}(\underline{h}_Y, G) \simeq Map(Y, Z) \simeq Holim_n Map(X_n, Z) \simeq Holim_n \mathbb{R}\underline{Hom}(F_*, G),$$

showing that $G$ is a stack. □

The corollary 1.3.2.5 implies that $\mathbb{R}\underline{h}$ provides a fully faithful functor

$$\mathbb{R}\underline{h} : \mathrm{Ho}(Aff_{\mathcal{C}}) \longrightarrow \mathrm{St}(\mathcal{C}, \tau).$$

Objects in the essential image of $\mathbb{R}\underline{h}$ will be called *representable objects*. If such an object corresponds to a commutative monoid $A \in \mathrm{Ho}(Comm(\mathcal{C}))$, it will also be denoted by $\mathbb{R}\underline{Spec}\, A \in \mathrm{St}(\mathcal{C}, \tau)$. In formula

$$\mathbb{R}\underline{Spec}\, A := \mathbb{R}\underline{h}_{Spec\, A},$$

for any $A \in Comm(\mathcal{C})$ corresponding to $Spec\, A \in Aff_{\mathcal{C}}$. As $\mathbb{R}\underline{h}$ commutes with $\mathbb{U}$-small homotopy limits, we see that the subcategory of representable stacks is stable by $\mathbb{U}$-small homotopy limits. The reader should be careful that a $\mathbb{V}$-small homotopy limit of representable stacks is not representable in general. Lemma 1.3.2.3 (1) also implies that a finite coproduct of representable stacks is a representable stack, and we have

$$\coprod_i \mathbb{R}\underline{h}_{U_i} \simeq \mathbb{R}\underline{h}_{\coprod_i^{\mathbb{L}} U_i}.$$

Also, by identifying the category $\mathrm{Ho}(Comm(\mathcal{C}))^{op}$ with the full subcategory of $\mathrm{St}(\mathcal{C},\tau)$ consisting of representable stacks, one can extend the notions of morphisms defined in §1.2 (e.g. (formally) étale, Zariski open immersion, flat, smooth ...) to morphisms between representable stacks. Indeed, they are all invariant by equivalences and therefore are properties of morphisms in $\mathrm{Ho}(Comm(\mathcal{C}))$. We will often use implicitly these extended notions. In particular, we will use the expression *$\tau$-covering families of representable stacks* to denote families of morphisms of representable stacks corresponding in $\mathrm{Ho}(Comm(\mathcal{C}))^{op}$ to $\tau$-covering families.

We will use the same terminology for the morphisms in $\mathrm{Ho}(Comm(\mathcal{C}))$ and for the corresponding morphisms of representable stacks, except for the notion of epimorphism (Def. 1.2.6.1), which for obvious reasons will be replaced by *monomorphism* in the context of stacks. This is justified since a morphism $A \longrightarrow B$ is an epimorphism in the sense of Def. 1.2.6.1 if and only if the induced morphism of stacks

$$\mathbb{R}\underline{Spec}\,B \longrightarrow \mathbb{R}\underline{Spec}\,A$$

is a monomorphism in the model category $Aff_{\mathcal{C}}^{\sim,\tau}$ (see Remark 1.2.6.2).

REMARK 1.3.2.6. The reader should be warned that we will also use the expression *epimorphism of stacks*, which will refer to a morphism of stacks that induces an epimorphism on the sheaves $\pi_0$ (see Def. 1.3.1.2 or [**HAGI**], where they are also called *coverings*). It is important to notice that a $\tau$-covering family of representable stacks $\{X_i \longrightarrow X\}$ induces an epimorphism of stacks $\coprod X_i \longrightarrow X$. On the contrary, there might very well exist families of morphisms of representable stacks $\{X_i \longrightarrow X\}$ such that $\coprod X_i \longrightarrow X$ is an epimorphism of stacks, but which are not $\tau$-covering families (e.g. a morphism between representable stacks that admits a section).

COROLLARY 1.3.2.7. *Let $\{u_i : X_i = Spec\,A_i \longrightarrow X = Spec\,A\}_{i\in I}$ be a covering family in $Aff_{\mathcal{C}}$. Then, the family of base change functors*

$$\{\mathbb{L}u_i^* : \mathrm{Ho}(A-Mod) \longrightarrow \mathrm{Ho}(A_i-Mod)\}_{i\in I}$$

*is conservative. In other words, a $\tau$-covering family in $Aff_{\mathcal{C}}$ is a formal covering in the sense of Def. 1.2.5.1.*

PROOF. By the quasi-compactness assumption on $\tau$ we can assume that the covering family is finite. Also, the morphism $A \longrightarrow \prod_i^h A_i = B$ is a covering. Therefore, the descent assumption 1.3.2.2 (3) implies that the base change functor

$$B \otimes_A^{\mathbb{L}} - : \mathrm{Ho}(A-Mod) \longrightarrow \mathrm{Ho}(B-Mod)$$

is conservative. Finally, we have seen during the proof of 1.3.2.3 (1) that the product of the base change functors

$$\prod_i A_i \otimes_B^{\mathbb{L}} - : \mathrm{Ho}(B-Mod) \longrightarrow \prod_i \mathrm{Ho}(A_i-Mod)$$

is an equivalence. Therefore, the composition

$$\prod_i A_i \otimes_A^{\mathbb{L}} - : \mathrm{Ho}(A-Mod) \longrightarrow \prod_i \mathrm{Ho}(A_i-Mod)$$

is a conservative functor. □

We also recall the Yoneda lemma for stacks, stating that for any $A \in Comm(\mathcal{C})$, and any fibrant object $F \in Aff_{\mathcal{C}}^{\sim,\tau}$, there is a natural equivalence of simplicial sets

$$\mathbb{R}\underline{Hom}(\mathbb{R}\underline{Spec}\,A, F) \simeq \mathbb{R}_\tau\underline{Hom}(\mathbb{R}\underline{Spec}\,A, F) \simeq F(A).$$

For an object $F \in \mathrm{St}(\mathcal{C}, \tau)$, and $A \in Comm(\mathcal{C})$ we will use the following notation

$$\mathbb{R}F(A) := \mathbb{R}_\tau \underline{Hom}(\mathbb{R}\underline{Spec}\, A, F).$$

Note that $\mathbb{R}F(A) \simeq RF(A)$, where $R$ is a fibrant replacement functor on $Aff_{\mathcal{C}}^{\sim,\tau}$. Note that there is always a natural morphism $F(A) \longrightarrow \mathbb{R}F(A)$, which is an equivalence precisely when $F$ is a stack.

Finally, another important consequence of assumption 1.3.2.2 is the local character of representable stacks.

PROPOSITION 1.3.2.8. *Let $G$ be a representable stack and $F \longrightarrow G$ be any morphism. Assume there exists a $\tau$-covering family of representable stacks*

$$\{G_i \longrightarrow G\},$$

*such that each stack $F \times^h_G G_i$ is representable. Then $F$ is a representable stack.*

PROOF. Let $X \in Aff_{\mathcal{C}}$ be a fibrant object such that $G \simeq \underline{h}_X$. We can of course assume that $G = \underline{h}_X$. We can also assume that $F \longrightarrow G$ is a fibration, and therefore that $G$ and $F$ are fibrant objects.

By choosing a refinement of the covering family $\{G_i \longrightarrow G\}$, one can suppose that the covering family is finite and that each morphism $G_i \longrightarrow G$ is the image by $\underline{h}$ of a fibration $U_i \longrightarrow X$ in $Aff_{\mathcal{C}}$. Finally, considering the coproduct $U = \coprod^{\mathbb{L}} U_i \longrightarrow X$ in $Aff_{\mathcal{C}}$ and using lemma 1.3.2.3 (1) one can suppose that the family $\{G_i \longrightarrow G\}$ has only one element and is the image by $\underline{h}$ of a fibration $U \longrightarrow X$ in $Aff_{\mathcal{C}}$.

We consider the augmented simplicial object $U_* \longrightarrow X$ in $Aff_{\mathcal{C}}$, which is the nerve of the morphism $U \to X$, and the corresponding augmented simplicial object $\underline{h}_{U_*} \longrightarrow \underline{h}_X$ in $Aff_{\mathcal{C}}^{\sim,\tau}$. We form the pullback in $Aff_{\mathcal{C}}^{\sim,\tau}$

$$\begin{array}{ccc} F & \longrightarrow & \underline{h}_X \\ \uparrow & & \uparrow \\ F_* & \longrightarrow & \underline{h}_{U_*} \end{array}$$

which is a homotopy pullback because of our choices. In particular, for any $n$, $F_n$ is a representable stack.

Clearly $F_*$ is the nerve of the fibration $F \times_{\underline{h}_X} \underline{h}_U \longrightarrow F$. As this last morphism is an epimorphism in $Aff_{\mathcal{C}}^{\sim,\tau}$ the natural morphism

$$|F_*| := Hocolim_n F_n \longrightarrow F$$

is an isomorphism in $\mathrm{St}(\mathcal{C}, \tau)$. Therefore, it remains to show that $|F_*|$ is a representable stack.

We will consider the category $sAff_{\mathcal{C}}^{\sim,\tau}$ of simplicial objects in $Aff_{\mathcal{C}}^{\sim,\tau}$, endowed with its Reedy model structure (see [**HAGI**, §4.4] for details and notations). In the same way, we will consider the Reedy model structure on $sAff_{\mathcal{C}}$, the category of simplicial objects in $Aff_{\mathcal{C}}$.

LEMMA 1.3.2.9. *There exists a simplicial object $V_*$ in $Aff_{\mathcal{C}}$ and an isomorphism $\mathbb{R}\underline{h}_{V_*} \simeq F_*$ in $\mathrm{Ho}(sAff_{\mathcal{C}}^{\sim,\tau})$.*

PROOF. First of all, our functor $\underline{h}$ extends in the obvious way to a functor on the categories of simplicial objects

$$\underline{h} : sAff_{\mathcal{C}} \longrightarrow sAff_{\mathcal{C}}^{\sim,\tau},$$

by the formula

$$(\underline{h}_{X_*})_n := \underline{h}_{X_n}$$

for any $X_* \in sAff_\mathcal{C}$. This functor possesses a right derived functor

$$\mathbb{R}\underline{h} : \mathrm{Ho}(sAff_\mathcal{C}) \longrightarrow \mathrm{Ho}(sAff_\mathcal{C}^{\sim,\tau}),$$

which is easily seen to be fully faithful.

We claim that the essential image of $\mathbb{R}\underline{h}$ consists of all simplicial objects $F_* \in \mathrm{Ho}(sAff_\mathcal{C}^{\sim,\tau})$ such that for any $n$, $F_n$ is a representable stack. This will obviously imply the lemma. Indeed, as $\mathbb{R}\underline{h}$ commutes with homotopy limits, one sees that this essential image is stable by ($\mathbb{U}$-small) homotopy limits. Also, any object $F_* \in sAff_\mathcal{C}^{\sim,\tau}$ can be written as a homotopy limit

$$F_* \simeq Holim_n \mathbb{R}Cosk_n(F_*)$$

of its derived coskeleta (see [**HAGI**, §4.4]). Recall that for a fibrant object $F_*$ in $sAff_\mathcal{C}^{\sim,\tau}$ one has

$$Cosk_n(F_*)_p \simeq \mathbb{R}Cosk_n(F_*)_p \simeq (F_*)^{Sk_n\Delta^p},$$

where, for a simplicial set $K$, $(F_*)^K$ is defined to be the equalizer of the two natural maps

$$\prod_{[n]}(F_n)^{K_n} \rightrightarrows \prod_{[n]\to[m]}(F_m)^{K_n}.$$

Now, for any simplicial set $K$, and any integer $n \geq 0$, there is a homotopy push out square

$$\begin{array}{ccc} Sk_nK & \longrightarrow & Sk_{n+1}K \\ \uparrow & & \uparrow \\ \coprod_{K^{\partial\Delta^{n+1}}} \partial\Delta^{n+1} & \longrightarrow & \coprod_{K^{\Delta^{n+1}}} \Delta^{n+1}. \end{array}$$

Using this, and the fact that $K \mapsto F_*^K$ sends homotopy push outs to homotopy pullbacks when $F_*$ is fibrant, we see that for any fibrant object $F_* \in sAff_\mathcal{C}^{\sim,\tau}$, and any $n$, we have a homotopy pullback diagram in $sAff_\mathcal{C}^{\sim,\tau}$

$$\begin{array}{ccc} \mathbb{R}Cosk_n(F_*) & \longrightarrow & \mathbb{R}Cosk_{n-1}(F_*) \\ \uparrow & & \uparrow \\ A_*^n & \longrightarrow & B_*^n. \end{array}$$

Here, $A_*^n$ and $B_*^n$ are defined by the following formulas

$$\begin{array}{rccc} A_*^n : & \Delta^{op} & \longrightarrow & Aff_\mathcal{C}^{\sim,\tau} \\ & [p] & \mapsto & \prod_{(\Delta^p)^{\Delta^{n+1}}} F_*^{\Delta^{n+1}} \end{array}$$

$$\begin{array}{rccc} B_*^n : & \Delta^{op} & \longrightarrow & Aff_\mathcal{C}^{\sim,\tau} \\ & [p] & \mapsto & \prod_{(\Delta^p)^{\partial\Delta^{n+1}}} F_*^{\partial\Delta^{n+1}}. \end{array}$$

Therefore, by induction on $n$, it is enough to see that if $F_*$ is fibrant then $Cosk_0(F_*)$, $A_*^n$ and $B_*^n$ all belongs to the essential image of $\mathbb{R}\underline{h}$.

The simplicial object $Cosk_0(F_*)$ is isomorphic to the nerve of $F_0 \longrightarrow *$, and therefore is the image by $\mathbb{R}\underline{h}$ of the nerve of a fibration $X \longrightarrow *$ in $Aff_\mathcal{C}$ representing $F_0$. This shows that $Cosk_0(F_*)$ is in the image of $\mathbb{R}\underline{h}$.

We have $F_*^{\Delta^{n+1}} = F_{n+1}$, which is a representable stack. Let $X_{n+1} \in Aff_\mathcal{C}$ be a fibrant object such that $F_{n+1}$ is equivalent to $\underline{h}_{X_{n+1}}$. Then, as $\underline{h}$ commutes with limits, we see that $A_*^n$ is equivalent the image by $\underline{h}$ of the simplicial object

$$[p] \mapsto \prod_{(\Delta^p)^{\Delta^{n+1}}} X_{n+1}.$$

In the same way, $F_*^{\partial\Delta^{n+1}}$ can be written as a finite homotopy limit of $F_i$'s, and therefore is a representable stack. Let $Y_{n+1}$ be a fibrant object in $Aff_{\mathcal{C}}$ such that $F_*^{\partial\Delta^{n+1}}$ is equivalent to $\underline{h}_{Y_{n+1}}$. Then, $B_*^n$ is equivalent to the image by $\underline{h}$ of the simplicial object

$$[p] \mapsto \prod_{(\Delta^p)^{\partial\Delta^{n+1}}} Y_{n+1}.$$

This proves the lemma. □

We now finish the proof of Proposition 1.3.2.8. Let $V_*$ be a simplicial object in $sAff_{\mathcal{C}}$ such that $F_* \simeq \mathbb{R}\underline{h}_{V_*}$. The augmentation $F_* \longrightarrow \mathbb{R}\underline{h}_{U_*}$ gives rise to a well defined morphism in $\mathrm{Ho}(sAff_{\mathcal{C}}^{\sim,\tau})$

$$q : \mathbb{R}\underline{h}_{V_*} \longrightarrow \mathbb{R}\underline{h}_{U_*}.$$

We can of course suppose that $V_*$ is a cofibrant object in $sAff_{\mathcal{C}}$. As $U_*$ is the nerve of a fibration between fibrant objects it is fibrant in $sAff_{\mathcal{C}}$. Therefore, as $\mathbb{R}\underline{h}$ is fully faithful, we can represent $q$, up to an isomorphism, as the image by $\underline{h}$ of a morphism in $sAff_{\mathcal{C}}$

$$r : V_* \longrightarrow U_*.$$

On the level of commutative monoids, the morphism $r$ is given by a morphism $B_* \longrightarrow C_*$ of co-simplicial objects in $Comm(\mathcal{C})$. By construction, for any morphism $[n] \to [m]$ in $\Delta$, the natural morphism

$$F_m \longrightarrow F_n \times^h_{\mathbb{R}\underline{h}_{U_n}} \mathbb{R}\underline{h}_{U_m}$$

is an isomorphism in $\mathrm{St}(\mathcal{C}, \tau)$. This implies that the underlying co-simplicial $B_*$-module of $C_*$ is homotopy cartesian in the sense of Def. 1.2.12.1. Our assumption 1.3.2.2 (3) implies that if $Y := Hocolim_n V_* \in Aff_{\mathcal{C}}$, then the natural morphism

$$V_* \longrightarrow U_* \times^h_X Y$$

is an isomorphism in $\mathrm{Ho}(sAff_{\mathcal{C}})$. As homotopy colimits in $Aff_{\mathcal{C}}^{\sim,\tau}$ commute with homotopy pullbacks, this implies that

$$|F_*| \simeq |\mathbb{R}\underline{h}_{V_*}| \simeq |\mathbb{R}\underline{h}_{U_*}| \times^h_{\mathbb{R}\underline{h}_X} \mathbb{R}\underline{h}_Y.$$

But, as $\mathbb{R}\underline{h}_{U_*} \longrightarrow \mathbb{R}\underline{h}_X$ is the homotopy nerve of an epimorphism we have

$$|\mathbb{R}\underline{h}_{U_*}| \simeq \mathbb{R}\underline{h}_X,$$

showing finally that $|F_*|$ is isomorphic to $\mathbb{R}\underline{h}_Y$ and therefore is a representable stack. □

Finally, we finish this first section by the following description of the comma model category $Aff_{\mathcal{C}}^{\sim,\tau}/\underline{h}_X$, for some fibrant object $X \in Aff_{\mathcal{C}}$. This is not a completely trivial task as the model category $Aff_{\mathcal{C}}^{\sim,\tau}$ is not right proper.

For this, let $A \in Comm(\mathcal{C})$ such that $X = Spec\, A$, so that $A$ is a cofibrant object in $Comm(\mathcal{C})$. We consider the comma model category $A - Comm(\mathcal{C}) = (Aff_{\mathcal{C}}/X)^{op}$, which is also the model category $Aff_{A-Mod}^{op} = Comm(A - Mod)$. The model pre-topology $\tau$ on $Aff_{\mathcal{C}}$ induces in a natural way a model pre-topology $\tau$ on $Aff_{\mathcal{C}}/X = Aff_{A-Mod}$. Note that there exists a natural equivalence of categories, compatible with the model structures, between $(Aff_{\mathcal{C}}/X)^{\sim,\tau}$ and $Aff_{\mathcal{C}}^{\sim,\tau}/h_X$. We consider the natural morphism $h_X \longrightarrow \underline{h}_X$. It gives rise to a Quillen adjunction

$$Aff_{\mathcal{C}}^{\sim,\tau}/h_X \longrightarrow Aff_{\mathcal{C}}^{\sim,\tau}/\underline{h}_X \qquad Aff_{\mathcal{C}}^{\sim,\tau}/h_X \longleftarrow Aff_{\mathcal{C}}^{\sim,\tau}/\underline{h}_X$$

where the right adjoint sends $F \longrightarrow \underline{h}_X$ to $F \times_{\underline{h}_X} h_X \longrightarrow h_X$.

PROPOSITION 1.3.2.10. *The above Quillen adjunction induces a Quillen equivalence between* $Aff_{\mathcal{C}}^{\sim,\tau}/h_X \simeq (Aff_{\mathcal{C}}/X)^{\sim,\tau}$ *and* $Aff_{\mathcal{C}}^{\sim,\tau}/\underline{h}_X$.

PROOF. For $F \longrightarrow \underline{h}_X$ a fibrant object, and $Y \in Aff_{\mathcal{C}}/X$, there is a homotopy cartesian square

$$\begin{array}{ccc} (F \times_{\underline{h}_X} h_X)(Y) & \longrightarrow & h_X(Y) \\ \downarrow & & \downarrow \\ F(Y) & \longrightarrow & \underline{h}_X(Y). \end{array}$$

As the morphism $h_X(Y) \longrightarrow \underline{h}_X(Y)$ is always surjective up to homotopy when $Y$ is cofibrant, this implies easily that the derived pullback functor

$$\mathrm{Ho}(Aff_{\mathcal{C}}^{\sim,\tau}/\underline{h}_X) \longrightarrow \mathrm{Ho}(Aff_{\mathcal{C}}^{\sim,\tau}/h_X)$$

is conservative. Therefore, it is enough to show that the forgetful functor

$$\mathrm{Ho}(Aff_{\mathcal{C}}/X^{\sim,\tau}) \longrightarrow \mathrm{Ho}(Aff_{\mathcal{C}}^{\sim,\tau}/\underline{h}_X)$$

is fully faithful.

Using the Yoneda lemma for $Aff_{\mathcal{C}}/X$, we have

$$Map_{(Aff_{\mathcal{C}}/X)^{\sim,\tau}}(h_Y, h_Z) \simeq Map_{Aff_{\mathcal{C}}/X}(Y, Z)$$

for any two objects $Y$ and $Z$ in $Aff_{\mathcal{C}}/X$. Therefore, there exists a natural fibration sequence

$$Map_{(Aff_{\mathcal{C}}/X)^{\sim,\tau}}(h_Y, h_Z) \longrightarrow Map_{Aff_{\mathcal{C}}}(Y, Z) \longrightarrow Map_{Aff_{\mathcal{C}}}(Z, X).$$

In the same way, the Yoneda lemma for $Aff_{\mathcal{C}}$ implies that there exists a fibration sequence

$$Map_{Aff_{\mathcal{C}}^{\sim,\tau}/\underline{h}_X}(h_Y, h_Z) \longrightarrow Map_{Aff_{\mathcal{C}}}(Y, Z) \longrightarrow Map_{Aff_{\mathcal{C}}}(Z, X).$$

This two fibration sequences implies that the forgetful functor induces equivalences of simplicial sets

$$Map_{Aff_{\mathcal{C}}/X^{\sim,\tau}}(h_Y, h_Z) \simeq Map_{Aff_{\mathcal{C}}^{\sim,\tau}/\underline{h}_X}(Y, Z).$$

In other words, the functor

$$\mathrm{Ho}(Aff_{\mathcal{C}}/X^{\sim,\tau}) \longrightarrow \mathrm{Ho}(Aff_{\mathcal{C}}^{\sim,\tau}/\underline{h}_X)$$

is fully faithful when restricted to the full subcategory of representable stacks. But, any object in $\mathrm{Ho}(Aff_{\mathcal{C}}/X^{\sim,\tau})$ is a homotopy colimit of representable stacks. Furthermore, as the derived pullback

$$\mathrm{Ho}(Aff_{\mathcal{C}}^{\sim,\tau}/\underline{h}_X) \longrightarrow \mathrm{Ho}(Aff_{\mathcal{C}}^{\sim,\tau}/h_X)$$

commutes with homotopy colimits (as homotopy pullbacks of simplicial sets do), this implies that the functor

$$\mathrm{Ho}(Aff_{\mathcal{C}}/X^{\sim,\tau}) \longrightarrow \mathrm{Ho}(Aff_{\mathcal{C}}^{\sim,\tau}/\underline{h}_X)$$

is fully faithful on the whole category. □

The important consequence of Prop. 1.3.2.10 comes from the fact that it allows to see objects in $\mathrm{Ho}(Aff_{\mathcal{C}}^{\sim,\tau}/\underline{h}_X)$ as functors

$$A - Comm(\mathcal{C}) \longrightarrow SSet.$$

This last fact will be used implicitly in the sequel of this work.

As explained in the introduction to this chapter, we will need to fix a class **P** of morphisms in $Aff_{\mathcal{C}}$. Such a class will be then used to glue representable stacks to get a *geometric stack*. In other words, geometric stacks will be the objects obtained by taking some quotient of representable stacks by equivalence relations whose structural morphisms are in **P**. Of course, different choices of **P** will lead to different notions of geometric stacks. To fix his intuition the reader may think of **P** as being the class of smooth morphisms (though in some applications **P** can be something different).

From now, and all along this section, we fix a class **P** of morphism in $Aff_{\mathcal{C}}$, that is stable by equivalences. As the Yoneda functor

$$\mathbb{R}\underline{h} : \mathrm{Ho}(Aff_{\mathcal{C}})^{op} \longrightarrow \mathrm{St}(\mathcal{C}, \tau)$$

is fully faithful we can extend the notion of morphisms belonging to **P** to its essential image. So, a morphism of representable objects in $\mathrm{Ho}(SPr(Aff_{\mathcal{C}}^{\sim,\tau}))$ is in **P** if by definition it correspond to a morphism in $\mathrm{Ho}(Aff_{\mathcal{C}})$ which is in **P**. We will make the following assumtions on morphisms of **P** with respect to the topology $\tau$, making "being in **P**" into a $\tau$-local property.

ASSUMPTION 1.3.2.11. (1) *Covering families consist of morphisms in* ***P*** *i.e. for any $\tau$-covering family $\{U_i \longrightarrow X\}_{i \in I}$ in $Aff_{\mathcal{C}}$, the morphism $U_i \longrightarrow X$ is in* ***P*** *for all $i \in I$.*
(2) *Morphisms in* ***P*** *are stable by compositions, equivalences and homotopy pull-backs.*
(3) *Let $f : X \longrightarrow Y$ be a morphism in $Aff_{\mathcal{C}}$. If there exists a $\tau$-covering family*

$$\{U_i \longrightarrow X\}$$

*such that each composite morphism $U_i \longrightarrow Y$ lies in* ***P****, then $f$ belongs to* ***P****.*
(4) *For any two objects $X$ and $Y$ in $Aff_{\mathcal{C}}$, the two natural morphisms*

$$X \longrightarrow X \coprod^{\mathbb{L}} Y \qquad Y \longrightarrow X \coprod^{\mathbb{L}} Y$$

*are in* ***P****.*

The reader will notice that assumptions 1.3.2.2 and 1.3.2.11 together imply the following useful fact.

LEMMA 1.3.2.12. *Let $\{X_i \longrightarrow X\}$ be a finite family of morphisms in* ***P****. The total morphism*

$$\coprod_i^{\mathbb{L}} X_i \longrightarrow X$$

*is also in* ***P****.*

PROOF. We consider the family of natural morphisms

$$\{X_j \longrightarrow \coprod_i^{\mathbb{L}} X_i\}_j.$$

According to our assumption 1.3.2.2 it is a $\tau$-covering family in $Aff_{\mathcal{C}}$. Moreover, each morphism $X_j \longrightarrow \coprod_i^{\mathbb{L}} X_i$ and $X_j \longrightarrow X$ is in **P**, so assumption 1.3.2.11 (3) implies that so is $\coprod_i^{\mathbb{L}} X_i \longrightarrow X$. □

We finish this section by the following definition.

DEFINITION 1.3.2.13. *A* Homotopical Algebraic Geometry context *(or simply* HAG context*) is a 5-tuple $(\mathcal{C}, \mathcal{C}_0, \mathcal{A}, \tau, \boldsymbol{P})$, where $(\mathcal{C}, \mathcal{C}_0, \mathcal{A})$ is a HA context in the sense of Def. 1.1.0.11, $\tau$ is a model pre-topology on $Aff_{\mathcal{C}}$, $\boldsymbol{P}$ is a class of morphism in $Aff_{\mathcal{C}}$, and such that assumptions 1.3.2.2 and 1.3.2.11 are satisfied.*

### 1.3.3. Main definitions and standard properties

From now on, we fix a HAG context $(\mathcal{C}, \mathcal{C}_0, \mathcal{A}, \tau, \mathbf{P})$ in the sense of Def. 1.3.2.13. We will consider the model category of stacks $Aff_{\mathcal{C}}^{\sim,\tau}$ as described in the previous section, and introduce the notion of *geometric* and *n-geometric stacks*, which are objects in $Aff_{\mathcal{C}}^{\sim,\tau}$ satisfying certain properties.

The basic geometric idea is that a stack is geometric if it is obtained by taking the quotient of a representable stack $X$ (or more generally of a disjoint union of representable stacks) by the action of a groupoid object $X_1$ acting on $X$, such that $X_1$ is itself representable, and such that the source morphism $X_1 \longrightarrow X$ is a morphism in the chosen class $\mathbf{P}$. If one thinks of $\mathbf{P}$ as being the class of certain smooth morphisms, being geometric is thus equivalent of being a quotient by a smooth groupoid action.

It turns out that this notion is not enough for certain applications, as some naturally arising stacks are obtained as quotients by an action of a groupoid in geometric stacks rather than in representable stacks (e.g. the quotients by a group-stack action). We will therefore also introduce the notion of $n$-geometric stack, which is defined inductively as a stack obtained as a quotient by an action of a groupoid object in $(n-1)$-geometric stacks whose source morphism is in $\mathbf{P}$. Of course, for this definition to make sense one must also explain, inductively on $n$, what are the morphisms in $\mathbf{P}$ between $n$-geometric stacks.

The inductive definition we give below uses a different (though equivalent) point of view, closer to the original definition of algebraic stacks due to Deligne-Mumford and M. Artin. It says that a stack $F$ is $n$-geometric if for any pair of points of $F$, the stack of equivalences between them is $(n-1)$-geometric, and if moreover it receives a morphism, which is surjective and in $\mathbf{P}$, from a representable stack (or from a disjoint union of representable stacks). The equivalence of this definition with the previously mentioned quotient-by-groupoids point of view will be established in the next section (see Prop. 1.3.4.2).

DEFINITION 1.3.3.1.
(1) *A stack is* $(-1)$-geometric *if it is representable.*
(2) *A morphism of stacks $f : F \longrightarrow G$ is* $(-1)$-representable *if for any representable stack $X$ and any morphism $X \longrightarrow G$, the homotopy pullback $F \times^h_G X$ is a representable stack.*
(3) *A morphism of stacks $f : F \longrightarrow G$ is* in $(-1)$-$\mathbf{P}$ *if it is $(-1)$-representable, and if for any representable stack $X$ and any morphism $X \longrightarrow G$, the induced morphism*

$$F \times^h_G X \longrightarrow X$$

*is a $\boldsymbol{P}$-morphism between representable stacks.*

*Now let $n \geq 0$.*

(1) *Let $F$ be any stack. An* $n$-atlas for $F$ *is a $\mathbb{U}$-small family of morphisms $\{U_i \longrightarrow F\}_{i\in I}$ such that*
   (a) *Each $U_i$ is representable.*
   (b) *Each morphism $U_i \longrightarrow F$ is in $(n-1)$-$\boldsymbol{P}$.*

(c) *The total morphism*

$$\coprod_{i\in I} U_i \longrightarrow F$$

*is an epimorphism.*

(2) *A stack $F$ is* $n$-geometric *if it satisfies the following two conditions.*
(a) *The diagonal morphism $F \longrightarrow F \times^h F$ is $(n-1)$-representable.*
(b) *The stack $F$ admits an $n$-atlas.*

(3) *A morphism of stacks $F \longrightarrow G$ is* $n$-representable *if for any representable stack $X$ and any morphism $X \longrightarrow G$, the homotopy pullback $F \times^h_G X$ is $n$-geometric.*

(4) *A morphism of stacks $F \longrightarrow G$ is* in $n$-**P** *(or* has the property $n$-**P**, *or* is a $n$-**P**-morphism*) if it is $n$-representable and if for any representable stack $X$, any morphism $X \longrightarrow G$, there exists an $n$-atlas $\{U_i\}$ of $F \times^h_G X$, such that each composite morphism $U_i \longrightarrow X$ is in* ***P****.*

REMARK 1.3.3.2. In the above definition, condition (2a) follows from condition (2b). This is not immediate now but will be an easy consequence of the description of geometric stacks as quotients by groupoids given in the next section. We prefer to keep the definition of n-geometric stacks with the two conditions (2a) and (2b) as it is very similar to the usual definition of algebraic stacks found in the literature (e.g. in [**La-Mo**]).

The next Proposition gives the fundamental properties of geometric $n$-stacks.

PROPOSITION 1.3.3.3.
(1) *Any $(n-1)$-representable morphism is $n$-representable.*
(2) *Any $(n-1)$-**P**-morphism is a $n$-**P**-morphism.*
(3) *$n$-representable morphisms are stable by isomorphisms, homotopy pullbacks and compositions.*
(4) *$n$-**P**-morphisms are stable by isomorphisms, homotopy pullbacks and compositions.*

PROOF. We use a big induction on $n$. All the assertions are easily verified for $n = -1$ using our assumptions 1.3.2.11 on the morphisms in **P**. So, we fix an integer $n \geq 0$ and suppose that *all* the assertions are true for any $m < n$; let's prove that they all remain true at the level $n$.

(1) By definition 1.3.3.1 it is enough to check that a $(n-1)$-geometric stack $F$ is $n$-geometric. But an $(n-1)$-atlas for $F$ is a $n$-atlas by induction hypothesis (which tells us in particular that a $(n-2)$-**P**-morphism is a $(n-1)$-**P**-morphism), and moreover the diagonal of $F$ is $(n-2)$-representable thus $(n-1)$-representable, again by induction hypothesis. Therefore $F$ is $n$-geometric.

(2) Let $f : F \longrightarrow G$ be an $(n-1)$-**P**-morphism. By definition it is $(n-1)$-representable, hence $n$-representable by (1). Let $X \longrightarrow G$ be a morphism with $X$ representable. Since $f$ is in $(n-1)$-**P**, there exists an $(n-1)$-atlas $\{U_i\}$ of $F \times^h_G X$ such that each $U_i \longrightarrow X$ is in **P**. But, as already observed in (1), our inductive hypothesis, shows that any $(n-1)$-atlas is an $n$-atlas, and we conclude that $f$ is also in $n$-**P**.

(3) Stability by isomorphisms and homotopy pullbacks is clear by definition. To prove the stability by composition, it is enough to prove that if $f : F \longrightarrow G$ is an $n$-representable morphism and $G$ is $n$-geometric then so is $F$.

Let $\{U_i\}$ be an $n$-atlas of $G$, and let $F_i := F\times^h_G U_i$. The stacks $F_i$ are $n$-geometric, so we can find an $n$-atlas $\{V_{i,j}\}_j$ for $F_i$, for any $i$. By induction hypothesis (telling us in particular that $(n-1)$-**P**-morphisms are closed under composition) we see that the family of morphisms $\{V_{i,j} \longrightarrow F\}$ is an $n$-atlas for $F$. It remains to show that the diagonal of $F$ is $(n-1)$-representable.

There is a homotopy cartesian square

$$\begin{array}{ccc} F\times^h_G F & \longrightarrow & F\times^h F \\ \downarrow & & \downarrow \\ G & \longrightarrow & G\times^h G. \end{array}$$

As $G$ is $n$-geometric, the stability of $(n-1)$-representable morphisms under homotopy pullbacks (true by induction hypothesis) implies that $F\times^h_G F \longrightarrow F\times^h F$ is $(n-1)$-representable. Now, the diagonal of $F$ factors has $F \longrightarrow F\times^h_G F \longrightarrow F\times^h F$, and therefore by stability of $(n-1)$-representable morphisms by composition (true by induction hypothesis), we see that it is enough to show that $F \longrightarrow F\times^h_G F$ is $(n-1)$-representable. Let $X$ be a representable stack and $X \longrightarrow F\times^h_G F$ be any morphism. Then, we have

$$F\times^h_{F\times^h_G F} X \simeq X\times^h_{(F\times^h_G X)\times^h(F\times^h_G X)} (F\times^h_G X).$$

As by hypothesis $F\times^h_G X$ is $n$-geometric this shows that $F\times^h_{F\times^h_G F} X$ is $(n-1)$-geometric, showing that $F \longrightarrow F\times^h_G F$ is $(n-1)$-representable.

(4) Stability by isomorphisms and homotopy pullbacks is clear by definition. Let $F \longrightarrow G \longrightarrow H$ be two $n$-**P**-morphisms of stacks. By (3) we already know the composite morphism to be $n$-representable. By definition of being in $n$-**P** one can assume that $H$ is representable. Then, there exists an $n$-atlas $\{U_i\}$ for $G$ such that each morphism $U_i \longrightarrow H$ is in **P**. Let $F_i := F\times^h_G U_i$, and let $\{V_{i,j}\}_j$ be an $n$-atlas for $F_i$ such that each $V_{i,j} \longrightarrow U_i$ is in **P**. Since by induction hypothesis $(n-1)$-**P**-morphisms are closed under composition, we see that $\{V_{i,j}\}$ is indeed an $n$-atlas for $F$ such that each $V_{i,j} \longrightarrow H$ is in **P**. □

An important consequence of our descent assumption 1.3.2.2 and Prop. 1.3.2.8 is the following useful proposition.

PROPOSITION 1.3.3.4. *Let $f : F \longrightarrow G$ be any morphism such that $G$ is an $n$-geometric stack. We suppose that there exists a $n$-atlas $\{U_i\}$ of $G$ such that each stack $F\times^h_G U_i$ is $n$-geometric. Then $F$ is $n$-geometric.*

*If furthermore each projection $F\times^h_X U_i \longrightarrow U_i$ is in $n$-**P**, then so is $f$.*

PROOF. Using the stability of $n$-representable morphisms by composition (see Prop. 1.3.3.3 (3)) we see that it is enough to show that $f$ is $n$-representable. The proof goes by induction on $n$. For $n=-1$ this is our corollary Prop. 1.3.2.8. Let us assume $n\geq 0$ and the proposition proved for rank less than $n$. Using Prop. 1.3.3.3 (3) it is enough to suppose that $G$ is a representable stack $X$.

Let $\{U_i\}$ be an $n$-atlas of $X$ as in the statement, and let $\{V_{i,j}\}$ be an $n$-atlas for $F_i := F\times^h_X U_i$. Then, the composite family $V_{i,j} \longrightarrow F$ is clearly an $n$-atlas for $F$. It remains to prove that $F$ has an $(n-1)$-representable diagonal.

The diagonal of $F$ factors as

$$F \longrightarrow F\times^h_X F \longrightarrow F\times^h F.$$

The last morphism being the homotopy pullback

$$\begin{array}{ccc} F\times^h_X F & \longrightarrow & F\times^h F \\ \downarrow & & \downarrow \\ X & \longrightarrow & X\times^h X \end{array}$$

is representable and therefore $(n-1)$-representable. Finally, let $Z$ be any representable stack and $Z \longrightarrow F\times^h_X F$ be a morphism. Then, the morphism

$$Z\times^h_{F\times^h_X F} F \longrightarrow X$$

satisfies the conditions of the proposition 1.3.3.4 for the rank $(n-1)$. Indeed, for any $i$ we have

$$(Z\times^h_{F\times^h_X F} F)\times^h_X U_i \simeq (Z\times^h_X U_i)\times^h_{F_i\times^h_{U_i} F_i} F_i.$$

Therefore, using the induction hypothesis we deduce that the stack $Z\times^h_{F\times^h_X F} F$ is $(n-1)$-geometric, proving that $F\longrightarrow F\times^h_X F$ is $(n-1)$-representable.

The last part of the proposition follows from the fact that any $n$-atlas $\{V_{i,j}\}$ of $F_i$ is such that each morphism $V_{i,j}\longrightarrow X$ is in $n$-**P** by construction. □

COROLLARY 1.3.3.5. *The full subcategory of $n$-geometric stacks in* $\mathrm{St}(\mathcal{C},\tau)$ *is stable by homotopy pullbacks, and by $\mathbb{U}$-small disjoint union if $n\geq 0$.*

PROOF. Let $F\longrightarrow H \longleftarrow G$ be a diagram of stacks. There are two homotopy cartesian squares

$$\begin{array}{ccc} F\times^h G & \longrightarrow & G \\ \downarrow & & \downarrow \\ F & \longrightarrow & \bullet \end{array} \qquad \begin{array}{ccc} F\times^h_H G & \longrightarrow & H \\ \downarrow & & \downarrow \\ F\times^h G & \longrightarrow & H\times^h H, \end{array}$$

showing that the stability under homotopy pullbacks follows from the stability of $n$-representable morphisms under compositions and homotopy pullbacks.

Let us prove the second part of the corollary, concerning $\mathbb{U}$-small disjoint union. Suppose now tha $n\geq 0$ and let $F$ be $\coprod_i F_i$ with each $F_i$ an $n$-geometric stack. Then, we have

$$F\times^h F\simeq \coprod_{i,j} F_i\times^h F_j.$$

For any representable stack $X$, and any morphism $X\longrightarrow F\times^h F$, there exists a 0-atlas $\{U_k\}$ of $X$, and commutative diagrams of stacks

$$\begin{array}{ccc} U_k & \longrightarrow & F_{i(k)}\times^h F_{j(k)} \\ \downarrow & & \downarrow \\ X & \longrightarrow & \coprod_{i,j} F_i\times^h F_j. \end{array}$$

We apply Prop. 1.3.3.4 to the morphism

$$G := F\times^h_{F\times^h F} X \longrightarrow X$$

and for the covering $\{U_k\}$. We havc

$$G\times^h_X U_k\simeq \emptyset \; if \; i(k)\neq j(k)$$

and

$$G\times^h_X U_k \simeq F_{i(k)}\times^h_{F_{i(k)}\times^h F_{i(k)}} U_k$$

otherwise. Prop. 1.3.3.4 implies that $G$ is $(n-1)$-geometric, and therefore that the diagonal of $F$ is $(n-1)$-representable. Finally, the same argument and assumption 1.3.2.11 show that the disjoint union of $n$-atlases of the $F_i$'s will form an $n$-atlas for $F$. □

Finally, let us mention the following important additional property.

PROPOSITION 1.3.3.6. *Let $f : F \longrightarrow G$ be an $n$-representable morphism. If $f$ is in $m$-$\mathbf{P}$ for $m > n$ then it is in $n$-$\mathbf{P}$.*

PROOF. By induction on $m$ it is enough to treat the case $m = n+1$. The proof goes then by induction on $n$. For $n = -1$ this is our assumption 1.3.2.11 (3). For $n \geq 0$ we can by definition assume $G$ is a representable stack and therefore that $F$ is $n$-geometric. Then, there exists an $(n+1)$-atlas $\{U_i\}$ for $F$ such that each $U_i \longrightarrow G$ is in $\mathbf{P}$. By induction, $\{U_i\}$ is also an $n$-atlas for $F$, which implies that the morphism $f$ is in fact in $\mathbf{P}$. □

The last proposition implies in particular that for an $n$-representable morphism of stacks the property of being in $n$-$\mathbf{P}$ does not depend on $n$. We will therefore give the following definition.

DEFINITION 1.3.3.7. *A morphism in* $\mathrm{St}(\mathcal{C}, \tau)$ *is in* $\mathbf{P}$ *if it is in $n$-$\mathbf{P}$ for some integer $n$.*

### 1.3.4. Quotient stacks

We will now present a characterization of geometric stacks in terms of quotient by groupoid actions. This point of view is very much similar to the presentation of manifolds by charts, and much less intrinsic than definition Def. 1.3.3.1. However, it is sometimes more easy to handle as several stacks have natural presentations as quotients by groupoids.

Let $X_*$ be a Segal groupoid object in a model category $M$ (Def. 1.3.1.6). Inverting the equivalence

$$X_2 \longrightarrow X_1 \times^h_{X_0} X_1$$

and composing with $d_1 : X_2 \longrightarrow X_1$ gives a well defined morphism in $\mathrm{Ho}(M)$

$$\mu : X_1 \times^h_{X_0} X_1 \longrightarrow X_1,$$

that is called *composition*. In the same way, inverting the equivalence

$$X_2 \longrightarrow X_1 \times^h_{d_0, X_0, d_0} X_1,$$

and composing with

$$Id \times^h s_0 : X_1 \longrightarrow X_1 \times^h_{d_0, X_0, d_0} X_1$$

$$d_2 : X_2 \longrightarrow X_1$$

gives a well defined isomorphism in $\mathrm{Ho}(M)$

$$i : X_1 \longrightarrow X_1,$$

called *inverse*. It is easy to check that $d_1 \circ i = d_0$ as morphisms in $\mathrm{Ho}(M)$, showing that the two morphisms $d_0$ and $d_1$ are always isomorphic in $\mathrm{Ho}(M)$. Finally, using condition (1) of Def. 1.3.1.6 we see that for any $i > 0$, all the face morphisms

$$X_i \longrightarrow X_{i-1}$$

of a Segal groupoid object are isomorphic in $\mathrm{Ho}(M)$.

DEFINITION 1.3.4.1. *A Segal groupoid object* $X_*$ *in* $\mathrm{Ho}(SPr(Aff_{\mathcal{C}}^{\sim,\tau}))$ *is* an $n$-**P** Segal groupoid *if it satisfies the following two conditions.*

1. *The stacks* $X_0$ *and* $X_1$ *are disjoint unions of* $n$*-geometric stacks.*
2. *The morphism* $d_0 : X_1 \longrightarrow X_0$ *is in* $n$-$\boldsymbol{P}$.

As $n$-geometric stacks are stable by homotopy pullbacks, $X_i$ is a disjoint union of $n$-geometric stacks for any $i$ and any $n$-**P** Segal groupoid $X_*$. Furthermore, the condition (2) of Def. 1.3.1.6 implies that the two morphisms

$$d_0, d_1 : X_1 \longrightarrow X_0$$

are isomorphic as morphisms in $\mathrm{St}(\mathcal{C}, \tau)$. Therefore, for any $n$-**P** Segal groupoid $X_*$, all the faces $X_i \longrightarrow X_{i-1}$ are in **P**.

PROPOSITION 1.3.4.2. *Let* $F \in \mathrm{Ho}(SPr(Aff_{\mathcal{C}}^{\sim,\tau}))$ *be a stack and* $n \geq 0$. *The following conditions are equivalent.*

1. *The stack* $F$ *is* $n$*-geometric.*
2. *There exists an* $(n-1)$-$\boldsymbol{P}$ *Segal groupoid object* $X_*$ *in* $SPr(Aff_{\mathcal{C}}^{\sim,\tau})$*, such that* $X_0$ *is a disjoint union of representable stacks, and an isomorphism in* $\mathrm{Ho}(SPr(Aff_{\mathcal{C}}^{\sim,\tau}))$
$$F \simeq |X_*| := Hocolim_{[n]\in\Delta} X_n.$$
3. *There exists an* $(n-1)$-$\boldsymbol{P}$ *Segal groupoid object* $X_*$ *in* $SPr(Aff_{\mathcal{C}}^{\sim,\tau})$*, and an isomorphism in* $\mathrm{Ho}(SPr(Aff_{\mathcal{C}}^{\sim,\tau}))$
$$F \simeq |X_*| := Hocolim_{[n]\in\Delta} X_n.$$

*If these conditions are satisfied we say that* $F$ is the quotient stack of the $(n-1)$-**P** Segal groupoid $X_*$.

PROOF. We have already seen that a 0-geometric stack is $n$-geometric for any $n$. Therefore, (2) $\Rightarrow$ (3). It remains to show that (1) implies (2) and (3) implies (1).

(1) $\Rightarrow$ (2) Let $F$ be an $n$-geometric stack, and $\{U_i\}$ be an $n$-atlas for $F$. We let

$$p : X_0 := \coprod_i U_i \longrightarrow F$$

be the natural projection. Up to an equivalence, we can represent $p$ by a fibration $X_0 \longrightarrow F$ between fibrant objects in $SPr(Aff_{\mathcal{C}}^{\sim,\tau})$. We define a simplicial object $X_*$ to be the nerve of $p$

$$X_n := \underbrace{X_0 \times_F X_0 \times_F \cdots \times_F X_0}_{n\ times}.$$

Clearly, $X_*$ is a groupoid object in $SPr(Aff_{\mathcal{C}}^{\sim,\tau})$ in the usual sense, and as $p$ is a fibration between fibrant objects it follows that it is also a Segal groupoid object in the sense of Def. 1.3.1.6. Finally, as $F$ is $n$-geometric, one has $X_1 \simeq \coprod_{i,j} U_i \times^h_F U_j$ which is therefore an $(n-1)$-geometric stack by Cor. 1.3.3.5. The morphism $d_0 :$ $X_1 \longrightarrow X_0 \simeq \coprod_i U_i$ is then given by the projections $U_i \times^h_F U_j \longrightarrow U_i$ which are in $(n-1)$-**P** as $\{U_i\}$ is an $n$-atlas. This implies that $X_*$ is an $(n-1)$-**P** Segal groupoid such that $X_0$ is a disjoint union of representable.

LEMMA 1.3.4.3. *The natural morphism*

$$|X_*| \longrightarrow F$$

*is an isomorphism in* $\mathrm{Ho}(SPr(Aff_{\mathcal{C}}^{\sim,\tau}))$.

PROOF. For any fibration of simplicial sets $f : X \longrightarrow Y$, we know that the natural morphism from the geometric realization of the nerve of $f$ to $Y$ is equivalent to an inclusion of connected components. This implies that the morphism $|X_*| \longrightarrow F$ induces an isomorphisms on the sheaves $\pi_i$ for $i > 0$ and an injection on $\pi_0$. As the morphism $p : X_0 \longrightarrow F$ is an epimorphism and factors as $X_0 \to |X_*| \to F$, the morphism $|X_*| \longrightarrow F$ is also an epimorphism. This shows that it is surjective on the sheaves $\pi_0$ and therefore is an isomorphism in $\mathrm{Ho}(SPr(Aff_{\mathcal{C}}^{\sim,\tau}))$. $\square$

The previous lemma finishes the proof of (1) $\Rightarrow$ (2).

(3) $\Rightarrow$ (1) Let $X_*$ be an $(n-1)$-**P** Segal groupoid and $F = |X_*|$. First of all, we recall the following important fact.

LEMMA 1.3.4.4. *Let $M$ be a $\mathbb{U}$-model topos, and $X_*$ be a Segal groupoid object in $M$ with homotopy colimit $|X_*|$. Then, for any $n > 0$, the natural morphism*

$$X_n \longrightarrow \underbrace{X_0 \times^h_{|X_*|} X_0 \times^h_{|X_*|} \cdots \times^h_{|X_*|} X_0}_{n \text{ times}}$$

*is an isomorphism in* $\mathrm{Ho}(M)$.

PROOF. This is one of the standard properties of model topoi. See Thm. 1.3.1.7. $\square$

Let $\{U_i\}$ be an $(n-1)$-atlas for $X_0$, and let us consider the composed morphisms

$$f_i : U_i \longrightarrow X_0 \longrightarrow F.$$

Clearly, $\coprod_i U_i \longrightarrow F$ is a composition of epimorphism, and is therefore a epimorphism. In order to prove that $\{U_i\}$ form an $n$-atlas for $F$ it is enough to prove that the morphism $X_0 \longrightarrow F$ is in $(n-1)$-**P**.

Let $X$ be any representable stack, $X \longrightarrow F$ be a morphism, and let $G$ be $X_0 \times^h_F X$. As the morphism $X_0 \longrightarrow F$ is an epimorphism, we can find a covering family $\{Z_j \longrightarrow X\}$, such that each $Z_j$ is representable, and such that there exists a commutative diagram in $\mathrm{St}(\mathcal{C}, \tau)$

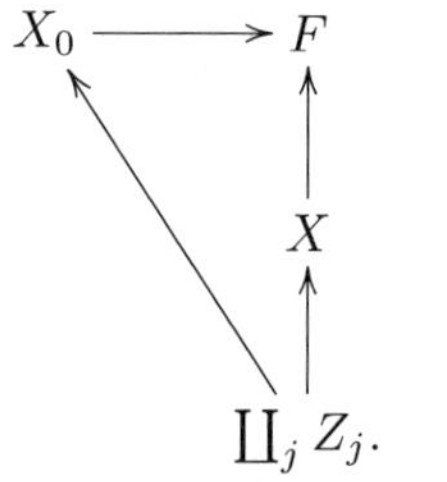


By assumption 1.3.2.11 (1) we can also assume that each morphism $Z_j \longrightarrow X$ is in **P**, and therefore that $\{Z_j \longrightarrow X\}$ is a 0-atlas of $X$.

In order to prove that $G \longrightarrow X$ is in $(n-1)$-**P** it is enough by Prop. 1.3.3.4 to prove that each stack $G_j := G \times^h_X Z_j \simeq X_0 \times^h_F Z_j$ is $(n-1)$-geometric, and furthermore that each projection $G_j \longrightarrow Z_j$ is in **P**. We have

$$G_j \simeq (X_0 \times^h_F X_0) \times^h_{X_0} Z_j.$$

Therefore, lemma 1.3.4.4 implies that $G_j \simeq X_1 \times^h_{X_0} Z_j$, showing that each $G_j$ is $(n-1)$-geometric and finally that $G$ is $(n-1)$-geometric. Furthermore, this also shows that each projection $G_j \longrightarrow Z_j$ is of the form $X_1 \times^h_{X_0} Z_j \longrightarrow Z_j$ which is in **P** by hypothesis on the Segal groupoid object $X_*$. $\square$

COROLLARY 1.3.4.5. *Let $f : F \longrightarrow G$ be an epimorphism of stacks and $n \geq 0$. If $F$ is $n$-geometric and $f$ is $(n-1)$-representable and in $\mathbf{P}$, then $G$ is $n$-geometric.*

PROOF. Indeed, let $U \longrightarrow F$ be an $n$-atlas, and let $g : U \longrightarrow G$ be the composition. The morphism $g$ is still an epimorphism and $(n-1)$-representable, and thus we can assume that $F$ is 0-representable. The morphism $f$ being an epimorphism, $G$ is equivalent to the quotient stack of the Segal groupoid $X_*$ which is the homotopy nerve of $f$. By assumption this Segal groupoid is an $(n-1)$-Segal groupoid and thus $G$ is $n$-geometric by Prop. 1.3.4.2 (1). □

## 1.3.5. Quotient stacks and torsors

Writing a stack as a quotient stack of a Segal groupoid is also useful in order to describe associated stacks to certain pre-stacks. Indeed, it often happens that a pre-stack is defined as the quotient of a Segal groupoid, and we are going to show in this section that the associated stack of such quotients can be described using an adequate notion of torsor over a Segal groupoid. This section is in fact completely independent of the notion of geometricity and concern pure stacky statements that are valid in any general model topos. We have decided to include this section as it helps understanding the quotient stack construction presented previously. However, it is not needed for the rest of this work and proofs will be more sketchy than usual.

We let $M$ be a general $\mathbb{U}$-model topos in the sense of Def. 1.3.1.5. The main case of application will be $M = Aff_{\mathcal{C}}^{\sim,\tau}$ but we rather prefer to state the results in the most general setting (in particular we do not even assume that $M$ is $t$-complete).

Let $X_*$ be a Segal groupoid object in $M$ in the sense of Def. 1.3.1.6, and we assume that each $X_n$ is a fibrant object in $M$. We will consider $sM$, the category of simplicial objects in $M$, which will be endowed with its levelwise projective model structure, for which fibrations and equivalences are defined levelwise. We consider $sM/X_*$, the model category of simplicial objects over $X_*$. Finally, we let $Z$ be a fibrant object in $M$, and $X_* \longrightarrow Z$ be a morphism in $sM$ (where $Z$ is considered as a constant simplicial object), and we assume that the induced morphism

$$|X_*| \longrightarrow Z$$

is an isomorphism in $\mathrm{Ho}(M)$.

We define a Quillen adjunction

$$\phi : sM/X_* \longrightarrow M/Z \qquad sM/X_* \longleftarrow M/Z : \psi$$

in the following way. For any $Y \longrightarrow Z$ in $M$, we set

$$\psi(Y) := Y \times_Z X_* \in sM,$$

or in other words $\psi(Y)_n = Y \times_Z X_n$ and with the obvious transitions morphisms. The left adjoint to $\psi$ sends a simplicial object $Y_* \longrightarrow X_*$ to its colimit in $M$

$$\phi(Y_*) := Colim_{n \in \Delta^{op}} Y_n \longrightarrow Colim_n X_{n \in \Delta^{op}} \longrightarrow Z.$$

It is easy to check that $(\phi, \psi)$ is a Quillen adjunction.

PROPOSITION 1.3.5.1. *The functor*

$$\mathbb{R}\psi : \mathrm{Ho}(M/Z) \longrightarrow \mathrm{Ho}(sM/X_*)$$

*is fully faithful. Its essential image consists of all* $Y_* \longrightarrow X_*$ *such that for any morphism* $[n] \to [m]$ *in* $\Delta$ *the square*

$$\begin{array}{ccc} Y_m & \longrightarrow & Y_n \\ \downarrow & & \downarrow \\ X_m & \longrightarrow & X_n \end{array}$$

*is homotopy cartesian.*

PROOF. Let $Y \to Z$ in $M/Z$. Proving that $\mathbb{R}\psi$ is fully faithful is equivalent to prove that the natural morphism

$$|Y \times^h_Z X_*| \longrightarrow Y$$

is an isomorphism in $\mathrm{Ho}(M)$. But, using the standard properties of model topoi (see Thm. 1.3.1.7), we have

$$|Y \times^h_Z X_*| \simeq Y \times^h_Z |X_*| \simeq Y$$

as $|X_*| \simeq Z$. This shows that $\mathbb{R}\psi$ is fully faithful.

By definition of the functors $\phi$ and $\psi$, it is clear that $\mathbb{R}\psi$ takes its values in the subcategory described in the statement of the proposition. Conversely, let $Y_* \longrightarrow X_*$ be an object in $\mathrm{Ho}(M)$ satisfying the condition of the proposition. As $X_*$ is a Segal groupoid object, we know that $X_*$ is naturally equivalent to the homotopy nerve of the augmentation morphism $X_0 \longrightarrow Z$ (see Thm. 1.3.1.7). Therefore, the object $\mathbb{R}\psi\mathbb{L}\phi(Y_*)$ is by definition the homotopy nerve of the morphism

$$|Y_*| \times^h_Z X_0 \longrightarrow |Y_*|.$$

But, we have

$$|Y_*| \times^h_Z X_0 \simeq |Y_* \times^h_Z X_0| \simeq Y_0$$

by hypothesis on $Y_*$. Therefore, the object $\mathbb{R}\psi\mathbb{L}\phi(Y_*)$ is naturally isomorphic in $\mathrm{Ho}(sM/X_*)$ to the homotopy nerve of the natural

$$Y_0 \longrightarrow |Y_*|.$$

Finally, as $X_*$ is a Segal groupoid object, so is $Y_*$ by assumption. The standard properties of model topoi (see Thm. 1.3.1.7) then tell us that $Y_*$ is naturally equivalent to the homotopy nerve of $Y_0 \longrightarrow |Y_*|$, and thus to $\mathbb{R}\psi\mathbb{L}\phi(Y_*)$ by what we have just done. □

The model category $sM/X_*$, or rather its full subcategory of objects satisfying the conditions of Prop. 1.3.5.1, can be seen as the category of objects in $M$ together with an *action of the Segal groupoid* $X_*$. Proposition 1.3.5.1 therefore says that the homotopy theory of stacks over $|X_*|$ is equivalent to the homotopy theory of stacks together with an action of $X_*$. This point of view will now help us to describe the stack associated to $|X_*|$.

For this, let $F$ be a fixed fibrant object in $M$. We define a new model category $sM/(X_*, F)$ in the following way. Its objects are pairs $(Y_*, f)$, where $Y_* \to X_*$ is an object in $sM/X_*$ and $f : Colim_n Y_n \longrightarrow F$ is a morphism in $M$. Morphisms $(Y_*, f) \to (Y'_*, f')$ are given by morphisms $Y_* \longrightarrow Y'_*$ in $sM/X_*$, such that

$$\begin{array}{ccc} Colim_n Y_n & \longrightarrow & Colim_n Y'_n \\ & {\scriptstyle f}\searrow \quad \swarrow{\scriptstyle f'} & \\ & F & \end{array}$$

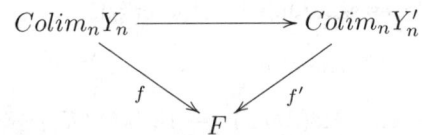

commutes in $M$. The model structure on $sM/(X_*, F)$ is defined in such a way that fibrations and equivalences are defined on the underlying objects in $sM$. The model category $sM/(X_*, F)$ is also the comma model category $sM/(X_* \times F)$ where $F$ is considered as a constant simplicial object in $M$.

DEFINITION 1.3.5.2. *An object $Y_* \in sM/(X_*, F)$ is a $X_*$-*torsor on $F$ *if it satisfies the following two conditions.*

(1) *For all morphism $[n] \to [m]$ in $\Delta$, the square*

$$\begin{array}{ccc} Y_m & \longrightarrow & Y_n \\ \downarrow & & \downarrow \\ X_m & \longrightarrow & X_n \end{array}$$

*is homotopy cartesian.*

(2) *The natural morphism*

$$|Y_*| \longrightarrow F$$

*is an isomorphism in* Ho$(M)$.

*The* space of $X_*$-torsors over $F$, *denoted by $Tors_{X_*}(F)$, is the nerve of the sub category of fibrant objects $sM/(X_*, F)^f$, consisting of equivalences between $X_*$-torsors on $F$.*

Suppose that $f : F \longrightarrow F'$ is a morphism between fibrant objects in $M$. We get a pullback functor

$$sM/(X_*, F') \longrightarrow sM/(X_*, F),$$

which is right Quillen, and such that the induced functor on fibrant objects

$$sM/(X_*, F')^f \longrightarrow sM/(X_*, F)^f$$

sends $X_*$-torsors over $F$ to $X_*$-torsors over $F'$ (this uses the commutation of homotopy colimits with homotopy pullbacks). Therefore, restricting to the sub categories of equivalences, we get a well defined morphism between spaces of torsors

$$f^* : Tors_{X_*}(F') \longrightarrow Tors_{X_*}(F).$$

By applying the standard strictification procedure, we can always suppose that $(f \circ g)^* = g^* \circ f^*$. This clearly defines a functor from $(M^f)^{op}$, the opposite full subcategory of fibrant objects in $M^f$, to $SSet$

$$Tors_{X_*} : (M^f)^{op} \longrightarrow SSet.$$

This functor sends equivalences in $M^f$ to equivalences of simplicial sets, and therefore induces a Ho$(SSet)$-enriched functor (using for example [**D-K**1])

$$Tors_{X_*} : \mathrm{Ho}(M^f)^{op} \simeq \mathrm{Ho}(M)^{op} \longrightarrow \mathrm{Ho}(SSet).$$

In other words, there are natural morphisms in Ho$(SSet)$

$$Map_M(F, F') \longrightarrow Map_{SSet}(Tors_{X_*}(F'), Tors_{X_*}(F)),$$

compatible with compositions.

The main classification result is the following. It gives a way to describe the stack associated to $|X_*|$ for some Segal groupoid object $X_*$ in $M$.

PROPOSITION 1.3.5.3. *Let $X_*$ be a Segal groupoid object in $M$ and $Z$ be a fibrant model for $|X_*|$ in $M$. Then, there exists an element $\alpha \in \pi_0(Tors_{X_*}(Z))$, such that for any fibrant object $F \in M$, the evaluation at $\alpha$*

$$\alpha^* : Map_M(F, Z) \longrightarrow Tors_{X_*}(F)$$

*is an isomorphism in* Ho$(SSet)$.

PROOF. We can of course assume that $X_*$ is cofibrant in $sM$, as everything is invariant by changing $X_*$ with something equivalent. Let $Colim_n X_n \longrightarrow Z$ be a morphism in $M$ such that $Z$ is fibrant and such that the induced morphism

$$|X_*| \longrightarrow Z$$

is an equivalence. Such a morphism exists as $Colim_n X_n$ is cofibrant in $M$ and computes the homotopy colimit $|X_*|$. The element $\alpha \in \pi_0(Tors_{X_*}(Z))$ is defined to be the pair $(X_*, p) \in \mathrm{Ho}(sM/(X_*, Z))$, consisting of the identity of $X_*$ and the natural augmentation $p : Colim_n X_n \longrightarrow Z$. Clearly, $\alpha$ is a $X_*$-torsor over $Z$.

Applying Cor. A.0.5 of Appendix A, we see that there exists a homotopy fiber sequence

$$Map_M(F, Z) \longrightarrow N((M/Z)_W^f) \longrightarrow N(M_W^f)$$

where the $(M/Z)_W^f$ (resp. $M_W^f$) is the subcategory of equivalences between fibrant objects in $M/Z$ (resp. in $M$), and the morphism on the right is the forgetful functor and the fiber is taken at $F$. In the same way, there exists a homotopy fiber sequence

$$Tors_{X_*}(F) \longrightarrow N(((sM/X_*)_W^f)^{cart}) \longrightarrow N(M_W^f)$$

where $(sM/X_*)_W^f)^{cart}$ is the subcategory of $sM/X_*$ consisting of equivalence between fibrant objects satisfying condition (1) of Def. 1.3.5.2, and where the morphism on the right is induced my the homotopy colimit functor of underlying simplicial objects and the fiber is taken at $F$. There exists a morphism of homotopy fiber sequences of simplicial sets

$$\begin{array}{ccccc}
Map_M(F, Z) & \longrightarrow & N((M/Z)_W^f) & \longrightarrow & N(M_W^f) \\
\downarrow{\scriptstyle \alpha^*} & & \downarrow & & \downarrow{\scriptstyle Id} \\
Tors_{X_*}(F) & \longrightarrow & N(((sM/X_*)_W^f)^{cart}) & \longrightarrow & N(M_W^f)
\end{array}$$

and the arrow in the middle is an equivalence because of our proposition 1.3.5.1. □

### 1.3.6. Properties of morphisms

We fix another class **Q** of morphisms in $Aff_{\mathcal{C}}$, which is stable by equivalences, compositions and homotopy pullbacks. Using the Yoneda embedding the notion of morphisms in **Q** (or simply ***Q**-morphisms*) is extended to a notion of morphisms between representable stacks.

DEFINITION 1.3.6.1. *We say that* morphisms in **Q** are compatible with $\tau$ and **P** *(or equivalently that* morphisms in **Q** are local with respect to $\tau$ and **P**) *if the following two conditions are satisfied:*

1. *If $f : X \longrightarrow Y$ is a morphism in $Aff_{\mathcal{C}}$ such that there exists a covering family*
$$\{U_i \longrightarrow X\}$$
*with each composite morphism $U_i \longrightarrow Y$ in **Q**, then $f$ belongs to **Q**.*
2. *If $f : X \longrightarrow Y$ is a morphism in $Aff_{\mathcal{C}}$ and there exists a covering family*
$$\{U_i \longrightarrow Y\}$$
*such that each homotopy pullback morphism*
$$X \times_Y^h U_i \longrightarrow U_i$$
*is in **Q**, then $f$ belongs to **Q**.*

For a class of morphism $\mathbf{Q}$, compatible with $\tau$ and $\mathbf{P}$ in the sense above we can make the following definition.

DEFINITION 1.3.6.2. *Let* $\boldsymbol{Q}$ *be a class of morphisms in* $Aff_{\mathcal{C}}$, *stable by equivalences, homotopy pullbacks and compositions, and which is compatible with* $\tau$ *and* $\boldsymbol{P}$ *in the sense above. A morphism of stacks* $f : F \longrightarrow G$ *is* in $\mathbf{Q}$ *(or equivalently* is a $\mathbf{Q}$-morphism*) if it is* $n$*-representable for some* $n$*, and if for any representable stack* $X$ *and any morphism* $X \longrightarrow G$ *there exists an* $n$*-atlas* $\{U_i\}$ *of* $F \times^h_G X$ *such that each morphism* $U_i \longrightarrow X$ *between representable stacks is in* $\boldsymbol{Q}$.

Clearly, because of our definition 1.3.6.1, the notion of morphism in $\mathbf{Q}$ of definition 1.3.6.2 is compatible with the original notion. Furthermore, it is easy to check, as it was done for $\mathbf{P}$-morphisms, the following proposition.

PROPOSITION 1.3.6.3. 
(1) *Morphisms in* $\boldsymbol{Q}$ *are stable by equivalences, compositions and homotopy pullbacks.*
(2) *Let* $f : F \longrightarrow G$ *be any morphism between* $n$*-geometric stacks. We suppose that there exists a* $n$*-atlas* $\{U_i\}$ *of* $G$ *such that each projection* $F \times^h_X U_i \longrightarrow U_i$ *is in* $\boldsymbol{Q}$*. Then* $f$ *is in* $\boldsymbol{Q}$.

PROOF. Exercise. □

We can also make the following two general definitions of morphisms of stacks.

DEFINITION 1.3.6.4. *Let* $f : F \longrightarrow G$ *be a morphism of stacks.*
(1) *The morphism is* categorically locally finitely presented *if for any representable stack* $X = \mathbb{R}\underline{Spec}\, A$*, any morphism* $X \longrightarrow G$*, and any* $\mathbb{U}$*-small filtered system of commutative* $A$*-algebras* $\{B_i\}$*, the natural morphism*

$$Hocolim_i Map_{Aff_{\mathcal{C}}^{\sim,\tau}/X}(\mathbb{R}\underline{Spec}\, B_i, F\times^h_G X) \longrightarrow Map_{Aff_{\mathcal{C}}^{\sim,\tau}/X}(\mathbb{R}\underline{Spec}\,(Hocolim_i B_i), F\times^h_G X)$$

*is an isomorphism in* $\mathrm{Ho}(SSet)$.
(2) *The morphism* $f$ *is* quasi-compact *if for any representable stack* $X$ *and any morphism* $X \longrightarrow G$ *there exists a finite family of representable stacks* $\{X_i\}$ *and an epimorphism*

$$\coprod_i X_i \longrightarrow F \times^h_G X.$$

(3) *The morphism* $f$ *is* categorically finitely presented *if it is categorically locally finitely presented and quasi-compact.*
(4) *The morphism* $f$ *is a* monomorphism *if the natural morphism*

$$F \longrightarrow F \times^h_G F$$

*is an isomorphism in* $\mathrm{St}(\mathcal{C}, \tau)$.
(5) *Assume that the class* $\boldsymbol{Q}$ *of finitely presented morphism in* $Aff_{\mathcal{C}}$ *is compatible with the model topology* $\tau$ *in the sense of Def. 1.3.6.1. The morphism* $f$ *is* locally finitely presented *if it is a* $\boldsymbol{Q}$*-morphism in the sense of Def. 1.3.6.2. It is a* finitely presented morphism *if it is quasi-compact and locally finitely presented.*

Clearly, the above notions are compatible with the one of definition 1.2.6.1, in the sense that a morphism of commutative monoids $A \longrightarrow B$ has a certain property in the sense of Def. 1.2.6.1 if and only if the corresponding morphism of stacks $\mathbb{R}\underline{Spec}\, B \longrightarrow \mathbb{R}\underline{Spec}\, A$ has the same property in the sense of Def. 1.3.6.4.

PROPOSITION 1.3.6.5. *Quasi-compact morphisms, categorically (locally) finitely presented morphisms, (locally) finitely presented morphisms and monomorphisms are stable by equivalences, composition and homotopy pullbacks.*

PROOF. Exercise. □

REMARK 1.3.6.6. When the class of finitely presented morphisms in $Aff_{\mathcal{C}}$ is compatible with the topology $\tau$, Def. 1.3.6.4 gives us two different notions of locally finitely presented morphism which are a priori rather difficult to compare. Giving precise conditions under which they coincide is however not so important as much probably they are already different in some of our main examples (e.g. for each example for which the representable stacks are not truncated, as in the complicial, or brave new algebraic geometry presented in §2.3 and §2.4). In this work we have chosen to use only the non categorical version of locally finitely presented morphisms, the price to pay being of course that they do not have easy functorial characterization.

### 1.3.7. Quasi-coherent modules, perfect modules and vector bundles

For a commutative monoid $A$ in $\mathcal{C}$, we define a category $A-QCoh$, of quasi-coherent modules on $A$ (or equivalently on $Spec\, A$) in the following way. Its objects are the data of a $B$-module $M_B$ for any commutative $A$-algebra $B \in A-Comm(\mathcal{C})$, together with an isomorphism

$$\alpha_u : M_B \otimes_B B' \longrightarrow M_{B'}$$

for any morphism $u : B \longrightarrow B'$ in $A-Comm(\mathcal{C})$, such that one has $\alpha_v \circ (\alpha_u \otimes_{B'} B'') = \alpha_{v \circ u}$ for any pair of morphisms

$$B \xrightarrow{u} B' \xrightarrow{v} B''$$

in $A-Comm(\mathcal{C})$. Such data will be denoted by $(M, \alpha)$. A morphism in $A-QCoh$, from $(M, \alpha)$ to $(M', \alpha')$ is given by a family of morphisms of $B$-modules $f_B : M_B \longrightarrow M'_B$, for any $B \in A-Comm(\mathcal{C})$, such that for any $u : B \to B'$ in $A-Comm(\mathcal{C})$ the diagram

$$\begin{array}{ccc} M_B \otimes_B B' & \xrightarrow{\alpha_u} & M_{B'} \\ {\scriptstyle f_B \otimes_B B'} \downarrow & & \downarrow {\scriptstyle f_{B'}} \\ (M'_B) \otimes_B B' & \xrightarrow[\alpha'_u]{} & M'_{B'} \end{array}$$

commutes. As the categories $A - Mod$ and $Comm(\mathcal{C})$ are all $\mathbb{V}$-small, so are the categories $A - QCoh$.

There exists a natural projection $A - QCoh \longrightarrow A - Mod$, sending $(M, \alpha)$ to $M_A$, and it is straightforward to check that it is an equivalence of categories. In particular, the model structure on $A-Mod$ will be transported naturally on $A-QCoh$ through this equivalence. Fibrations (resp. equivalences) in $A-QCoh$ are simply the morphisms $f : (M, \alpha) \longrightarrow (M', \alpha')$ such that $f_A : M_A \longrightarrow M'_A$ is a fibration (resp. an equivalence).

Let now $f : A \longrightarrow A'$ be a morphism of commutative monoids in $\mathcal{C}$. There exists a pullback functor

$$f^* : A - QCoh \longrightarrow A' - QCoh$$

defined by $f^*(M, \alpha)_B := M_B$ for any $B \in A - Comm(\mathcal{C})$, and for $u : B \longrightarrow B'$ in $A - Comm(\mathcal{C})$ the transition morphism

$$f^*(M, \alpha)_B \otimes_B B' = M_B \otimes_B B' \longrightarrow f^*(M, \alpha)_{B'} = M_{B'}$$

is given by $\alpha_u$. By definition of the model structure on $A - QCoh$, the functor

$$f^* : A - QCoh \longrightarrow A' - QCoh$$

is clearly a left Quillen functor, as the natural diagram

$$\begin{array}{ccc} A - QCoh & \xrightarrow{f^*} & A' - QCoh \\ \downarrow & & \downarrow \\ A - Mod & \xrightarrow[f^*]{} & A' - Mod \end{array}$$

commutes up to a natural isomorphism. Furthermore, for any pair of morphisms

$$A \xrightarrow{f} A' \xrightarrow{g} A''$$

in $Comm(\mathcal{C})$, there is an equality $(g \circ f)^* = g^* \circ f^*$. In other words, the rule

$$A \mapsto A - QCoh \qquad (f : A \to A') \mapsto f^*$$

defines a $\mathbb{U}$-cofibrantly generated left Quilllen presheaf on $Aff_{\mathcal{C}} = Comm(\mathcal{C})^{op}$ in the sense of Appendix B.

We now consider for any $A \in Comm(\mathcal{C})$, the subcategory $A - QCoh^c_W$ of $A - QCoh$, consisting of equivalences between cofibrant objects. As these are preserved by the pullback functors $f^*$, we obtain this way a new presheaf of $\mathbb{V}$-small categories

$$\begin{array}{cccc} QCoh^c_W : & Comm(\mathcal{C}) = Aff^{op}_{\mathcal{C}} & \longrightarrow & Cat_{\mathbb{V}} \\ & A & \mapsto & A - QCoh^c_W. \end{array}$$

Composing with the nerve functor

$$N : Cat_{\mathbb{V}} \longrightarrow SSet_{\mathbb{V}}$$

we get a simplicial presheaf

$$\begin{array}{cccc} N(QCoh^c_W) : & Comm(\mathcal{C}) = Aff^{op}_{\mathcal{C}} & \longrightarrow & SSet_{\mathbb{V}} \\ & A & \mapsto & N(A - QCoh^c_W). \end{array}$$

DEFINITION 1.3.7.1. *The simplicial presheaf of* quasi-coherent modules *is* $N(QCoh^c_W)$ *defined above. It is denoted by* **QCoh**, *and is considered as an object in* $Aff^{\sim,\tau}_{\mathcal{C}}$.

It is important to note that for any $A \in Comm(\mathcal{C})$, the simplicial set $\mathbf{QCoh}(A)$ is naturally equivalent to the nerve of $A - Mod^c_W$, the subcategory of equivalences between cofibrant objects in $A-Mod$, and therefore also to the nerve of $A-Mod_W$, the subcategory of equivalences in $A - Mod$. In particular, $\pi_0(\mathbf{QCoh}(A))$ is in bijection with isomorphisms classes of objects in $\mathrm{Ho}(A - Mod)$ (i.e. equivalence classes of objects in $A - Mod$). Furthermore, by [**D-K**3] (see also Appendix A), for any $x \in \mathbf{QCoh}(A)$, corresponding to an equivalence class of $M \in A - Mod$, the connected component of $\mathbf{QCoh}(A)$ containing $x$ is naturally equivalent to $BAut(M)$, where $Aut(M)$ is the simplicial monoid of self equivalences of $M$ in $A - Mod$. In particular, we have

$$\pi_1(\mathbf{QCoh}(A), x) \simeq Aut_{\mathrm{Ho}(A-Mod)}(M)$$

$$\pi_{i+1}(\mathbf{QCoh}(A), x) \simeq [S^i M, M]_{A-Mod} \ \forall\, i > 1.$$

The main result of this section is the following.

THEOREM 1.3.7.2. *The simplicial presheaf* **QCoh** *is a stack.*

PROOF. This is a direct application of the strictification theorem B.0.7 recalled in Appendix B.

More precisely, we use our lemma 1.3.2.3 (2). Concerning finite direct sums, we have already seen (during the proof of lemma 1.3.2.3) that the natural functor

$$(-\otimes_{A\times^h B} A)\times(-\otimes_{A\times^h B} B):(A\times^h B)-Mod\longrightarrow A-Mod\times B-Mod$$

is a Quillen equivalence. This implies that for any two objects $X, Y \in Aff_{\mathcal{C}}$ the natural morphism

$$\mathbf{QCoh}(X\coprod^{\mathbb{L}} Y)\longrightarrow \mathbf{QCoh}(X)\times\mathbf{QCoh}(Y)$$

is an equivalence. It only remains to show that **QCoh** has the descent property with respect to hypercovers of the type described in lemma 1.3.2.3 (2). But this is nothing else than our assumption 1.3.2.2 (3) together with Cor. B.0.8. □

An important consequence of theorem 1.3.7.2 is the following.

COROLLARY 1.3.7.3. *Let $A \in Comm(\mathcal{C})$ be a commutative monoid, $A - Mod_W$ the subcategory of equivalences in $A - Mod$, and $N(A - Mod_W)$ be its nerve. Then, there exists natural isomorphisms in* $\mathrm{Ho}(SSet)$

$$\mathbb{R}\mathbf{QCoh}(A)\simeq\mathbb{R}_{\tau}\underline{Hom}(\mathbb{R}\underline{Spec}\,A,\mathbf{QCoh})\simeq N(A-Mod_W).$$

To finish this section, we will describe two important sub-stacks of **QCoh**, namely the stack of *perfect modules* and the stack of *vector bundles*.

For any commutative monoid $A$ in $\mathcal{C}$, we let $\mathbf{Perf}(A)$ be the sub-simplicial set of $\mathbf{QCoh}(A)$ consisting of all connected components corresponding to perfect objects in $\mathrm{Ho}(A-Mod)$ (in the sense of Def. 1.2.3.6). More precisely, if $Iso(D)$ denotes the set of isomorphisms classes of a category $D$, the simplicial set $\mathbf{Perf}(A)$ is defined as the pullback

$$\begin{array}{ccc}\mathbf{Perf}(A) & \longrightarrow & Iso(\mathrm{Ho}(A-Mod)^{perf})\\ \downarrow & & \downarrow\\ \mathbf{QCoh}(A) & \longrightarrow & \pi_0\mathbf{QCoh}(A)\simeq Iso(\mathrm{Ho}(A-Mod))\end{array}$$

where $\mathrm{Ho}(A-Mod)^{perf}$ is the full subcategory of $\mathrm{Ho}(A-Mod)$ consisting of perfect $A$-modules in the sense of Def. 1.2.3.6.

We say that an $A$-module $M \in \mathrm{Ho}(A - Mod)$ is a *rank $n$ vector bundle*, if there exists a covering family $A \longrightarrow A'$ such that $M \otimes^{\mathbb{L}}_A A'$ is isomorphic in $\mathrm{Ho}(A' - Mod)$ to $(A')^n$. As we have defined the sub simplicial set $\mathbf{Perf}(A)$ of $\mathbf{QCoh}(A)$ we define $\mathbf{Vect}_n(A)$ to be the sub simplicial set of $\mathbf{QCoh}(A)$ consisting of connected components corresponding to rank $n$ vector bundles.

For any morphism of commutative monoids $u : A \longrightarrow A'$, the base change functor

$$\mathbb{L}u^*:\mathrm{Ho}(A-Mod)\longrightarrow\mathrm{Ho}(A'-Mod)$$

preserves perfect modules as well as rank $n$ vector bundles. Therefore, the sub simplicial sets $\mathbf{Vect}_n(A)$ and **Perf** form in fact full sub simplicial presheaves

$$\mathbf{Vect}_n\subset\mathbf{QCoh}\qquad\mathbf{Perf}\subset\mathbf{QCoh}.$$

The simplicial presheaves $\mathbf{Vect}_n$ and **Perf** then define objects in $Aff_{\mathcal{C}}^{\sim,\tau}$.

COROLLARY 1.3.7.4. *The simplicial presheaves* **Perf** *and* $\mathbf{Vect}_n$ *are stacks.*

PROOF. Indeed, as they are full sub-simplicial presheaves of the stack **QCoh**, it is clearly enough to show that being a perfect module and being a vector bundle or rank $n$ is a local condition for the topology $\tau$. For vector bundles this is obvious from the definition.

Let $A \in Comm(\mathcal{C})$ be a commutative monoid, and $P$ be an $A$-module, such that there exists a $\tau$-covering $A \longrightarrow B$ such that $P \otimes^{\mathbb{L}}_A B$ is a perfect $B$-module. Assume that $A \to B$ is a cofibration, and let $B_*$ be its co-nerve, considered as a co-simplicial object in $A - Comm(\mathcal{C})$. Let $Q$ be any $A$-module, and define two objects in $\mathrm{Ho}(csB_* - Mod)$ by

$$\mathbb{R}\underline{Hom}_A(P,Q)_* := \mathbb{R}\underline{Hom}_A(P,Q) \otimes^{\mathbb{L}}_A B_*$$

$$\mathbb{R}\underline{Hom}_A(P,Q_*) := \mathbb{R}\underline{Hom}_A(P,Q \otimes^{\mathbb{L}}_A B_*).$$

There is a natural morphism

$$\mathbb{R}\underline{Hom}_A(P,Q)_* \longrightarrow \mathbb{R}\underline{Hom}_A(P,Q_*).$$

These co-simplicial objects are both cartesian, and by assumption 1.3.2.2 (3) applied to the $A$-modules $Q$ and $\mathbb{R}\underline{Hom}_A(P,Q)$ the induced morphism in $\mathrm{Ho}(A - Mod)$

$$\int \mathbb{R}\underline{Hom}_A(P,Q)_* \simeq \mathbb{R}\underline{Hom}_A(P,Q) \simeq$$

$$\simeq Holim_{n\in\Delta}\mathbb{R}\underline{Hom}_A(P,Q \otimes^{\mathbb{L}}_A B_n) \simeq \int \mathbb{R}\underline{Hom}_A(P,Q \otimes^{\mathbb{L}}_A B_*)$$

is an isomorphism. Therefore, assumption 1.3.2.2 (3) implies that the natural morphism

$$\mathbb{R}\underline{Hom}_A(P,Q)_0 \longrightarrow \mathbb{R}\underline{Hom}_A(P,Q_0)$$

is an isomorphism. By definition this implies that

$$\mathbb{R}\underline{Hom}_A(P,Q) \otimes^{\mathbb{L}}_A B \longrightarrow \mathbb{R}\underline{Hom}_A(P,Q \otimes^{\mathbb{L}}_A B)$$

is an isomorphism in $\mathrm{Ho}(A - Mod)$. In particular, when $Q = A$ we find that the natural morphism

$$P^{\vee} \otimes^{\mathbb{L}}_A B \longrightarrow \mathbb{R}\underline{Hom}_B(P \otimes^{\mathbb{L}}_A B, B)$$

is an isomorphism in $\mathrm{Ho}(B - Mod)$. As $P \otimes^{\mathbb{L}}_A B$ is by assumption a perfect $B$-module, we find that the natural morphism

$$P^{\vee} \otimes^{\mathbb{L}}_A Q \longrightarrow \mathbb{R}\underline{Hom}(P,Q)$$

becomes an isomorphism after base changing to $B$, and this for any $A$-module $Q$. As $A \longrightarrow B$ is a $\tau$-covering, we see that this implies that

$$P^{\vee} \otimes^{\mathbb{L}}_A Q \longrightarrow \mathbb{R}\underline{Hom}(P,Q)$$

is always an isomorphism in $\mathrm{Ho}(A - Mod)$, for any $Q$, and thus that $P$ is a perfect $A$-module. □

DEFINITION 1.3.7.5. *The* stack of vector bundles of rank $n$ *is* $\mathbf{Vect}_n$. *The* stack of perfect modules *is* **Perf**.

The same construction can also been done in the stable context. For a commutative monoid $A$ in $\mathcal{C}$, we define a category $A - QCoh^{Sp}$, of stable quasi-coherent modules on $A$ (or equivalently on $Spec\,A$) in the following way. Its objects are the data of a stable $B$-module $M_B \in Sp(B - Mod)$ for any commutative $A$-algebra $B \in A - Comm(\mathcal{C})$, together with an isomorphism

$$\alpha_u : M_B \otimes_B C \longrightarrow M_{B'}$$

for any morphism $u : B \longrightarrow B'$ in $A-Comm(\mathcal{C})$, such that one has $\alpha_v \circ (\alpha_u \otimes_{B'} B'') = \alpha_{v\circ u}$ for any pair of morphisms

$$B \xrightarrow{u} B' \xrightarrow{v} B''$$

in $A-Comm(\mathcal{C})$. Such data will be denoted by $(M, \alpha)$. A morphism in $A-QCoh^{Sp}$, from $(M, \alpha)$ to $(M', \alpha')$ is given by a family of morphisms of stable $B$-modules $f_B : M_B \longrightarrow M'_B$, for any $B \in A - Comm(\mathcal{C})$, such that for any $u : B \to B'$ in $A - Comm(\mathcal{C})$ the diagram

$$\begin{array}{ccc} M_B \otimes_B B' & \xrightarrow{\alpha_u} & M_{B'} \\ {\scriptstyle f_B \otimes_B B'} \downarrow & & \downarrow {\scriptstyle f_{B'}} \\ (M'_B) \otimes_B B' & \xrightarrow[\alpha'_u]{} & M'_{B'} \end{array}$$

commutes. As the categories $Sp(A - Mod)$ and $Comm(\mathcal{C})$ are all $\mathbb{V}$-small, so are the categories $A - QCoh^{Sp}$.

There exists a natural projection $A - QCoh^{Sp} \longrightarrow Sp(A - Mod)$, sending $(M, \alpha)$ to $M_A$, and it is straightforward to check that it is an equivalence of categories. In particular, the model structure on $Sp(A - Mod)$ will be transported naturally on $A - QCoh^{Sp}$ through this equivalence.

Let now $f : A \longrightarrow A'$ be a morphism of commutative monoids in $\mathcal{C}$. There exists a pullback functor

$$f^* : A - QCoh^{Sp} \longrightarrow A' - QCoh^{Sp}$$

defined by $f(M, \alpha)_B := M_B$ for any $B \in A - Comm(\mathcal{C})$, and for $u : B \longrightarrow B'$ in $A - Comm(\mathcal{C})$ the transition morphism

$$f(M, \alpha)_B \otimes_B B' = M_B \otimes_B B' \longrightarrow f(M, \alpha)_{B'} = M_{B'}$$

is given by $\alpha_u$. By definition of the model structure on $A - QCoh^{Sp}$, the functor

$$f^* : A - QCoh^{Sp} \longrightarrow A' - QCoh^{Sp}$$

is clearly a left Quillen functor. Furthermore, for any pair of morphisms

$$A \xrightarrow{f} A' \xrightarrow{g} A''$$

in $Comm(\mathcal{C})$, there is an equality $(g \circ f)^* = g^* \circ f^*$. In other words, the rule

$$A \mapsto A - QCoh^{Sp} \qquad (f : A \to A') \mapsto f^*$$

defines a $\mathbb{U}$-cofibrantly generated left Quilllen presheaf on $Aff_{\mathcal{C}} = Comm(\mathcal{C})^{op}$ in the sense of Appendix B.

We now consider for any $A \in Comm(\mathcal{C})$, the subcategory $A - QCoh_W^{Sp,c}$ of $A - QCoh^{Sp}$, consisting of equivalences between cofibrant objects. As these are preserved by the pullback functors $f^*$, one gets this way a new presheaf of $\mathbb{V}$-small categories

$$\begin{array}{cccc} QCoh_W^{Sp,c} : & Comm(\mathcal{C}) = Aff_{\mathcal{C}}^{op} & \longrightarrow & Cat_{\mathbb{V}} \\ & A & \mapsto & A - QCoh_W^{Sp,c}. \end{array}$$

Composing with the nerve functor

$$N : Cat_{\mathbb{V}} \longrightarrow SSet_{\mathbb{V}}$$

one gets a simplicial presheaf

$$\begin{array}{cccc} N(QCoh_W^{Sp,c}) : & Comm(\mathcal{C}) = Aff_{\mathcal{C}}^{op} & \longrightarrow & SSet_{\mathbb{V}} \\ & A & \mapsto & N(A - QCoh_W^{Sp,c}). \end{array}$$

DEFINITION 1.3.7.6. *The simplicial presheaf of* stable quasi-coherent modules *is* $N(QCoh_W^{Sp,c})$ *defined above. It is denoted by* $\mathbf{QCoh}^{Sp}$, *and is considered as an object in* $Aff_{\mathcal{C}}^{\sim,\tau}$.

It is important to note that for any $A \in Comm(\mathcal{C})$, the simplicial set $\mathbf{QCoh}^{Sp}(A)$ is naturally equivalent to the nerve of $A - Mod_W^{Sp,c}$, the subcategory of equivalences between cofibrant objects in $Sp(A - Mod)$, and therefore also to the nerve of $Sp(A - Mod)_W$, the subcategory of equivalences in $Sp(A - Mod)$. In particular, $\pi_0(\mathbf{QCoh}^{Sp}(A))$ is in bijection with isomorphisms classes of objects in $\mathrm{Ho}(Sp(A - Mod))$ (i.e. equivalence classes of objects in $Sp(A - Mod)$). Furthermore, by [**D-K**3] (see also Appendix A), for any $x \in \mathbf{QCoh}^{Sp}(A)$, corresponding to an equivalence class of $M \in Sp(A - Mod)$, the connected component of $\mathbf{QCoh}^{Sp}(A)$ containing $x$ is naturally equivalent to $BAut(M)$, where $Aut(M)$ is the simplicial monoid of self equivalences of $M$ in $Sp(A - Mod)$. In particular, we have

$$\pi_1(\mathbf{QCoh}^{Sp}(A), x) \simeq Aut_{\mathrm{Ho}(Sp(A-Mod))}(M)$$

$$\pi_{i+1}(\mathbf{QCoh}^{Sp}(A), x) \simeq [S^i M, M]_{Sp(A-Mod)} \ \forall\, i > 1.$$

The same proof as theorem 1.3.7.2, but based on Proposition 1.2.12.5 gives the following stable version.

THEOREM 1.3.7.7. *Assume that the two conditions are satisfied.*

1. *The suspension functor* $S : \mathrm{Ho}(\mathcal{C}) \longrightarrow \mathrm{Ho}(\mathcal{C})$ *is fully faithful.*
2. *For all* $\tau$*-covering family* $\{U_i \to X\}$ *in* $Aff_{\mathcal{C}}$, *each morphism* $U_i \to X$ *is flat in the sense of Def. 1.2.4.1.*

*Then, the simplicial presheaf* $\mathbf{QCoh}^{Sp}$ *is a stack.*

For any commutative monoid $A$ in $\mathcal{C}$, we let $\mathbf{Perf}^{Sp}(A)$ be the sub-simplicial set of $\mathbf{QCoh}^{Sp}(A)$ consisting of all connected components corresponding to perfect objects in $\mathrm{Ho}(Sp(A - Mod))$ (in the sense of Def. 1.2.3.6). This defines a full sub-prestack $\mathbf{Perf}^{Sp}$ of $\mathbf{QCoh}^{Sp}$. Then, the same argument as for Cor. 1.3.7.4 gives the following corollary.

COROLLARY 1.3.7.8. *Assume that the two conditions are satisfied.*

1. *The suspension functor* $S : \mathrm{Ho}(\mathcal{C}) \longrightarrow \mathrm{Ho}(\mathcal{C})$ *is fully faithful.*
2. *For all* $\tau$*-covering family* $\{U_i \to X\}$ *in* $Aff_{\mathcal{C}}$, *each morphism* $U_i \to X$ *is flat in the sense of Def. 1.2.4.1.*

*The simplicial presheaf* $\mathbf{Perf}^{Sp}$ *is a stack.*

This justifies the following definition.

DEFINITION 1.3.7.9. *Under the condition of Cor. 1.3.7.8, the* stack of stable perfect modules *is* $\mathbf{Perf}^{Sp}$.

We finish this section by the standard description of $\mathbf{Vect}_n$ as the classifying stack of the group stack $\mathbf{Gl}_n$, of invertible $n$ by $n$ matrices. For this, we notice that the natural morphism of stacks $* \longrightarrow \mathbf{Vect}_n$, pointing the trivial rank $n$ vector bundle, induces an isomorphism of sheaves of sets $* \simeq \pi_0(\mathbf{Vect}_n)$. Therefore, the stack $\mathbf{Vect}_n$ can be written as the geometric realization of the homotopy nerve of the morphism $* \longrightarrow \mathbf{Vect}_n$. In other words, we can find a Segal groupoid object $X_*$, such that $X_0 = *$, and with $|X_*| \simeq \mathbf{Vect}_n$. Furthermore, the object $X_1$ is naturally equivalent to the loop stack $\Omega_* \mathbf{Vect}_n := * \times^h_{\mathbf{Vect}_n} *$. By construction and by Prop. A.0.6, this loop stack can be described as the simplicial presheaf

$$\begin{array}{rccc} \Omega_* \mathbf{Vect}_n : & Comm(\mathcal{C}) & \longrightarrow & SSet \\ & A & \mapsto & Map'_{\mathcal{C}}(1^n, A^n), \end{array}$$

where $Map'_{\mathcal{C}}(1^n, A^n)$ is the sub simplicial set of the mapping space $Map_{\mathcal{C}}(1^n, A^n)$ consisting of all connected components corresponding to automorphisms in

$$\pi_0 Map_{\mathcal{C}}(1^n, A^n) \simeq \pi_0 Map_{A-Mod}(A^n, A^n) \simeq [A^n, A^n]_{A-Mod}.$$

The important fact concerning the stack $\Omega_*\mathbf{Vect}_n$ is the following result.

PROPOSITION 1.3.7.10. (1) *The stack $\Omega_*\mathbf{Vect}_n$ is representable and the morphism $\Omega_*\mathbf{Vect}_n \longrightarrow *$ is formally smooth.*

(2) *If Moreover $\mathbf{1}$ is finitely presented in $\mathcal{C}$, then the morphism $\Omega_*\mathbf{Vect}_n \longrightarrow *$ is finitely presented (and thus smooth by (1)).*

PROOF. (1) We start by defining a larger stack $\mathbf{M}_n$, of $n$ by $n$ matrices. We set

$$\begin{array}{cccc} \mathbf{M}_n: & Comm(\mathcal{C}) & \longrightarrow & SSet \\ & A & \mapsto & Map_{\mathcal{C}}(1^n, A^n). \end{array}$$

This stack is representable, as it is isomorphic in $\mathrm{St}(\mathcal{C}, \tau)$ to $\mathbb{R}\underline{Spec}\, B$, where $B = \mathbb{L}F(\mathbf{1}^{n^2})$ is the free commutative monoid generated by the object $\mathbf{1}^{n^2} \in \mathcal{C}$. We claim that the natural inclusion morphism

$$\Omega_*\mathbf{Vect}_n \longrightarrow \mathbf{M}_n$$

is $(-1)$-representable and a formally étale morphism. Indeed, let $A$ be any commutative monoid,

$$x : X := \mathbb{R}\underline{Spec}\, A \longrightarrow \mathbf{M}_n$$

be a morphism of stacks, and let us consider the stack

$$F := \Omega_*\mathbf{Vect}_n \times^h_{\mathbf{M}_n} X \longrightarrow X.$$

The point $x$ corresponds via the Yoneda lemma to a morphism $u : A^n \longrightarrow A^n$ in $\mathrm{Ho}(A - Mod)$. Now, for any commutative monoid $A'$, the natural morphism

$$\mathbb{R}F(A') \longrightarrow (\mathbb{R}\underline{Spec}\, A)(A')$$

identifies $\mathbb{R}F(A')$ with the sub simplicial set of $(\mathbb{R}\underline{Spec}\, A)(A') \simeq Map_{Comm(\mathcal{C})}(A, A')$ consisting of all connected components corresponding to morphisms $A \longrightarrow A'$ in $\mathrm{Ho}(Comm(\mathcal{C}))$ such that

$$u \otimes^{\mathbb{L}}_A A' : (A')^n \longrightarrow (A')^n$$

is an isomorphism in $\mathrm{Ho}(A' - Mod)$. Considering $u$ as an element of $[A^n, A^n] \simeq M_n(\pi_0(A))$, we can consider its determinant $d(u) \in \pi_0(A)$. Then, using notations from Def. 1.2.9.2, we clearly have an isomorphism of stacks

$$F \simeq \mathbb{R}\underline{Spec}\,(A[d(u)^{-1}]).$$

By Prop. 1.2.9.5 this shows that the morphism

$$F \longrightarrow \mathbb{R}\underline{Spec}\, A$$

is a formally étale morphism between representable stacks. As $\mathbf{M}_n$ is representable this implies that $\Omega_*\mathbf{Vect}_n$ is a representable stack and that the morphism

$$\Omega_*\mathbf{Vect}_n \longrightarrow \mathbf{M}_n$$

is formally étale (it is also a flat monomorphism by 1.2.9.4).

Moreover, we have $\mathbf{M}_n \simeq \mathbb{R}\underline{Spec}\, B$, where $B := \mathbb{L}F(\mathbf{1}^{n^2})$ is the derived free commutative monoid over $\mathbf{1}^{n^2}$. This implies that $\mathbb{L}_B$ is a free $B$-module of rank $n^2$, and therefore that the morphism $\mathbf{1} \longrightarrow B$ is formally smooth in the sense of Def. 1.2.7.1. By composition, we find that $\Omega_*\mathbf{Vect}_n \longrightarrow *$ is a formally smooth morphism as required.

(2) This follows from (1), Prop. 1.2.9.4 (2) and the fact that $B = \mathbb{L}F(\mathbf{1}^{n^2})$ is a finitely presented object in $Comm(\mathcal{C})$. □

DEFINITION 1.3.7.11. (1) *The stack* $\Omega_*\mathbf{Vect}_n$ *is denoted by* $\mathbf{Gl}_n$*, and is called the* linear group stack of rank $n$. *The stack* $\mathbf{Gl}_1$ *is denoted by* $\mathbb{G}_m$*, and is called the* multiplicative group stack.

(2) *The stack* $\mathbf{M}_n$ *defined during the proof of 1.3.7.10* (1) *is called the* stack of $n \times n$ matrices. *The stack* $\mathbf{M}_1$ *is denoted by* $\mathbb{G}_a$*, and is called the* additive group stack.

Being a stack of loops, the stack $\mathbf{Gl}_n = \Omega_*\mathbf{Vect}_n$ has a natural group structure, encoded in the fact that it is the $X_1$ of a Segal groupoid object $X_*$ with $X_0 = *$. Symbolically, we will simply write

$$B\mathbf{Gl}_n := |X_*|.$$

Our conclusion is that the stack $\mathbf{Vect}_n$ can be written as $B\mathbf{Gl}_n$, where $\mathbf{Gl}_n$ is a *formally smooth representable group stack.* Furthermore this group stack is smooth when the unit $\mathbf{1}$ is finitely presented. As a corollary we get the following geometricity result on $\mathbf{Vect}_n$.

COROLLARY 1.3.7.12. *Assume that the unit* 1 *is a finitely presented object in* $\mathcal{C}$. *Assume that all smooth morphisms in* $Comm(\mathcal{C})$ *belong to* $\mathbf{P}$. *Then, the stack* $\mathbf{Vect}_n$ *is* 1*-geometric, the morphism* $\mathbf{Vect}_n \longrightarrow *$ *is in* $\mathbf{P}$ *and finitely presented, and furthermore its diagonal is a* $(-1)$*-representable morphism.*

PROOF. The 1-geometricity statement is a consequence of Prop. 1.3.4.2 and the fact that the natural morphism $* \longrightarrow \mathbf{Vect}_n$ is a 1-$\mathbf{P}$-atlas. That $*$ is a 1-$\mathbf{P}$-atlas also implies that $\mathbf{Vect}_n \longrightarrow *$ is in $\mathbf{P}$ and finitely presented. The statement concerning the diagonal follows from the fact that $\mathbf{Gl}_n$ is a representable stack and the locality of representable objects Prop. 1.3.2.8. □

We finish with an analogous situation for perfect modules. We let $K$ be a perfect object in $\mathcal{C}$, and we define a stack $\mathbb{R}\underline{End}(K)$ in the following way. We chose a cofibrant replacement $QK$ of $K$, and $\Gamma_*$ a simplicial resolution functor on $\mathcal{C}$. One sets

$$\begin{array}{rccc} \mathbb{R}\underline{End}(K): & Comm(\mathcal{C}) & \longrightarrow & SSet_{\mathbb{V}} \\ & A & \mapsto & Hom(QK, \Gamma_*(QK \otimes A)). \end{array}$$

Note that for any $A$ the simplicial set $\mathbb{R}\underline{End}(K)(A)$ is naturally equivalent to

$$Map_{A-Mod}(K \otimes^{\mathbb{L}} A, K \otimes^{\mathbb{L}} A).$$

LEMMA 1.3.7.13. *The simplicial presheaf* $\mathbb{R}\underline{End}(K)(A) \in Aff_{\mathcal{C}}^{\sim,\tau}$ *is a stack. It is furthermore representable.*

PROOF. This is clear as $K$ being perfect one sees that there exists an isomorphism in $\mathrm{Ho}(SPr(Aff_{\mathcal{C}}))$

$$\mathbb{R}\underline{End}(K) \simeq \mathbb{R}\underline{Spec}\, B,$$

where $B := \mathbb{L}F(K \otimes^{\mathbb{L}} K^{\vee})$ is the derived free commutative monoid on the object $K \otimes^{\mathbb{L}} K^{\vee}$. □

We now define a sub-stack $\mathbb{R}\underline{Aut}(K)$ of $\mathbb{R}\underline{End}(K)$. For a commutative monoid $A \in Comm(\mathcal{C})$, we define $\mathbb{R}\underline{Aut}(K)(A)$ to be the union of connected components of $\mathbb{R}\underline{End}(K)(A)$ corresponding to isomorphisms in

$$\pi_0(\mathbb{R}\underline{End}(K)(A)) \simeq [K \otimes^{\mathbb{L}} A, K \otimes^{\mathbb{L}} A]_{A-Mod}.$$

This clearly defines a full sub-simplicial presheaf $\mathbb{R}\underline{Aut}(K)$ of $\mathbb{R}\underline{End}(K)$, which is a sub-stack as one can see easily using Cor. 1.3.2.7.

PROPOSITION 1.3.7.14. *Assume that $\mathcal{C}$ is a stable model category. Then, the stack $\mathbb{R}\underline{Aut}(K)$ is representable. Furthermore the morphism $\mathbb{R}\underline{Aut}(K) \longrightarrow \mathbb{R}\underline{End}(K)$ is a formal Zariski open immersion, and $\mathbb{R}\underline{Aut}(K) \longrightarrow *$ is fp. If furthermore $\mathbf{1}$ is finitely presented in $\mathcal{C}$ then $\mathbb{R}\underline{Aut}(K) \longrightarrow *$ is a perfect morphism.*

PROOF. It is the same as 1.3.7.10 but using the construction $A_K$ and Prop. 1.2.10.1, instead of the standard localization $A[a^{-1}]$. More precisely, for a representable stack $X := \mathbb{R}\underline{Spec}\, A$ and a morphism $x : X \longrightarrow \mathbb{R}\underline{End}(K)$, the homotopy pullback

$$\mathbb{R}\underline{Aut}(K) \times^h_{\mathbb{R}\underline{End}(K)} X \longrightarrow X$$

is isomorphic in $\mathrm{Ho}(Aff_{\mathcal{C}}^{\sim,\tau}/X)$ to

$$\mathbb{R}\underline{Spec}\, A_E \longrightarrow \mathbb{R}\underline{Spec}\, A,$$

where $E$ is the homotopy cofiber of the endomorphism $x : K \otimes^{\mathbb{L}} A \longrightarrow K \otimes^{\mathbb{L}} A$ corresponding to the point $x$. □

CHAPTER 1.4

# Geometric stacks: Infinitesimal theory

As in the previous chapter, we fix once for all a HAG context $(\mathcal{C}, \mathcal{C}_0, \mathcal{A}, \tau, \mathbf{P})$.

In this chapter we will assume furthermore that the suspension functor

$$S : \mathrm{Ho}(\mathcal{C}) \longrightarrow \mathrm{Ho}(\mathcal{C})$$

is fully faithful. In particular, for any commutative monoid $A \in Comm(\mathcal{C})$, the stabilization functor

$$S_A : \mathrm{Ho}(A - Mod) \longrightarrow \mathrm{Ho}(Sp(A - Mod))$$

is fully faithful (see 1.2.11.2). We will therefore forget to mention the functor $S_A$ and simply consider $A$-modules as objects in $\mathrm{Ho}(Sp(A - Mod))$, corresponding to 0-connective objects.

## 1.4.1. Tangent stacks and cotangent complexes

We consider the initial commutative monoid $\mathbf{1} \in Comm(\mathcal{C})$. It can be seen as a module over itself, and gives rise to a trivial square zero extension $\mathbf{1} \oplus \mathbf{1}$ (see §1.2.1).

DEFINITION 1.4.1.1. *The* dual numbers over $\mathcal{C}$ *is the commutative monoid* $\mathbf{1} \oplus \mathbf{1}$*, and is denoted by* $\mathbf{1}[\epsilon]$*. The corresponding representable stack is denoted by*

$$\mathbb{D}_\epsilon := \mathbb{R}\underline{Spec}\,(\mathbf{1}[\epsilon])$$

*and is called the* infinitesimal disk.

Of course, as every trivial square zero extension the natural morphism $\mathbf{1} \longrightarrow \mathbf{1}[\epsilon]$ possesses a natural retraction $\mathbf{1}[\epsilon] \longrightarrow \mathbf{1}$. On the level of representable stacks this defines a natural global point

$$* \longrightarrow \mathbb{D}_\epsilon.$$

We recal that $\mathrm{Ho}(Aff_{\mathcal{C}}^{\sim,\tau})$ being the homotopy category of a model topos has internal $Hom$'s objects $\mathbb{R}_\tau\underline{\mathcal{H}om}(-,-)$. They satisfy the usual adjunction isomorphisms

$$Map_{Aff_{\mathcal{C}}^{\sim,\tau}}(F, \mathbb{R}_\tau\underline{\mathcal{H}om}(G, H)) \simeq Map_{Aff_{\mathcal{C}}^{\sim,\tau}}(F \times^h G, H).$$

DEFINITION 1.4.1.2. *Let* $F \in \mathrm{St}(\mathcal{C}, \tau)$ *be a stack. The* tangent stack *of* $F$ *is defined to be*

$$TF := \mathbb{R}_\tau\underline{\mathcal{H}om}(\mathbb{D}_\epsilon, F).$$

*The natural morphism* $* \longrightarrow \mathbb{D}_\epsilon$ *induces a well defined projection*

$$\pi : TF \longrightarrow F,$$

*and the projection* $\mathbb{D}_\epsilon \longrightarrow *$ *induces a natural section of* $\pi$

$$e : F \longrightarrow TF.$$

For any commutative monoid $A \in Comm(\mathcal{C})$, it is clear that there is a natural equivalence

$$A \otimes^{\mathbb{L}} (\mathbf{1}[\epsilon]) \simeq A \oplus A,$$

where $A \oplus A$ is the trivial square zero extension of $A$ by itself. We will simply denote $A \oplus A$ by $A[\epsilon]$, and

$$\mathbb{D}_{\epsilon}^{A} := \mathbb{R}\underline{Spec}\,(A[\epsilon]) \simeq \mathbb{R}\underline{Spec}\,(A) \times \mathbb{D}_{\epsilon}$$

the infinitesimal disk over $A$.

With these notations, and for any fibrant object $F \in Aff_{\mathcal{C}}^{\sim,\tau}$, the stack $TF$ can be described as the following simplicial presheaf

$$\begin{array}{rccc} TF: & Comm(\mathcal{C}) = Aff_{\mathcal{C}}^{op} & \longrightarrow & SSet_{\mathbb{V}} \\ & A & \mapsto & F(A[\epsilon]). \end{array}$$

Note that if $F$ is fibrant, then so is $TF$ as defined above. In other words for any $A \in Comm(\mathcal{C})$ there exists a natural equivalence of simplicial sets

$$\mathbb{R}TF(A) \simeq \mathbb{R}F(A[\epsilon]).$$

Proposition 1.4.1.3. *The functor $F \mapsto TF$ commutes with $\mathbb{V}$-small homotopy limits.*

Proof. This is clear as $\mathbb{R}_{\tau}\underline{\mathcal{H}om}(H, -)$ always commutes with homotopy limits for any $H$. ☐

Let $F \in \mathrm{St}(\mathcal{C}, \tau)$ be a stack, $A \in Comm(\mathcal{C})$ a commutative monoid and

$$x : \mathbb{R}\underline{Spec}\, A \longrightarrow F$$

be a $A$-point. Let $M$ be an $A$-module, and let $A \oplus M$ be the trivial square zero extension of $A$ by $M$. Let us fix the following notations

$$X := \mathbb{R}\underline{Spec}\, A \qquad X[M] := \mathbb{R}\underline{Spec}\,(A \oplus M).$$

The natural augmentation $A \oplus M \longrightarrow A$ gives rise to a natural morphism of stacks $X \longrightarrow X[M]$.

Definition 1.4.1.4. *Let $x : X \longrightarrow F$ be as above. The simplicial set of* derivations from $F$ to $M$ at the point $x$ *is defined by*

$$\mathbb{D}er_F(X, M) := Map_{X/Aff_{\mathcal{C}}^{\sim,\tau}}(X[M], F) \in \mathrm{Ho}(SSet_{\mathbb{V}}).$$

*It will also denoted by $\mathbb{D}er_F(x, M)$.*

As the construction $M \mapsto X[M]$ is functorial in $M$, we get this way a well defined functor

$$\mathbb{D}er_F(X, -) : \mathrm{Ho}(A - Mod) \longrightarrow \mathrm{Ho}(SSet_{\mathbb{V}}).$$

This functor is furthermore naturally compatible with the $\mathrm{Ho}(SSet)$-enrichment, in the sense that there exists natural morphisms

$$Map_{A-Mod}(M, N) \longrightarrow Map_{SSet}(\mathbb{D}er_F(X, M), \mathbb{D}er_F(X, N))$$

which are compatible with compositions. Note that the morphism $X \longrightarrow X[M]$ has a natural retraction $X[M] \longrightarrow X$, and therefore that the simplicial set $\mathbb{D}er_F(X, M)$ has a distinguished base point, the trivial derivation. The functor $\mathbb{D}er_F(X, -)$ takes its values in the homotopy category of pointed simplicial sets.

We can also describe the functor $\mathbb{D}er_F(X, M)$ using functors of points in the following way. Let $F \in Aff_{\mathcal{C}}^{\sim,\tau}$ be a fibrant object. For any commutative monoid $A$, any $A$-module $M$ and any point $x \in F(A)$, we consider the standard homotopy fiber[1] of

$$F(A \oplus M) \longrightarrow F(A)$$

at the point $x$. This clearly defines a functor

$$\begin{array}{ccc} A - Mod & \longrightarrow & SSet_{\mathbb{V}} \\ M & \mapsto & Hofiber\,(F(A \oplus M) \to F(A)) \end{array}$$

which is a lift of the functor considered above

$$\mathbb{D}er_F(X, -) : \mathrm{Ho}(A - Mod) \longrightarrow \mathrm{Ho}(SSet_{\mathbb{V}}),$$

where $x \in F(A)$ corresponds via the Yoneda lemma to a morphism $X = \mathbb{R}\underline{Spec}\, A \longrightarrow F$. The fact that the functor $\mathbb{D}er_F(X, -)$ has a natural lift as above is important, as it then makes sense to say that it commutes with homotopy limits or homotopy colimits. The functor $\mathbb{D}er_F(X, -)$ can be considered as an object in $\mathrm{Ho}((A - Mod^{op})^{\wedge})$, the homotopy category of pre-stacks on the model category $A - Mod^{op}$, as defined in [**HAGI**, §4.1] (see also §1.3.1). Restricting to the subcategory $A - Mod_0$ we get an object

$$\mathbb{D}er_F(X, -) \in \mathrm{Ho}((A - Mod_0^{op})^{\wedge}).$$

In the sequel the functor $\mathbb{D}er_F(X, -)$ will always be considered as an object in $\mathrm{Ho}((A - Mod_0^{op})^{\wedge})$.

DEFINITION 1.4.1.5. *Let $F$ be any stack and let $A \in \mathcal{A}$.*

(1) *Let $x : X := \mathbb{R}\underline{Spec}\, A \longrightarrow F$ be an $A$-point. We say that $F$* has a cotangent complex at $x$ *if there exists an integer $n$, an $(-n)$-connective stable $A$-module $\mathbb{L}_{F,x} \in \mathrm{Ho}(Sp(A - Mod))$, and an isomorphism in $\mathrm{Ho}((A - Mod_0^{op})^{\wedge})$*

$$\mathbb{D}er_F(X, -) \simeq \mathbb{R}\underline{h}_s^{\mathbb{L}_{F,x}}.$$

(2) *If $F$ has a cotangent complex at $x$, the stable $A$-module $\mathbb{L}_{F,x}$ is then called the* cotangent complex of $F$ at $x$.

(3) *If $F$ has a cotangent complex at $x$, the* tangent complex of $F$ at $x$ *is then the stable $A$-module*

$$\mathbb{T}_{F,x} := \mathbb{R}\underline{Hom}_A^{Sp}(\mathbb{L}_{F,x}, A) \in \mathrm{Ho}(Sp(A - Mod)).$$

In other words, the existence of a cotangent complex of $F$ at $x : X := \mathbb{R}\underline{Spec}\, A \longrightarrow F$ is equivalent to the co-representability of the functor

$$\mathbb{D}er_F(X, -) : A - Mod_0 \longrightarrow SSet_{\mathbb{V}},$$

by some $(-n)$-connective object $\mathbb{L}_{F,x} \in \mathrm{Ho}(Sp(A - Mod))$. The fact that $\mathbb{L}_{F,x}$ is well defined is justified by our Prop. 1.2.11.3.

The first relation between cotangent complexes and the tangent stack is given by the following proposition.

PROPOSITION 1.4.1.6. *Let $F$ be a stack and $x : X := \mathbb{R}\underline{Spec}\, A \longrightarrow F$ be an $A$-point with $A \in \mathcal{A}$. If $F$ has a cotangent complex $\mathbb{L}_{F,x}$ at the point $x$ then there exists a natural isomorphism in* $\mathrm{Ho}(SSet_{\mathbb{V}})$

$$\mathbb{R}\underline{Hom}_{Aff_C^{\sim,\tau}/F}(X, TF) \simeq Map_{Sp(A-Mod)}(A, \mathbb{T}_{F,x}) \simeq Map_{Sp(A-Mod)}(\mathbb{L}_{F,x}, A).$$

---

[1] The standard homotopy fiber product of a diagram $X \longrightarrow Z \longleftarrow Y$ of fibrant simplicial sets is defined for example by

$$X \times_Z^h Y := (X \times Y) \times_{Z \times Z} Z^{\Delta^1}.$$

PROOF. By definition of $TF$, there is a natural isomorphism

$$\mathbb{R}\underline{Hom}_{Aff_{\mathcal{C}}^{\sim,\tau}/F}(X, TF) \simeq \mathbb{R}\underline{Hom}_{Aff_{\mathcal{C}}^{\sim,\tau}/X}(X[A], TF) \simeq Map_{Sp(A-Mod)}(\mathbb{L}_{F,x}, A).$$

Moreover, by definition of the tangent complex we have

$$Map_{Sp(A-Mod)}(\mathbb{L}_{F,x}, A) \simeq Map_{Sp(A-Mod)}(A, \mathbb{T}_{F,x}).$$

□

Now, let $F$ be a stack, and $u$ a morphism in $\mathrm{Ho}(Aff_{\mathcal{C}}^{\sim,\tau}/F)$

$$\begin{array}{ccc} Y & \xrightarrow{u} & X \\ & {\scriptstyle y}\searrow \quad \swarrow{\scriptstyle y} & \\ & F, & \end{array}$$

with $X = \mathbb{R}\underline{Spec}\, A$ and $Y = \mathbb{R}\underline{Spec}\, B$ belonging to $\mathcal{A}$. Let $M \in B - Mod_0$, which is also an $A$-module by the forgetful functor $A - Mod_0 \longrightarrow B - Mod_0$. There is a commutative diagram of commutative monoids

$$\begin{array}{ccc} A \oplus M & \longrightarrow & B \oplus M \\ \downarrow & & \downarrow \\ A & \longrightarrow & B, \end{array}$$

inducing a commutative square of representable stacks

$$\begin{array}{ccc} X[M] & \longleftarrow & Y[M] \\ \uparrow & & \uparrow \\ X & \longleftarrow & Y. \end{array}$$

This implies the existence of a natural morphism in $\mathrm{Ho}(SSet_{\mathbb{V}})$

$$\mathbb{D}er_F(Y, M) \longrightarrow \mathbb{D}er_F(X, M).$$

If the stack $F$ has cotangent complexes at both points $x$ and $y$, Prop. 1.2.11.3 induces a well defined morphism in $\mathrm{Ho}(Sp(B - Mod))$

$$u^* : \mathbb{L}_{F,x} \otimes_A^{\mathbb{L}} B \longrightarrow \mathbb{L}_{F,y}.$$

Of course, we have $(u \circ v)^* = v^* \circ u^*$ whenever this formula makes sense.

In the same way, the construction of $\mathbb{L}_{F,x}$ is functorial in $F$. Let $f : F \longrightarrow F'$ be a morphism of stacks, $A \in \mathcal{A}$, and $x : \mathbb{R}\underline{Spec}\, A \longrightarrow F$ be an $A$-point with image $f(x) : \mathbb{R}\underline{Spec}\, A \longrightarrow F'$. Then, for any $A$-module $M$, there is a natural morphism

$$\mathbb{D}er_F(X, M) \longrightarrow \mathbb{D}er_{F'}(X, M).$$

Therefore, if $F$ has a cotangent complex at $x$ and $F'$ has a cotangent complex at $x'$, we get a natural morphism in $\mathrm{Ho}(Sp(A - Mod))$

$$df_x : \mathbb{L}_{F',f(x)} \longrightarrow \mathbb{L}_{F,x},$$

called the *differential of $f$ at $x$*. Once again, we have $d(f \circ g)_x = dg_x \circ df_x$ each time this formula makes sense (this is the chain rule). Dually, we also get by duality the *derivative of $f$ at $x$*

$$Tf_x : \mathbb{T}_{F,x} \longrightarrow \mathbb{T}_{F',f(x)}.$$

DEFINITION 1.4.1.7. *A stack $F$ has a global cotangent complex relative to the* HA context $(\mathcal{C}, \mathcal{C}_0, \mathcal{A})$ *(or simply* has a cotangent complex *when there is no ambiguity on the context) if the following two conditions are satisfied.*

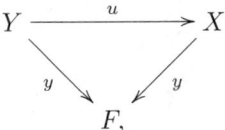

(1) *For any* $A \in \mathcal{A}$, *and any point* $x : \mathbb{R}\underline{Spec}\,A \longrightarrow F$, *the stack* $F$ *has a cotangent complex* $\mathbb{L}_{F,x}$ *at* $x$.
(2) *For any morphism* $u : A \longrightarrow B$ *in* $\mathcal{A}$, *and any morphism in* $\mathrm{Ho}(Aff_{\mathcal{C}}^{\sim,\tau}/F)$

$$\begin{array}{ccc} Y := \mathbb{R}\underline{Spec}\,B & \xrightarrow{\quad u \quad} & X := \mathbb{R}\underline{Spec}\,A \\ & {}_{y}\searrow \quad \swarrow_{x} & \\ & F, & \end{array}$$

*the induced morphism*

$$u^* : \mathbb{L}_{F,x} \otimes^{\mathbb{L}}_A B \longrightarrow \mathbb{L}_{F,y}$$

*is an isomorphism in* $\mathrm{Ho}(Sp(B - Mod))$.

As a corollary of Prop. 1.2.1.2 and the standard properties of derivations any representable stack has a cotangent complex.

PROPOSITION 1.4.1.8. *Any representable stack* $F = \mathbb{R}\underline{Spec}A$ *has a global cotangent complex.*

PROOF. This is nothing else than the existence of a universal derivation as proved in Prop. 1.2.1.2. □

If $X = \mathbb{R}\underline{Spec}\,A$ is a representable stack in $\mathcal{A}$, and $x : X \to X$ is the identity, then the stable $A$-module $\mathbb{L}_{X,x}$ is naturally isomorphic in $\mathrm{Ho}(Sp(A - Mod))$ to $\mathbb{L}_A$. More generally, for any morphism $A \longrightarrow B$ with $B \in \mathcal{A}$, corresponding to $y : \mathbb{R}\underline{Spec}\,B \longrightarrow X$, the $B$-module $\mathbb{L}_{X,y}$ is naturally isomorphic in $\mathrm{Ho}(Sp(B - Mod))$ to $\mathbb{L}_A \otimes^{\mathbb{L}}_A B$.

The next proposition explains the relation between the tangent stack and the global cotangent complex when it exists. It is a globalization of Prop.1.4.1.6.

PROPOSITION 1.4.1.9. *Let* $F$ *be a stack having a cotangent complex. Let* $x : X = \mathbb{R}\underline{Spec}\,A \longrightarrow F$ *be any morphism, and*

$$TF_x := TF \times^h_F X \longrightarrow X$$

*the natural projection. Let* $A \longrightarrow B$ *be a morphism with* $B \in \mathcal{A}$, *corresponding to a morphism of representable stacks* $X = \mathbb{R}\underline{Spec}\,A \longrightarrow Y = \mathbb{R}\underline{Spec}\,B$. *Then, there exists a natural isomorphism in* $\mathrm{Ho}(SSet)$

$$\mathbb{R}TF_x(B) \simeq Map_{Aff_{\mathcal{C}}^{\sim,\tau}/F}(Y, TF) \simeq Map_{Sp(A-Mod)}(\mathbb{L}_{F,x}, B).$$

PROOF. This is a reformulation of Prop. 1.4.1.6, and the fact that

$$Map_{Sp(A-Mod)}(\mathbb{L}_{F,x}, B) \simeq Map_{Sp(B-Mod)}(\mathbb{L}_{F,x} \otimes^{\mathbb{L}}_A B, B).$$

□

REMARK 1.4.1.10. Of course, the isomorphism of proposition Prop. 1.4.1.9 is functorial in $F$.

PROPOSITION 1.4.1.11. *Let* $F$ *be an* $n$*-geometric stack. We assume that for any* $A \in \mathcal{A}$, *and any point* $x : X := \mathbb{R}\underline{Spec}\,A \longrightarrow F$, *and any* $A$*-module* $M \in A - Mod_0$, *the natural morphism*

$$\mathbb{D}er_F(X, M) \simeq \mathbb{D}er_F(X, \Omega S(M)) \longrightarrow \Omega \mathbb{D}er_F(X, S(M))$$

*is an isomorphism in* $\mathrm{Ho}(SSet)$. *Then* $F$ *has a global cotangent complex, which is furthermore* $(-n)$*-connective.*

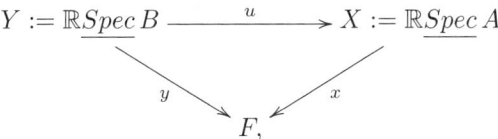

PROOF. The proof is by induction on $n$. For $n = -1$ this is Prop. 1.4.1.8 and does not use our exactness condition on the functor $\mathbb{D}er_F(X, -)$.

Let $n \geq 0$ be an integer and $F$ be an $n$-geometric stack. Let $X = \mathbb{R}\underline{Spec}\,A$ be a representable stack in $\mathcal{A}$ and $x : X \longrightarrow F$ be any morphism in $\mathrm{St}(\mathcal{C}, \tau)$. We consider the natural morphisms

$$d : X \longrightarrow X \times^h X \qquad d_F : X \longrightarrow X \times^h_F X.$$

By induction on $n$, we see that the stacks $X \times^h X$ and $X \times^h_F X$ both have cotangent complexes at the point $d$ and $d_F$, denoted respectively by $\mathbb{L}$ and $\mathbb{L}'$. There is moreover a natural morphism in $\mathrm{Ho}(Sp(A - Mod))$

$$f : \mathbb{L}' \longrightarrow \mathbb{L},$$

induced by

$$X \longrightarrow X \times^h_F X \longrightarrow X \times^h X.$$

We set $\mathbb{L}''$ as the homotopy cofiber of $f$ in $Sp(A - Mod)$. By construction, for any $A$-module $M \in A - Mod_0$, the simplicial set $Map_{Sp(A-Mod)}(\mathbb{L}'', M)$ is naturally equivalent to the homotopy fiber of

$$\mathbb{D}er_X(X, M) \times^h_{\mathbb{D}er_F(X,M)} \mathbb{D}er_X(X, M) \longrightarrow \mathbb{D}er_X(X, M) \times \mathbb{D}er_X(X, M),$$

and thus is naturally equivalent to $\Omega_* \mathbb{D}er_F(X, M)$. By assumption we have natural isomorphisms in $\mathrm{Ho}(SSet)$

$$Map_{Sp(A-Mod)}(\Omega(\mathbb{L}''), M) \simeq Map_{Sp(A-Mod)}(\mathbb{L}'', S(M))$$
$$\simeq \Omega_* \mathbb{D}er_F(X, S(M)) \simeq \mathbb{D}er_F(X, \Omega_* S(M)) \simeq \mathbb{D}er_F(X, M).$$

This implies that $\Omega_* \mathbb{L}'' \in \mathrm{Ho}(Sp(A - Mod))$ is a cotangent complex of $F$ at the point $x$. By induction on $n$ and by construction we also see that this cotangent complex is $(-n)$-connective.

Now, let

$$\begin{array}{ccc} Y = \mathbb{R}\underline{Spec}\,B & \xrightarrow{\ u\ } & X = \mathbb{R}\underline{Spec}\,A \\ & {\scriptstyle y}\searrow \quad \swarrow{\scriptstyle x} & \\ & F, & \end{array}$$

be a morphism in $\mathrm{Ho}(Aff^{\sim,\tau}_{\mathcal{C}}/F)$, with $A$ and $B$ in $\mathcal{A}$. We consider the commutative diagram with homotopy cartesian squares

$$\begin{array}{ccccccc} & & X & \longrightarrow & X \times^h_F X & \longrightarrow & X \times^h X \\ & \nearrow & \uparrow & & \uparrow & & \uparrow \\ Y & \longrightarrow & Y \times^h_X Y & \longrightarrow & Y \times^h_F Y & \longrightarrow & Y \times^h Y. \end{array}$$

By the above explicit construction and an induction on $n$, the fact that the natural morphism

$$u^* : \mathbb{L}_{F,x} \otimes^{\mathbb{L}}_A B \longrightarrow \mathbb{L}_{F,y}$$

is an isomorphism simply follows from the next lemma.

LEMMA 1.4.1.12. *Let*

$$\begin{array}{ccccc} X = \mathbb{R}\underline{Spec}\,A & \xrightarrow{\ x\ } & F & \longrightarrow & G \\ {\scriptstyle u}\uparrow & & \uparrow & & \uparrow \\ Y = \mathbb{R}\underline{Spec}\,B & \xrightarrow[\ y\ ]{} & F' & \longrightarrow & G' \end{array}$$

*be a commutative diagram with the right hand square being homotopy cartesian in $Aff_{\mathcal{C}}^{\sim,\tau}$. We assume that $A$ and $B$ are in $\mathcal{A}$ and that $F$ and $G$ have global cotangent complexes. Then the natural square*

$$\begin{array}{ccc} \mathbb{L}_{G',y} & \longrightarrow & \mathbb{L}_{F',y} \\ \uparrow & & \uparrow \\ \mathbb{L}_{G,x} \otimes_A^{\mathbb{L}} B & \longrightarrow & \mathbb{L}_{F,x} \otimes_A^{\mathbb{L}} B \end{array}$$

*is homotopy cartesian in $Sp(B-Mod)$.*

PROOF. This is immediate from the definition and the homotopy cartesian square

$$\begin{array}{ccc} \mathbb{D}er_F(Y,M) \simeq \mathbb{D}er_F(X,M) & \longrightarrow & \mathbb{D}er_G(Y,M) \simeq \mathbb{D}er_G(X,M) \\ \uparrow & & \uparrow \\ \mathbb{D}er_{F'}(Y,M) & \longrightarrow & \mathbb{D}er_{G'}(Y,M) \end{array}$$

for any $A$-module $M$. □

This finishes the proof of Prop. 1.4.1.11. □

In fact, the proof of Proposition 1.4.1.11 also proves the following

PROPOSITION 1.4.1.13. *Let $F$ be a stack such that the diagonal $F \longrightarrow F \times^h F$ is $(n-1)$-representable. We suppose that for an $A \in \mathcal{A}$, any point $x : X := \mathbb{R}\underline{Spec}\,A \longrightarrow F$, and any $A$-module $M \in A - Mod_0$ the natural morphism*

$$\mathbb{D}er_F(X,M) \simeq \mathbb{D}er_F(X,\Omega S(M)) \longrightarrow \Omega\mathbb{D}er_F(X,S(M))$$

*is an isomorphism in $\mathrm{Ho}(SSet)$. Then $F$ has a cotangent complex, which is furthermore $(-n)$-connective.*

We finish this section by the notion of relative cotangent complex and its relation with the absolute notion. Let $f : F \longrightarrow G$ be a morphism of stacks, $A \in \mathcal{A}$, and $X := \mathbb{R}\underline{Spec}\,A \longrightarrow F$ be a morphism. We define an object $\mathbb{D}er_{F/G}(X,-) \in \mathrm{Ho}((A - Mod_0^{op})^{\wedge})$, to be the standard homotopy fiber of the morphism of the natural morphism

$$df : \mathbb{D}er_F(X,-) \longrightarrow \mathbb{D}er_G(X,-).$$

In terms of functors the object $\mathbb{D}er_{F/G}(X,-)$ sends an $A$-module $M \in A - Mod_0$ to the simplicial set

$$\mathbb{D}er_{F/G}(X,M) = Map_{X/Aff_{\mathcal{C}}^{\sim,\tau}/G}(X[M],F).$$

DEFINITION 1.4.1.14. *Let $f : F \longrightarrow G$ be a morphism of stacks.*

(1) *Let $A \in \mathcal{A}$, and $x : X := \mathbb{R}\underline{Spec}\,A \longrightarrow F$ be an $A$-point. We say that $f$* has a (relative) cotangent complex at $x$ relative to $\mathcal{A}$ *(or simply $f$* has a (relative) cotangent complex at $x$ *when $\mathcal{A}$ is unambiguous) if there exists an integer $n$, and an $(-n)$-connective stable $A$-module $\mathbb{L}_{F/G,x} \in \mathrm{Ho}(Sp(A - Mod))$, and an isomorphism in $\mathrm{Ho}(A - Mod_0^{op})^{\wedge}$*

$$\mathbb{D}er_{F/G}(X,-) \simeq \mathbb{R}\underline{h}_s^{\mathbb{L}_{F/G,x}}.$$

(2) *If $f$ has a cotangent complex at $x$, the stable $A$-module $\mathbb{L}_{F,x}$ is then called the* (relative) cotangent complex of $f$ at $x$.

(3) *If $f$ has a cotangent complex at $x$, the* (relative) tangent complex of $f$ at $x$ *is then the stable $A$-module*

$$\mathbb{T}_{F/G,x} := \mathbb{R}\underline{Hom}_A^{Sp}(\mathbb{L}_{F/G,x}, A) \in \mathrm{Ho}(Sp(A-Mod)).$$

Let now be a morphism of stacks $f : F \longrightarrow G$, and a commutative diagram in $Aff_{\mathcal{C}}^{\sim,\tau}$

$$\begin{array}{ccc} Y & \xrightarrow{u} & X \\ & {\scriptstyle y}\searrow \quad \swarrow{\scriptstyle y} & \\ & F, & \end{array}$$

with $X = \mathbb{R}\underline{Spec}\,A$ and $Y = \mathbb{R}\underline{Spec}\,B$ belonging to $\mathcal{A}$. We have a natural morphism in $\mathrm{Ho}((A-\underline{Mod}_0^{op})^\wedge)$

$$\mathbb{D}er_{F/G}(Y, ) \longrightarrow \mathbb{D}er_{F/G}(X, -).$$

If the morphism $f : F \longrightarrow G$ has cotangent complexes at both points $x$ and $y$, Prop. 1.2.11.3 induces a well defined morphism in $\mathrm{Ho}(Sp(B-Mod))$

$$u^* : \mathbb{L}_{F/G,x} \otimes_A^{\mathbb{L}} B \longrightarrow \mathbb{L}_{F/G,y}.$$

Of course, we have $(u \circ v)^* = v^* \circ u^*$ when this formula makes sense.

DEFINITION 1.4.1.15. *A morphism of stacks $f : F \longrightarrow G$* has a relative cotangent complex relative to $(\mathcal{C}, \mathcal{C}_0, \mathcal{A})$ *(or simply* has a cotangent complex *when the HA context is clear) if the following two conditions are satisfied.*

(1) *For any $A \in \mathcal{A}$, and any point $x : \mathbb{R}\underline{Spec}\,A \longrightarrow F$, the morphism $f$ has a cotangent complex $\mathbb{L}_{F/G,x}$ at $x$.*

(2) *For any morphism $u : A \longrightarrow B$ in $\mathcal{A}$, and any morphism in* $\mathrm{Ho}(Aff_{\mathcal{C}}^{\sim,\tau}/F)$

$$\begin{array}{ccc} Y := \mathbb{R}\underline{Spec}\,B & \xrightarrow{u} & X := \mathbb{R}\underline{Spec}\,A \\ & {\scriptstyle y}\searrow \quad \swarrow{\scriptstyle x} & \\ & F, & \end{array}$$

*the induced morphism*

$$u^* : \mathbb{L}_{F/G,x} \otimes_A^{\mathbb{L}} B \longrightarrow \mathbb{L}_{F/G,y}$$

*is an isomorphism in* $\mathrm{Ho}(Sp(B-Mod))$.

The important remark is the following, relating absolute and relative notions of cotangent complexes.

LEMMA 1.4.1.16. *Let $f : F \longrightarrow G$ be a morphism of stacks.*

(1) *If both stacks $F$ and $G$ have cotangent complexes then the morphism $f$ has a cotangent complex. Furthermore, for any $A \in \mathcal{A}$, and any morphism of stacks $X = \mathbb{R}\underline{Spec}\,A \longrightarrow F$, there is a natural homotopy cofiber sequence of stable $A$-modules*

$$\mathbb{L}_{G,x} \longrightarrow \mathbb{L}_{F,x} \longrightarrow \mathbb{L}_{F/G,x}.$$

(2) *If the morphism $f$ has a cotangent complex then for any stack $H$ and any morphism $H \longrightarrow G$, the morphism $F \times_G^h H \longrightarrow H$ has a relative cotangent complex and furthermore we have*

$$\mathbb{L}_{F/G,x} \simeq \mathbb{L}_{F\times_G^h H/H,x}$$

*for any $A \in \mathcal{A}$, and any morphism of stacks $X = \mathbb{R}\underline{Spec}\,A \longrightarrow F \times_G^h H$.*

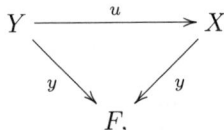

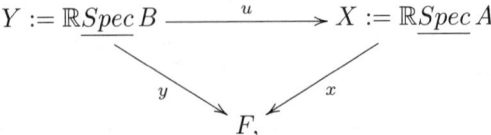

(3) *If for any $A \in \mathcal{A}$ and any morphism of stacks $x : X := \mathbb{R}\underline{Spec}\,A \longrightarrow F$, the morphism $F \times^h_G X \longrightarrow X$ has a relative cotangent complex, then the morphism $f$ has a relative cotangent complex. Furthermore, we have*

$$\mathbb{L}_{F/G,x} \simeq \mathbb{L}_{F\times^h_G X/X,x}.$$

(4) *If for any $A \in \mathcal{A}$ and any morphism of stacks $x : X := \mathbb{R}\underline{Spec}\,A \longrightarrow F$, the stack $F \times^h_G X$ has a cotangent complex, then the morphism $f$ has a relative cotangent complex. Furthermore we have a natural homotopy cofiber sequence*

$$\mathbb{L}_A \longrightarrow \mathbb{L}_{F\times^h_G X,x} \longrightarrow \mathbb{L}_{F/G,x}.$$

PROOF. (1) and (2) follow easily from the definition. Point (3) follows from (2). Finally, point (4) follows from (3), (1) and Prop. 1.4.1.8. □

## 1.4.2. Obstruction theory

Recall from 1.2.1.7 that for any commutative monoid $A$, any $A$-module $M$, and any derivation $d : A \longrightarrow A \oplus M$, we can form the square zero extension of $A$ by $\Omega M$, denoted by $A \oplus_d \Omega M$, as the homotopy cartesian square in $Comm(\mathcal{C})$

$$\begin{array}{ccc} A \oplus_d \Omega M & \xrightarrow{p} & A \\ \downarrow & & \downarrow d \\ A & \xrightarrow[s]{} & A \oplus M \end{array}$$

where $s : A \longrightarrow A \oplus M$ is the trivial derivation. In the sequel, the morphism $p : A \oplus_d \Omega M \longrightarrow A$ will be called the natural projection.

DEFINITION 1.4.2.1. (1) *A stack $F$ is* infinitesimally cartesian relative to the HA context $(\mathcal{C}, \mathcal{C}_0, \mathcal{A})$ *(or simply* inf-cartesian *when the HA context is unambiguous) if for any commutative monoid $A \in \mathcal{A}$, any $M \in A - Mod_1$, and any derivation $d \in \pi_0(\mathbb{D}er(A, M))$, corresponding to a morphism $d : A \longrightarrow A \oplus M$ in $\mathrm{Ho}(Comm(\mathcal{C})/A)$, the square*

$$\begin{array}{ccc} \mathbb{R}F(A \oplus_d \Omega M) & \longrightarrow & \mathbb{R}F(A) \\ \downarrow & & \downarrow d \\ \mathbb{R}F(A) & \xrightarrow[s]{} & \mathbb{R}F(A \oplus M) \end{array}$$

*is homotopy cartesian.*

(2) *A stack $F$* has an obstruction theory *(relative to $(\mathcal{C}, \mathcal{C}_0, \mathcal{A})$) if it has a (global) cotangent complex and if it is infinitesimally cartesian (relative to $(\mathcal{C}, \mathcal{C}_0, \mathcal{A})$).*

One also has a relative version.

DEFINITION 1.4.2.2. (1) *A morphism of stacks $F \longrightarrow G$ is* infinitesimally cartesian relative to $(\mathcal{C}, \mathcal{C}_0, \mathcal{A})$ *(or simply* inf-cartesian *if the HA context is clear) if for any commutative monoid $A \in \mathcal{A}$, any $A$-module $M \in A - Mod_1$, and any derivation $d \in \pi_0(\mathbb{D}er(A, M))$, corresponding to a morphism $d$ :*

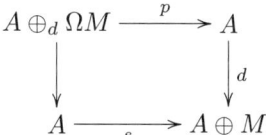

$A \longrightarrow A \oplus M$ *in* $\mathrm{Ho}(Comm(\mathcal{C})/A)$, *the square*

$$\begin{array}{ccc} \mathbb{R}F(A \oplus_d \Omega M) & \longrightarrow & \mathbb{R}G(A \oplus_d \Omega M) \\ \downarrow & & \downarrow \\ \mathbb{R}F(A) \times^h_{\mathbb{R}F(A\oplus M)} \mathbb{R}F(A) & \longrightarrow & \mathbb{R}G(A) \times^h_{\mathbb{R}G(A\oplus M)} \mathbb{R}G(A) \end{array}$$

*is homotopy cartesian.*

(2) *A morphism of stacks* $f : F \longrightarrow G$ has an obstruction theory relative to $(\mathcal{C}, \mathcal{C}_0, \mathcal{A})$ *(or simply* has an obstruction theory *if the HA context is clear) if it has a (global) cotangent complex and if it is infinitesimally cartesian relative* $(\mathcal{C}, \mathcal{C}_0, \mathcal{A})$.

As our HA context $(\mathcal{C}, \mathcal{C}_0, \mathcal{A})$ is fixed once for all we will from now avoid to mention the expression *relative to* $(\mathcal{C}, \mathcal{C}_0, \mathcal{A})$ when referring to the property of having an obstruction theory. The more precise terminology will only be used when two different HA contexts are involved (this will only happen in §2.3).

We have the following generalization of lemma 1.4.1.16.

LEMMA 1.4.2.3. *Let* $f : F \longrightarrow G$ *be a morphism of stacks.*

(1) *If both stacks* $F$ *and* $G$ *have an obstruction theory then the morphism* $f$ *has an obstruction theory.*
(2) *If the morphism* $f$ *has an obstruction theory then for any stack* $H$ *and any morphism* $H \longrightarrow G$, *the morphism* $F \times^h_G H \longrightarrow H$ *has a relative obstruction theory.*
(3) *If for any* $B \in Comm(\mathcal{C})$ *and any morphism of stacks* $y : Y := \mathbb{R}\underline{Spec}\, B \longrightarrow G$, *the stack* $F \times^h_G B$ *has an obstruction theory , then the morphism* $f$ *has a relative obstruction theory.*

PROOF. The existence of the cotangent complexes is done in Lem. 1.4.1.16, and it only remains to deal with the inf-cartesian property. The points (1) and (2) are clear by definition.

(3) Let $A \in \mathcal{A}$, $M \in A - Mod_1$ and $d \in \pi_0(\mathbb{D}er(A, M))$. We need to show that the square

$$\begin{array}{ccc} \mathbb{R}F(A \oplus_d \Omega M) & \longrightarrow & \mathbb{R}G(A \oplus_d \Omega M) \\ \downarrow & & \downarrow \\ \mathbb{R}F(A) \times^h_{\mathbb{R}F(A\oplus M)} \mathbb{R}F(A) & \longrightarrow & \mathbb{R}G(A) \times^h_{\mathbb{R}G(A\oplus M)} \mathbb{R}G(A) \end{array}$$

is homotopy cartesian. Let $z$ be a point in $\mathbb{R}G(A \oplus_d \Omega M)$. We need to prove that the morphism induced on the homotopy fibers of the two horizontal morphisms taken at $z$ is an equivalence. But this easily follows from the fact that the pullback of $f$ by the morphism corresponding to $z$,

$$F \times^h_G X_d[\Omega M] \longrightarrow X_d[\Omega M],$$

has an obstruction theory. □

PROPOSITION 1.4.2.4. (1) *Any representable stack has an obstruction theory.*
(2) *Any representable morphism has an obstruction theory.*

PROOF. (1) By Prop. 1.4.1.8 we already know that representable stacks have cotangent complexes. Using the Yoneda lemma, it is obvious to check that any representable stack is inf-cartesian. Indeed, for $F = \mathbb{R}\underline{Spec}\, B$ we have

$$\mathbb{R}F(A \oplus_d \Omega M) \simeq Map_{Comm(\mathcal{C})}(B, A \oplus_d \Omega M) \simeq$$

$$Map_{Comm(\mathcal{C})}(B, A) \times^h_{Map_{Comm(\mathcal{C})}(B, A\oplus M)} Map_{Comm(\mathcal{C})}(B, A) \simeq \mathbb{R}F(A) \times^h_{\mathbb{R}F(A\oplus M)} \mathbb{R}F(A).$$

(2) Follows from (1) and Lem. 1.4.2.3 (3). $\square$

In general, the expression *has an obstruction theory relative to* $(\mathcal{C}, \mathcal{C}_0, \mathcal{A})$ is justified by the following proposition.

PROPOSITION 1.4.2.5. *Let $F$ be a stack which has an obstruction theory. Let $A \in \mathcal{A}$, $M \in A - Mod_1$ and let $d \in \pi_0(\mathbb{D}er(A, M))$ be a derivation with $A \oplus_d \Omega M$ the corresponding square zero extension. Let us denote by*

$$X := \mathbb{R}\underline{Spec}\, A \longrightarrow X_d[\Omega M] := \mathbb{R}\underline{Spec}\,(A \oplus_d \Omega M)$$

*the morphism of representable stacks corresponding to the natural projection $A \oplus_d \Omega M \longrightarrow A$. Finally, let $x : X \longrightarrow F$ be an $A$-point of $F$.*

(1) *There exists a natural obstruction*

$$\alpha(x) \in \pi_0(Map_{Sp(A-Mod)}(\mathbb{L}_{F,x}, M)) = [\mathbb{L}_{F,x}, M]_{Sp(A-Mod)}$$

*vanishing if and only $x$ extends to a morphism $x'$ in* $\mathrm{Ho}(X/Aff_{\mathcal{C}}^{\sim,\tau})$

$$\begin{array}{ccc} X & \longrightarrow & X_d[\Omega M] \\ & {\scriptstyle x}\searrow \quad \swarrow{\scriptstyle x'} & \\ & F. & \end{array}$$

(2) *Let us suppose that $\alpha(x) = 0$. Then, the simplicial set of lifts of $x$*

$$\mathbb{R}\underline{Hom}_{X/Aff_{\mathcal{C}}^{\sim,\tau}}(X_d[\Omega M], F)$$

*is (non canonically) isomorphic in* $\mathrm{Ho}(SSet)$ *to the simplicial set*

$$Map_{Sp(A-Mod)}(\mathbb{L}_{F,x}, \Omega M) \simeq \Omega Map_{Sp(A-Mod)}(\mathbb{L}_{F,x}, M).$$

*More precisely, it is a simplicial torsor over the simplicial group*

$$\Omega Map_{Sp(A-Mod)}(\mathbb{L}_{F,x}, M).$$

PROOF. First of all, the space of lifts $x'$ is by definition $\mathbb{R}\underline{Hom}_{X/Aff_{\mathcal{C}}^{\sim,\tau}}(X_d[\Omega M], F)$, which is naturally equivalent to the homotopy fiber at $x$ of the natural morphism

$$\mathbb{R}_\tau \underline{Hom}(X_d[\Omega M], F) \longrightarrow \mathbb{R}_\tau \underline{Hom}(X, F).$$

Using that $F$ is inf-cartesian, we see that there exists a homotopy cartesian square

$$\begin{array}{ccc} \mathbb{R}_\tau \underline{Hom}(X_d[\Omega M], F) & \longrightarrow & \mathbb{R}_\tau \underline{Hom}(X, F) \\ \downarrow & & \downarrow{\scriptstyle d} \\ \mathbb{R}_\tau \underline{Hom}(X, F) & \xrightarrow{s} & \mathbb{R}_\tau \underline{Hom}(X[M], F). \end{array}$$

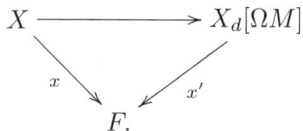

Therefore, the simplicial set $\mathbb{R}\underline{Hom}_{X/Aff_{\mathcal{C}}^{\sim,\tau}}(X_d[\Omega M], F)$ fits into a homotopy cartesian square

$$\begin{array}{ccc} \mathbb{R}\underline{Hom}_{X/Aff_{\mathcal{C}}^{\sim,\tau}}(X_d[\Omega M], F) & \longrightarrow & \bullet \\ \downarrow & & \downarrow{\scriptstyle d} \\ \bullet & \underset{0}{\longrightarrow} & \mathbb{R}\underline{Hom}_{X/Aff_{\mathcal{C}}^{\sim,\tau}}(X[M], F). \end{array}$$

As $F$ has a cotangent complex we have

$$\mathbb{R}\underline{Hom}_{X/Aff_{\mathcal{C}}^{\sim,\tau}}(X[M], F) \simeq Map_{Sp(A-Mod)}(\mathbb{L}_{F,x}, M),$$

and we see that the image of the right vertical arrow in the last diagram provides the element $\alpha(x) \in \pi_0(Map_{Sp(A-Mod)}(\mathbb{L}_{F,x}, M))$, which clearly vanishes if and only if $\mathbb{R}\underline{Hom}_{X/Aff_{\mathcal{C}}^{\sim,\tau}}(X_d[\Omega M], F)$ is non-empty. Furthermore, this last homotopy cartesian diagram also shows that if $\alpha(x) = 0$, then one has an isomorphism in $\mathrm{Ho}(SSet)$

$$\mathbb{R}\underline{Hom}_{X/Aff_{\mathcal{C}}^{\sim,\tau}}(X_d[\Omega M], F) \simeq \Omega Map_{Sp(A-Mod)}(\mathbb{L}_{F,x}, M).$$

□

One checks immediately that the obstruction of Prop. 1.4.2.5 is functorial in $F$ and $X$ in the following sense. If $f : F \longrightarrow F'$ be a morphism of stacks, then clearly

$$df(\alpha(x)) = \alpha(f(x)) \in \pi_0(\mathbb{D}er_{F'}(X, M)).$$

In the same way, if $A \longrightarrow B$ is a morphism in $\mathcal{A}$, corresponding to a morphism of representable stacks $Y \longrightarrow X$, and $y : Y \longrightarrow F$ be the composition, then we have

$$y^*(\alpha(x)) = \alpha_y \in \pi_0(\mathbb{D}er_F(Y, M \otimes_A^{\mathbb{L}} B)).$$

Proposition 1.4.2.5 also has a relative form, whose proof is essentially the same. We will also express it in a more precise way.

Proposition 1.4.2.6. *Let $f : F \longrightarrow G$ be a morphism of stacks which has an obstruction theory. Let $A \in \mathcal{A}$, $M \in A - Mod_1$, $d \in \pi_0(\mathbb{D}er(A, M))$ a derivation and $A \oplus_d \Omega M$ the corresponding square zero extension. Let $x$ be a point in $\mathbb{R}F(A) \times^h_{\mathbb{R}G(A \oplus_d \Omega M)} \mathbb{R}G(A)$ with projection $y \in \mathbb{R}F(A)$, and let $L(x)$ be the homotopy fiber, taken at $x$, of the morphism*

$$\mathbb{R}F(A \oplus_d \Omega M) \longrightarrow \mathbb{R}F(A) \times^h_{\mathbb{R}G(A \oplus_d \Omega M)} \mathbb{R}G(A).$$

*Then there exists a natural point $\alpha(x)$ in $Map_{A-Mod}(\mathbb{L}_{F/G,x}, M)$, and a natural isomorphism in* $\mathrm{Ho}(SSet)$

$$L(z) \simeq \Omega_{\alpha(x),0} Map_{A-Mod}(\mathbb{L}_{F/G,y}, M),$$

*where $\Omega_{\alpha(x),0} Map_{A-Mod}(\mathbb{L}_{F/G,y}, M)$ is the simplicial set of paths from $\alpha(x)$ to* 0.

Proof. Essentially the same as for Prop. 1.4.2.5. The point $x$ corresponds to a commutative diagram in $\mathrm{Ho}(Aff_{\mathcal{C}}^{\sim,\tau}/G)$

$$\begin{array}{ccc} X & \longrightarrow & F\,, \\ \downarrow & & \downarrow \\ X_d[\Omega M] & \longrightarrow & G \end{array}$$

where $X := \mathbb{R}\underline{Spec}\,A$ and $X_d[\Omega M] := \mathbb{R}\underline{Spec}\,(A \oplus_d \Omega M)$. Composing with the natural commutative diagram

$$\begin{array}{ccc} X[M] & \xrightarrow{d} & X \\ {\scriptstyle s}\downarrow & & \downarrow \\ X & \longrightarrow & X_d[\Omega M] \end{array}$$

we get a well defined commutative diagram in $\mathrm{Ho}(Aff_{\mathcal{C}}^{\sim,\tau}/G)$

$$\begin{array}{ccccc} X[M] & \xrightarrow{d} & X & \longrightarrow & F \\ {\scriptstyle s}\downarrow & & & & \downarrow \\ X & & \longrightarrow & & G, \end{array}$$

giving rise to a well defined point

$$\alpha(x) \in \mathbb{R}\underline{Hom}_{X/Aff_{\mathcal{C}}^{\sim,\tau}/G}(X[M], F) = \mathbb{D}er_{F/G}(X, M).$$

Using that the morphism $f$ is inf-cartesian, we easily see that the simplicial set $\Omega_{\alpha(x),0}\mathbb{D}er_{F/G}(X, M)$ is naturally equivalent to the space of lifts

$$L(x) = \mathbb{R}\underline{Hom}_{X/Aff_{\mathcal{C}}^{\sim,\tau}/G}(X_d[\Omega M], F) \simeq \Omega_{\alpha(x),0}\mathbb{D}er_{F/G}(X, M).$$

□

PROPOSITION 1.4.2.7. *Let $F$ be a stack whose diagonal $F \longrightarrow F \times^h F$ is $n$-representable for some $n$. Then $F$ has an obstruction theory if and only if it is inf-cartesian.*

PROOF. It is enough to show that a stack $F$ that is inf-cartesian satisfies the condition of proposition 1.4.1.11. But this follows easily from the following homotopy cartesian square

$$\begin{array}{ccc} A \oplus M & \longrightarrow & A \\ \downarrow & & \downarrow{\scriptstyle s} \\ A & \xrightarrow[s]{} & A \oplus S(M) \end{array}$$

for any commutative monoid $A \in \mathcal{A}$ and any $A$-module $M \in A - Mod_1$. □

### 1.4.3. Artin conditions

In this section we will give conditions on the topology $\tau$ and on the class of morphisms $\mathbf{P}$ ensuring that any stack which is geometric for such a $\tau$ and $\mathbf{P}$ will have an obstruction theory. We call these conditions *Artin's conditions*, though we warn the reader that these are not the rather famous conditions for a functor to be representable by an algebraic space. We refer instead to the fact that an algebraic stack in the sense of Artin (i.e. with a smooth atlas) has a good infinitesimal and obstruction theory. We think that this has been first noticed by M. Artin, since this is precisely one part of the easy direction of his representability criterion ([**Ar**, I, 1.6] or [**La-Mo**, Thm. 10.10]).

DEFINITION 1.4.3.1. *We will say that $\tau$* and $\mathbf{P}$ satisfy Artin's conditions relative to $(\mathcal{C}, \mathcal{C}_0, \mathcal{A})$ *(or simply* satisfy Artin's conditions *if the HA context is clear) if there exists a class $\mathbf{E}$ of morphisms in $Aff_{\mathcal{C}}$ such that the following conditions are satisfied.*

(1) *Any morphism in $\mathbf{P}$ is formally i-smooth in the sense of Def. 1.2.8.1.*

(2) *Morphisms in* $\boldsymbol{E}$ *are formally étale, are stable by equivalences, homotopy pullbacks and composition.*
(3) *For any morphism* $A \longrightarrow B$ *in* $\boldsymbol{E}$ *with* $A \in \mathcal{A}$, *we have* $B \in \mathcal{A}$.
(4) *For any epimorphism of representable stacks* $Y \longrightarrow X$, *which is a* $\boldsymbol{P}$-*morphism, there exists an epimorphism of representable stacks* $X' \longrightarrow X$, *which is in* $\boldsymbol{E}$, *and a commutative diagram in* $\mathrm{St}(\mathcal{C}, \tau)$

$$\begin{array}{ccc} & & Y \\ & \nearrow & \downarrow \\ X' & \longrightarrow & X. \end{array}$$

(5) *Let* $A \in \mathcal{A}$, $M \in A - Mod_1$, $d \in \pi_0(\mathbb{D}er(A, M))$ *be a derivation, and*

$$X := \mathbb{R}\underline{Spec}\, A \longrightarrow X_d[\Omega M] := \mathbb{R}\underline{Spec}, (A \oplus_d \Omega M)$$

*be the natural morphism in* $\mathrm{St}(\mathcal{C}, \tau)$. *Then, a formally étale morphism of representable stacks* $p : U \longrightarrow X_d[\Omega M]$ *is in* $\boldsymbol{E}$ *if and only if* $U \times^h_{X_d[\Omega M]} X \longrightarrow X$ *is so. Furthermore, if* $p$ *is in* $\boldsymbol{E}$, *then* $p$ *is an epimorphism of stacks if and only if* $U \times^h_{X_d[\Omega M]} X \longrightarrow X$ *is so.*

The above definition might seem a bit technical and somehow hard to follow. In order to fix his intuition, we suggest the reader to think in terms of standard algebraic geometry with $\tau$ being the etale topology, $\mathbf{P}$ the class of smooth morphisms, and $\mathbf{E}$ the class of étale morphisms. This is only meant to convey some classical geometric intuition because this classical situation in algebraic geometry does not really fit into the above definition; in fact in this case the base category is the category of $k$-modules and thus the suspension functor is not fully faithful.

In order to simplify notations we will say that a morphism $A \longrightarrow B$ in $Comm(\mathcal{C})$ is an $\boldsymbol{E}$*-covering* if it is in $\mathbf{E}$ and if the corresponding morphism of stacks

$$\mathbb{R}\underline{Spec}\, B \longrightarrow \mathbb{R}\underline{Spec}\, A$$

is an epimorphism of stacks.

THEOREM 1.4.3.2. *Assume* $\tau$ *and* $\boldsymbol{P}$ *satisfy Artin's conditions.*
(1) *Any* $n$*-representable morphism of stacks has an obstruction theory.*
(2) *Let* $f : F \longrightarrow G$ *be an* $n$*-representable morphism of stacks. If* $f$ *is in* $\boldsymbol{P}$ *then for any* $A \in \mathcal{A}$, *and any morphism* $x : X := \mathbb{R}\underline{Spec}\, A \longrightarrow F$ *there exists an* $E$*-covering*

$$x' : X' := \mathbb{R}\underline{Spec}\, A' \longrightarrow X$$

*such that for any* $M \in A' - Mod_1$ *the natural morphism*

$$[\mathbb{L}_{X'/G,x'}, M] \longrightarrow [\mathbb{L}_{F/G,x}, M]_{A-Mod}$$

*is zero.*

PROOF. Before starting the proof we will need the following general fact, that will be used all along the proof of the theorem.

LEMMA 1.4.3.3. *Let* $D$ *be a pointed model category for which the suspension functor*

$$S : \mathrm{Ho}(D) \longrightarrow \mathrm{Ho}(D)$$

*is fully faithful. Then, a homotopy co-cartesian square in* $D$ *is also homotopy cartesian.*

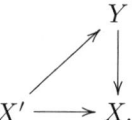

PROOF. When $D$ is a stable model category this is well known since homotopy fiber sequences are also homotopy cofiber sequences (see [**Ho1**]). The general case is proved in the same way. When $D$ is furthermore $\mathbb{U}$-cellular (which will be our case), one can even deduce the result from the stable case by using the left Quillen functor $D \longrightarrow Sp(D)$, from $D$ to the model category of spectra in $D$ as defined in [**Ho2**], and using the fact that it is homotopically fully faithful. □

The previous lemma can be applied to $\mathcal{C}$, but also to the model categories $B-Mod$ of modules over some commutative monoid $B$. In particular, homotopy cartesian square of $B$-modules which are also homotopy co-cartesian will remain homotopy cartesian after a derived tensor product by any $B$-module.

Let us now start the proof of theorem 1.4.3.2.

*Some topological invariance statements*

We start with several results concerning topological invariance of formally étale morphisms and morphisms in **E**.

LEMMA 1.4.3.4. *Let $A$ be a commutative monoid, $M$ an $A$-module and $A \oplus M \longrightarrow A$ the natural augmentation. Then, the homotopy push out functor*

$$A \otimes^{\mathbb{L}}_{A\oplus M} - : \mathrm{Ho}((A \oplus M) - Comm(\mathcal{C})) \longrightarrow \mathrm{Ho}(A - Comm(\mathcal{C}))$$

*induces an equivalence between the full sub-categories consisting of formally étale commutative $A \oplus M$-algebras and formally étale commutative $A$-algebras.*

PROOF. We see that the functor is essentially surjective as the morphism $A \oplus M \longrightarrow A$ possesses a section $A \longrightarrow A \oplus M$. Next, we show that the functor $A \otimes^{\mathbb{L}}_{A\oplus M} -$ is conservative. For this, let us consider the commutative square

$$\begin{array}{ccc} A \oplus M & \longrightarrow & A \\ \downarrow & & \downarrow \\ A & \longrightarrow & A \oplus S(M), \end{array}$$

which, as a commutative square in $\mathcal{C}$, is homotopy cocartesian and homotopy cartesian. Therefore, this square is homotopy cocartesian in $(A \oplus M) - Mod$, and thus lemma 1.4.3.3 implies that for any commutative $A \oplus M$-algebra $B$, the natural morphism

$$B \longrightarrow (A \otimes^{\mathbb{L}}_{A\oplus M} B) \times^h_{(A\oplus S(M))\otimes^{\mathbb{L}}_{A\oplus M} B} (A \otimes^{\mathbb{L}}_{A\oplus M} B) \simeq \left(A \times^h_{A\oplus S(M)} A\right) \otimes^{\mathbb{L}}_{A\oplus M} B$$

is then an isomorphism in $\mathrm{Ho}(Comm(\mathcal{C}))$. This clearly implies that the functor $A \otimes^{\mathbb{L}}_{A\oplus M} -$ is conservative.

Now, let $A \oplus M \longrightarrow B$ be a formally étale morphism of commutative monoids. The diagonal morphism $M \longrightarrow M \times^h M \simeq M \oplus M$ in $\mathrm{Ho}(A - Mod)$, induces a well defined morphism in $\mathrm{Ho}(Comm(\mathcal{C})/A \oplus M)$

$$\begin{array}{ccc} & & (A \oplus M) \oplus M \\ & \nearrow & \downarrow \\ A \oplus M & \xrightarrow[Id]{} & A \oplus M, \end{array}$$

and therefore a natural element in $\pi_0\mathbb{D}er(A\oplus M, M)$. Composing with $M \longrightarrow M\otimes^{\mathbb{L}}_A B$, we get a well defined element in $\pi_0\mathbb{D}er(A\oplus M, M\otimes^{\mathbb{L}}_A B)$. Using that $A\oplus M \longrightarrow B$ is formally étale, this element extends uniquely to an element in $\pi_0\mathbb{D}er(B, M\otimes^{\mathbb{L}}_A B)$. This last derivation gives rise to a well defined morphism in $\mathrm{Ho}((A\oplus M)-Comm(\mathcal{C}))$

$$u : B \longrightarrow (A\otimes^{\mathbb{L}}_{A\oplus M} B)\oplus M\otimes^{\mathbb{L}}_A B.$$

Furthermore, by construction, this morphism is sent to the identity of $A\otimes^{\mathbb{L}}_{A\oplus M} B$ by the functor $A\otimes^{\mathbb{L}}_{A\oplus M} -$, and as we have seen this implies that $u$ is an isomorphism in $\mathrm{Ho}((A\oplus M)-Comm(\mathcal{C}))$.

We now finish the proof of the lemma by showing that the functor $A\otimes^{\mathbb{L}}_{A\oplus M} -$ is fully faithful. For this, let $A\oplus M \longrightarrow B$ and $A\oplus M \longrightarrow B'$ be two formally étale morphisms of commutative monoids. As we have seen, $B'$ can be written as

$$B' \simeq (A\oplus M)\otimes^{\mathbb{L}}_A A' \simeq A'\oplus M'$$

where $A \longrightarrow A'$ is formally étale, and $M' := M\otimes^{\mathbb{L}}_A A'$. We consider the natural morphism

$$Map_{(A\oplus M)-Comm(\mathcal{C})}(B, B') \simeq Map_{(A\oplus M)-Comm(\mathcal{C})}(B, A'\oplus M') \longrightarrow$$

$$Map_{A-Comm(\mathcal{C})}(A\otimes^{\mathbb{L}}_{A\oplus M} B, A\otimes^{\mathbb{L}}_{A\oplus M} B') \simeq Map_{A-Comm(\mathcal{C})}(A\otimes^{\mathbb{L}}_{A\oplus M} B, A').$$

The homotopy fiber of this morphism at a point $B \to A'$ is identified with $\mathbb{D}er_{A\oplus M}(B, M')$, which is contractible as $B$ is formally étale over $A\oplus M$. This shows that the morphism

$$Map_{(A\oplus M)-Comm(\mathcal{C})}(B, B') \longrightarrow Map_{A-Comm(\mathcal{C})}(A\otimes^{\mathbb{L}}_{A\oplus M} B, A\otimes^{\mathbb{L}}_{A\oplus M} B')$$

has contractible homotopy fibers and therefore is an isomorphism in $\mathrm{Ho}(SSet)$, and finishes the proof of the lemma. □

LEMMA 1.4.3.5. *Let $A$ be a commutative monoid and $M$ an $A$-module. Then the homotopy push out functor*

$$A\otimes^{\mathbb{L}}_{A\oplus M} - : \mathrm{Ho}((A\oplus M)-Comm(\mathcal{C})) \longrightarrow \mathrm{Ho}(A-Comm(\mathcal{C}))$$

*induces an equivalence between the full sub-categories consisting of $\mathbf{E}$-coverings of $A\oplus M$ and of $\mathbf{E}$-coverings of $A$.*

PROOF. Using lemma 1.4.3.4 it is enough to show that a formally étale morphism $f : A\oplus M \longrightarrow B$ is in $\mathbf{E}$ (respectively an $\mathbf{E}$-covering) if and only if $A \longrightarrow A' := A\otimes^{\mathbb{L}}_{A\oplus M} B$ is in $\mathbf{E}$ (respectively an $\mathbf{E}$-covering). But, as $f$ is formally étale, lemma 1.4.3.4 implies that it can written as

$$A\oplus M \longrightarrow A'\oplus M' \simeq (A\oplus M)\otimes^{\mathbb{L}}_A A',$$

with $M' := M\otimes^{\mathbb{L}}_A A'$. Therefore the lemma simply follows from the stability of epimorphisms and morphisms in $\mathbf{E}$ by homotopy pullbacks. □

LEMMA 1.4.3.6. *Let $A \in Comm(\mathcal{C})$, $M \in A-Mod_1$, and $d \in \pi_0(\mathbb{D}er(A, M))$ be a derivation. Let $B := A\oplus_d \Omega M \longrightarrow A$ be the natural augmentation, and let us consider the base change functor*

$$A\otimes^{\mathbb{L}}_B - : \mathrm{Ho}(B-Comm(\mathcal{C})) \longrightarrow \mathrm{Ho}(A-Comm(\mathcal{C})).$$

*Then, $A\otimes^{\mathbb{L}}_B -$ restricted to the full subcategory consisting of formally étale commutative $B$-algebras is fully faithful.*

PROOF. Let

$$\begin{array}{ccc} B & \longrightarrow & A \\ \downarrow & & \downarrow d \\ A & \xrightarrow[s]{} & A \oplus M \end{array}$$

be the standard homotopy cartesian square of commutative monoids, which is also homotopy co-cartesian in $\mathcal{C}$ as $M \in A - Mod_1$. We represent it as a fibered square in $Comm(\mathcal{C})$

$$\begin{array}{ccc} B & \longrightarrow & A \\ \downarrow & & \downarrow d \\ A' & \xrightarrow[s']{} & A \oplus M, \end{array}$$

where $s' : A' \longrightarrow A \oplus M$ is fibrant replacement of the trivial section $s : A \longrightarrow A \oplus M$.

We define a model category $D$ whose objects are 5-plets $(B_1, B_2, B_3, a, b)$, where $B_1 \in A - Comm(\mathcal{C})$, $B_2 \in A' - Comm(\mathcal{C})$, $B_3 \in (A \oplus M) - Comm(\mathcal{C})$, and $a$ and $b$ are morphisms of commutative $(A \oplus M)$-algebras

$$(A \oplus M) \otimes_A B_1 \xrightarrow{a} B_3 \xleftarrow{b} (A \oplus M) \otimes_{A'} B_2$$

(where the co-base change on the left is taken with respect of the morphism $s' : A' \to A \oplus M$ and the one on the right with respect to $d : A \to A \oplus M$). For an object $(B_1, B_2, B_3, a, b)$ in $D$, the morphisms $a$ and $b$ can also be understood as $B_1 \longrightarrow B_3$ in $A - Comm(\mathcal{C})$ and $B_2 \longrightarrow B_3$ in $A' - Comm(\mathcal{C})$. The morphisms

$$(B_1, B_2, B_3, a, b) \longrightarrow (B'_1, B'_2, B'_3, a', b')$$

in $D$ are defined in the obvious way, as families of morphisms $\{B_i \to B'_i\}$ commuting with the $a$'s and $b$'s. A morphism in $D$ is defined to be an equivalence or a cofibration if each morphism $B_i \to B'_i$ is so. A morphism in $D$ is defined to be a fibration if each morphism $B_i \to B'_i$ is a fibration in $\mathcal{C}$, and if the natural morphisms

$$B_1 \longrightarrow B'_1 \times_{B'_3} B_3 \qquad B_2 \longrightarrow B'_2 \times_{B'_3} B_3$$

are fibrations in $\mathcal{C}$. This defines a model category structure on $D$ which is a Reedy type model structure. An important fact concerning $D$ is the description of its mapping spaces as the following homotopy cartesian square

$$\begin{array}{ccc} Map_D((\underline{B}, a, b), (\underline{B'}, a', b')) & \longrightarrow & Map_{A-Comm(\mathcal{C})}(B_1, B'_1) \times Map_{A'-Comm(\mathcal{C})}(B_2, B'_2) \\ \downarrow & & \downarrow \\ Map_{(A\oplus M)-Comm(\mathcal{C})}(B_3, B'_3) & \longrightarrow & Map_{A-Comm(\mathcal{C})}(B_1, B'_3) \times Map_{A'-Comm(\mathcal{C})}(B_2, B'_3) \end{array}$$

where we have denoted $\underline{B} := (B_1, B_2, B_3)$ and $\underline{B'} := (B'_1, B'_2, B'_3)$. There exists a natural functor

$$F : B - Comm(\mathcal{C}) \longrightarrow D$$

sending a commutative $B$-algebra $B'$ to the object

$$F(B') := (A \otimes_B B', A' \otimes_B B', (A \oplus M) \otimes_B B', a, b)$$

where

$$a : (A \oplus M) \otimes_A (A \otimes_B B') \simeq (A \oplus M) \otimes_B B' \qquad b : (A \oplus M) \otimes_{A'} (A' \otimes_B B') \simeq (A \oplus M) \otimes_B B'$$

are the two natural isomorphisms in $(A\oplus M)-Comm(\mathcal{C})$. The functor $F$ has a right adjoint $G$, sending an object $(B_1,B_2,B_3,a,b)$ to the pullback in $B-Comm(\mathcal{C})$

$$\begin{array}{ccc} G(B_1,B_2,B_3,a,b) & \longrightarrow & B_1 \\ \downarrow & & \downarrow a \\ B_2 & \xrightarrow[b]{} & B_3. \end{array}$$

Clearly the adjunction $(F,G)$ is a Quillen adjunction. Furthermore, lemma 1.4.3.3 implies that for any commutative $B$-algebra $B\to B'$ the adjunction morphism

$$B'\longrightarrow \mathbb{R}G\mathbb{L}F(B')=A\otimes^{\mathbb{L}}_B B'\times^h_{(A\oplus M)\otimes^{\mathbb{L}}_B B'}A'\otimes^{\mathbb{L}}_B B'\simeq (A\times^h_{A\oplus M}A')\otimes^{\mathbb{L}}_B B'$$

is an isomorphism in $\mathrm{Ho}(B-Comm(\mathcal{C}))$. This implies in particular that

$$\mathbb{L}F:\mathrm{Ho}(B-Comm(\mathcal{C}))\longrightarrow \mathrm{Ho}(D)$$

is fully faithful.

We now consider the functor

$$D\longrightarrow A-Comm(\mathcal{C})$$

sending $(B_1,B_2,B_3,a,b)$ to $B_1$. Using our lemma 1.4.3.4, and the description of the mapping spaces in $D$, it is not hard to see that the induced functor

$$\mathrm{Ho}(D)\longrightarrow \mathrm{Ho}(A-Comm(\mathcal{C}))$$

becomes fully faithful when restricted to the full subcategory of $\mathrm{Ho}(D)$ consisting of objects $(B_1,B_2,B_3,a,b)$ such that $A\to B_1$ and $A'\to B_2$ are formally étale and the induced morphism

$$a:(A\oplus M)\otimes^{\mathbb{L}}_A B_1\longrightarrow B_3 \qquad b:(A\oplus M)\otimes^{\mathbb{L}}_{A'} B_2\longrightarrow B_3$$

are isomorphisms in $\mathrm{Ho}((A\oplus M)-Comm(\mathcal{C}))$. Putting all of this together we deduce that the functor

$$A\otimes^{\mathbb{L}}_B - :\mathrm{Ho}(B-Comm(\mathcal{C}))\longrightarrow \mathrm{Ho}(A-Comm(\mathcal{C}))$$

is fully faithful when restricted to the full subcategory of formally étale morphisms. □

LEMMA 1.4.3.7. *Let $A$ be a commutative monoid, $M$ an $A$-module and $d:\mathbb{D}er(A,M)$ be a derivation. Let $B:=A\oplus_d\Omega M\longrightarrow A$ be the natural augmentation. Then, there exists a natural homotopy cofiber sequence of $A$-modules*

$$\mathbb{L}_B\otimes^{\mathbb{L}}_B A\longrightarrow \mathbb{L}_A\xrightarrow{d}\mathbb{L}QZ(M),$$

*where*

$$Q:A-Comm_{nu}(\mathcal{C})\longrightarrow A-Mod \qquad A-Comm_{nu}(\mathcal{C})\longleftarrow A-Mod:Z$$

*is the Quillen adjunction described during the proof of 1.2.1.2.*

PROOF. When $d$ is the trivial derivation, we know the lemma is correct as by Prop. 1.2.1.6 (4) we have

$$\mathbb{L}_{A\oplus M}\otimes^{\mathbb{L}}_{A\oplus M}A\simeq \mathbb{L}_A\coprod \mathbb{L}QZ(M).$$

For the general situation, we use our left Quillen functor

$$F:B-Comm(\mathcal{C})\longrightarrow D$$

defined during the proof of lemma 1.4.3.6. The commutative monoid $A$ is naturally an $(A \oplus M)$-algebra, and can be considered as a natural object $(A, A, A)$ in $D$ with the obvious transition morphisms

$$A \otimes_A (A \oplus M) \simeq A \oplus M \to A,$$

$$A \otimes_{A'} (A \oplus M) \to A \otimes_A (A \oplus M) \simeq A \oplus M \to A.$$

For any $A$-module $N$, one can consider $A \oplus N$ as a commutative $A$-algebra, and therefore as as an object $(A \oplus N, A \oplus N, A \oplus N)$ in $D$ (with the obvious transition morphisms). We will simply denote by $A$ the object $(A, A, A) \in D$, and by $A \oplus N$ the object $(A \oplus N, A \oplus N, A \oplus N) \in D$. The left Quillen property of $F$ implies that

$$\mathbb{D}er(B, N) \simeq Map_{D/A}(F(B), A \oplus N).$$

This shows that the morphism

$$\mathbb{D}er(A, N) \longrightarrow \mathbb{D}er(B, N)$$

is equivalent to the morphism

$$\mathbb{D}er(A, N) \longrightarrow \mathbb{D}er(A, N) \times^h_{\mathbb{D}er(A \oplus M, N)} \mathbb{D}er(A, N).$$

This implies the the morphism

$$\mathbb{L}_B \otimes^{\mathbb{L}}_B A \longrightarrow \mathbb{L}_A$$

is naturally equivalent to the morphism of $A$-modules

$$\mathbb{L}_A \coprod^{\mathbb{L}}_{\mathbb{L}_{A \oplus M} \otimes^{\mathbb{L}}_{A \oplus M} A} \mathbb{L}_A \longrightarrow \mathbb{L}_A.$$

Using the already known result for the trivial extension $A \oplus M$ we get the required natural cofiber sequence

$$\mathbb{L}_B \otimes^{\mathbb{L}}_B A \longrightarrow \mathbb{L}_A \longrightarrow \mathbb{L}QZ(M).$$

$\square$

LEMMA 1.4.3.8. *Let $A$ be a commutative monoid, $M \in A - Mod_1$ and $d \in \pi_0(\mathbb{D}er(A, M))$ be a derivation. Let $B := A \oplus_d \Omega M \longrightarrow A$ be the natural augmentation, and let us consider the base change functor*

$$A \otimes^{\mathbb{L}}_B - : \mathrm{Ho}(B - Comm(\mathcal{C})) \longrightarrow \mathrm{Ho}(A - Comm(\mathcal{C})).$$

*Then, $A \otimes^{\mathbb{L}}_B -$ induces an equivalence between the full sub-categories consisting of formally étale commutative $B$-algebras and of formally étale commutative $A$-algebras.*

PROOF. By lemma 1.4.3.6 we already know that the functor is fully faithful, and it only remains to show that any formally étale $A$-algebra $A \to A'$ is of the form $A \otimes^{\mathbb{L}}_B B'$ for some formally étale morphism $B \to B'$.

Let $A \longrightarrow A'$ be a formally étale morphism. The derivation $d \in \pi_0\mathbb{D}er(A, M)$ lifts uniquely to a derivation $d' \in \pi_0\mathbb{D}er(A', M')$ where $M' := M \otimes^{\mathbb{L}}_A A'$. We form the corresponding square zero extension

$$\begin{array}{ccc} B' := A' \oplus_d \Omega M' & \longrightarrow & A' \\ \downarrow & & \downarrow d' \\ A' & \longrightarrow & A' \oplus M' \end{array}$$

which comes equipped with a natural morphism $B \longrightarrow B'$ fitting in a homotopy commutative square

$$\begin{array}{ccc} B' & \longrightarrow & A' \\ \uparrow & & \uparrow \\ B & \longrightarrow & A. \end{array}$$

We claim that $B \longrightarrow B'$ is formally étale and that the natural morphism $A \otimes^{\mathbb{L}}_B B' \longrightarrow A'$ is an isomorphism in $\mathrm{Ho}(A - Comm(\mathcal{C}))$.

There are natural cofiber sequences in $\mathcal{C}$

$$B' \longrightarrow A' \longrightarrow M'$$

$$B \longrightarrow A \longrightarrow M,$$

as well as a natural morphism of cofiber sequences in $\mathcal{C}$

$$\begin{array}{ccccc} A \otimes^{\mathbb{L}}_B B' & \longrightarrow & A \otimes^{\mathbb{L}}_B A' & \longrightarrow & A \otimes^{\mathbb{L}}_B M' \\ \downarrow & & \downarrow & & \downarrow \\ B \otimes^{\mathbb{L}}_B A' \simeq A' & \longrightarrow & A \otimes^{\mathbb{L}}_B A' & \longrightarrow & M \otimes^{\mathbb{L}}_B A', \end{array}$$

which by our lemma 1.4.3.3 is also a morphism of fiber sequences. The two right vertical morphisms are equivalences and thus so is the arrow on the left. This shows that $A \otimes^{\mathbb{L}}_B B' \simeq A'$.

Finally, lemma 1.4.3.7 implies the existence of a natural morphism of cofiber sequences of $A'$-modules

$$\begin{array}{ccccc} \mathbb{L}_B \otimes^{\mathbb{L}}_B A & \longrightarrow & \mathbb{L}_A & \longrightarrow & \mathbb{L}QZ(M) \\ \downarrow & & \downarrow & & \downarrow \\ \mathbb{L}_{B'} \otimes^{\mathbb{L}}_{B'} A' & \longrightarrow & \mathbb{L}_{A'} & \longrightarrow & \mathbb{L}QZ(M') \simeq \mathbb{L}QZ(M) \otimes^{\mathbb{L}}_A A'. \end{array}$$

As $A \to A'$ is étale, this implies that the natural morphism

$$(\mathbb{L}_B \otimes^{\mathbb{L}}_B B') \otimes^{\mathbb{L}}_{B'} A' \simeq (\mathbb{L}_B \otimes^{\mathbb{L}}_B A) \otimes^{\mathbb{L}}_A A' \longrightarrow \mathbb{L}_{B'} \otimes^{\mathbb{L}}_{B'} A'$$

is an isomorphism in $\mathrm{Ho}(A' - Mod)$. This would show that $B \longrightarrow B'$ is formally étale if one knew that the base change functor

$$A' \otimes^{\mathbb{L}}_{B'} - : \mathrm{Ho}(B' - Mod) \longrightarrow \mathrm{Ho}(A' - Mod)$$

were conservative. However, this is the case as lemma 1.4.3.3 implies that for any $B'$-module $N$ we have a natural isomorphism in $\mathrm{Ho}(B' - Mod)$

$$N \simeq (A' \otimes^{\mathbb{L}}_{B'} N) \times^h_{(A' \oplus M') \otimes^{\mathbb{L}}_{B'} N} (A' \otimes^{\mathbb{L}}_{B'} N).$$

□

LEMMA 1.4.3.9. *Let $A \in \mathcal{A}$, $M \in A - Mod_1$ and $d \in \pi_0(\mathbb{D}er(A, M))$ be a derivation. Let*

$$A \oplus_d \Omega M \longrightarrow A$$

*be the natural morphism and let us consider the homotopy push-out functor*

$$(A \oplus_d \Omega M) \otimes^{\mathbb{L}}_A - : \mathrm{Ho}((A \oplus_d \Omega M) - Comm(\mathcal{C})) \longrightarrow \mathrm{Ho}(A - Comm(\mathcal{C})).$$

*Then, $(A \oplus_d \Omega M) \otimes^{\mathbb{L}}_A -$ induces an equivalence between the full sub-categories consisting of $\mathbf{E}$-covers of $A \oplus_d \Omega M$ and of $\mathbf{E}$-covers of $A$.*

PROOF. This is immediate from Lem. 1.4.3.8 and condition (4) of Artin's conditions 1.4.3.1. ☐

*Proof of Theorem 1.4.3.2*

We are now ready to prove that $F$ has an obstruction theory. To simplify notations we assume that $F$ is fibrant in $Aff_{\mathcal{C}}^{\sim,\tau}$, so $F(A) \simeq \mathbb{R}F(A)$ for any $A \in Comm(\mathcal{C})$. We then argue by induction on the integer $n$. For $n = -1$ Theorem 1.4.3.2 follows from Prop. 1.4.2.4, hypothesis Def. 1.4.3.1 (5) and Prop. 1.2.8.3. We now assume that $n \geq 0$ and that both statement of theorem 1.4.3.2 are true for all $m < n$.

We start by proving Thm. 1.4.3.2 (1) for rank $n$. For this, we use Lem. 1.4.2.3 and Prop. 1.4.2.7, which show that we only need to prove that any $n$-geometric stack is inf-cartesian.

LEMMA 1.4.3.10. *Let $F$ be an $n$-geometric stack. Then $F$ is inf-cartesian.*

PROOF. Let $A \in \mathcal{A}$, $M \in A-Mod_1$, and $d \in \pi_0(\mathbb{D}er(A,M))$. Let $x$ be a point in $\pi_0(F(A) \times^h_{F(A\oplus M)} F(A))$, whith projection $x_1 \in \pi_0(F(A))$ on the first factor. We need to show that the homotopy fiber, taken at $x$, of the morphism

$$F(A \oplus_d \Omega M) \longrightarrow F(A) \times^h_{F(A\oplus M)} F(A)$$

is contractible. For this, we replace the homotopy cartesian diagram

$$\begin{array}{ccc} A \oplus_d \Omega M & \longrightarrow & A \\ \downarrow & & \downarrow \\ A & \longrightarrow & A \oplus M \end{array}$$

be an equivalent commutative diagram in $Comm(\mathcal{C})$

$$\begin{array}{ccc} B & \longrightarrow & B_1 \\ \downarrow & & \downarrow \\ B_2 & \longrightarrow & B_3, \end{array}$$

in such a way that each morphism is a cofibration in $Comm(\mathcal{C})$. The point $x$ can be represented as a point in the standard homotopy pullback $F(B_1) \times^h_{F(B_3)} F(B_2)$. We then define a functor

$$S : B - Comm(\mathcal{C}) = (Aff_{\mathcal{C}}/Spec\, B)^{op} \longrightarrow SSet_{\mathbb{V}},$$

in the following way. For any morphism of commutative monoids $B \longrightarrow B'$, the simplicial set $S(B')$ is defined to be the standard homotopy fiber, taken at $x$, of the natural morphism

$$F(B') \longrightarrow F(B_1 \otimes_B B') \times^h_{F(B_3 \otimes_B B')} F(B_2 \otimes_B B').$$

Because of our choices on the $B_i$'s, it is clear that the simplicial presheaf $S$ is a stack on the comma model site $Aff_{\mathcal{C}}/Spec\, B$. Therefore, in order to show that $S(B)$ is contractible it is enough to show that $S(B')$ is contractible for some morphism $B \longrightarrow B'$ such that $\mathbb{R}\underline{Spec}\, B' \longrightarrow \mathbb{R}\underline{Spec}\, B$ is an epimorphism of stacks. In particular, we are allowed to homotopy base change by some **E**-covering of $B$. Also, using our lemma 1.4.3.8 (or rather its proof), we see that for any **E**-covering $B \to B'$, the homotopy cartesian square

$$\begin{array}{ccc} B' & \longrightarrow & B_1 \otimes_B B' \\ \downarrow & & \downarrow \\ B_2 \otimes_B B' & \longrightarrow & B_3 \otimes_B B', \end{array}$$

is in fact equivalent to some

$$\begin{array}{ccc} A' \oplus_{d'} \Omega M' & \longrightarrow & A' \\ \downarrow & & \downarrow d' \\ A' & \longrightarrow & A' \oplus M', \end{array}$$

for some **E**-covering $A \longrightarrow A'$ (and with $M' \simeq M \otimes_A^{\mathbb{L}} A'$, and where $d'$ is the unique derivation $d' \in \pi_0 \mathbb{D}er(A', M')$ extending $d$). This shows that we can always replace $A$ by $A'$, $d$ by $d'$ and $M$ by $M'$. In particular, for an $(n-1)$-atlas $\{U_i \longrightarrow F\}$ $F$, we can assume that the point $x_1 \in \pi_0(F(A))$, image of the point $x$, lifts to a point in $y_1 \in \pi_0(U_j(A))$ for some $j$. We will denote $U := U_j$.

SUB-LEMMA 1.4.3.11. *The point $x \in \pi_0(F(A) \times^h_{F(A\oplus M)} F(A))$ lifts to point $y \in \pi_0(U(A) \times^h_{U(A\oplus M)} U(A))$*

PROOF. We consider the commutative diagram of simplicial sets

$$\begin{array}{ccc} U(A) \times^h_{U(A\oplus M)} U(A) & \stackrel{f}{\longrightarrow} & F(A) \times^h_{F(A\oplus M)} F(A) \\ p \downarrow & & \downarrow q \\ U(A) & \longrightarrow & F(A) \end{array}$$

induced by the natural projection $A \oplus M \to A$. Let $F(p)$ and $F(q)$ be the homotopy fibers of the morphisms $p$ and $q$ taken at $y_1$ and $x_1$. We have a natural morphism $g : F(p) \longrightarrow F(q)$. Moreover, the homotopy fiber of the morphism $f$, taken at the point $x$, receives a natural morphism from the homotopy fiber of the morphism $g$. It is therefore enough to show that the homotopy fiber of $g$ is not empty. But, by definition of derivations, the morphism $g$ is equivalent to the morphism

$$\Omega_{d,0}\mathbb{D}er_U(X, M) \longrightarrow \Omega_{d,0}\mathbb{D}er_F(X, M),$$

where the derivation $d$ is given by the image of the point $y_1$ by $d : A \to A \oplus M$. Therefore, the homotopy fiber of the morphism $g$ is equivalent to

$$\Omega_{d,0}\mathbb{D}er_{U/F}(X, M) \simeq \Omega_{d,0}Map(\mathbb{L}_{U/F,y_1}, M).$$

But, using Thm. 1.4.3.2 (2) at rank $(n-1)$ and for the morphism $U \longrightarrow F$ we obtain that $\Omega_{d,0}\mathbb{D}er_{U/F}(X, M)$ is non empty. This finishes the proof of the sub-lemma. $\square$

We now consider the commutative diagram

$$\begin{array}{ccc} U(A \oplus_d \Omega M) & \stackrel{a}{\longrightarrow} & U(A) \times^h_{U(A\oplus M)} U(A) \\ b' \downarrow & & \downarrow b \\ F(A \oplus_d \Omega M) & \underset{a'}{\longrightarrow} & F(A) \times^h_{F(A\oplus M)} F(A). \end{array}$$

The morphism $a$ is an equivalence because $U$ is representable. Furthermore, by our inductive assumption the above square is homotopy cartesian. This implies that the homotopy fiber of $a'$ at $x$ is either contractible or empty. But, by the above sub-lemma the point $x$ lifts, up to homotopy, to a point in $U(A) \times^h_{U(A\oplus M)} U(A)$, showing that this homotopy fiber is non empty. $\square$

We have finished the proof of lemma 1.4.3.10 which implies that any $n$-geometric stack is inf-cartesian, and thus that any $n$-representable morphism has an obstruction theory. It only remain to show part (2) of Thm. 1.4.3.2 at rank $n$. For this, we use Lem. 1.4.1.16 (3) which implies that we can assume that $G = *$. Let $U \longrightarrow F$ be an $n$-atlas, $A \in \mathcal{A}$ and $x : X := \mathbb{R}\underline{Spec\,A} \longrightarrow F$ be a point. By passing to an epimorphism of representable stacks $X' \longrightarrow X$ which is in $\mathbf{E}$, we can suppose that the point $x$ factors through a point $u : X \longrightarrow U$, where $U$ is representable and $U \longrightarrow F$ is in $\mathbf{P}$. By composition and the hypothesis that morphisms in $\mathbf{P}$ are formally i-smooth, we see that $U \longrightarrow *$ is a formally i-smooth morphism. We then have a diagram

$$\mathbb{L}_{F,x} \longrightarrow \mathbb{L}_{U,u} \longrightarrow \mathbb{L}_{X,x},$$

which obviously implies that for any $M \in A-Mod_1$ the natural morphism

$$[\mathbb{L}_{X,x}, M] \longrightarrow [\mathbb{L}_{F,x}, M]$$

factors through the morphism

$$[\mathbb{L}_{X,x}, M] \longrightarrow [\mathbb{L}_{U,u}, M]$$

which is itself equal to zero by Prop. 1.2.8.3. $\square$

We also extract from Lem. 1.4.3.8 and its proof the following important corollary.

COROLLARY 1.4.3.12. *Let $A \in Comm(\mathcal{C})$, $M \in A-Mod$ and $d \in \pi_0(\mathbb{D}er(A, M))$ be a derivation. Assume that the square*

$$\begin{array}{ccc} B = A \oplus_d \Omega M & \longrightarrow & A \\ \downarrow & & \downarrow \\ A & \longrightarrow & A \oplus M \end{array}$$

*is homotopy co-cartesian in $\mathcal{C}$, then the base change functor*

$$A \otimes^{\mathbb{L}}_B - : \mathrm{Ho}(B - Comm(\mathcal{C})) \longrightarrow \mathrm{Ho}(A - Comm(\mathcal{C}))$$

*induces an equivalence between the full sub-categories of formally étale commutative $B$-algebras and formally étale commutative $A$-algebras. The same statement holds with* formally étale *replaced by* étale.

PROOF. Only the assertion with *formally étale* replaced by *étale* requires an argument. For this, we only need to prove that if a formally étale morphism $f : B \longrightarrow B'$ is such that $A \longrightarrow A \otimes^{\mathbb{L}}_B B' = A'$ is finitely presented, then so is $f$. For this we use the fully faithful functor

$$\mathbb{L}F : \mathrm{Ho}(B - Comm(\mathcal{C})) \longrightarrow \mathrm{Ho}(D)$$

defined during the proof of Lem. 1.4.3.6. Using the description of mappping spaces in $D$ in terms of a certain homotopy pullbacks, and using the fact that filtered homotopy colimits in $SSet$ commutes with homotopy pullbacks, we deduce the statement. $\square$

# Part 2

# Applications

# Introduction to Part 2

In this second part we apply the theory developed in the first part to study the geometry of stacks in various HAG contexts (Def. 1.3.2.13). In particular we will specialize our base symmetric monoidal model category $\mathcal{C}$ to the following cases:

- $\mathcal{C} = \mathbb{Z} - Mod$, the category of $\mathbb{Z}$-modules to get a theory of *geometric stacks in* (classical) *Algebraic Geometry* (§2.1);
- $\mathcal{C} = sMod_k$, the category of simplicial modules over an arbitrary base commutative ring $k$ to get a theory of *derived* or $D^-$*-geometric stacks* (§2.2);
- $\mathcal{C} = C(k)$, the category of unbounded cochain complexes of modules over a characteristic zero base commutative ring $k$ to get a theory of *geometric stacks in complicial algebraic geometry*, also called geometric $D$-stacks (§2.3);
- $\mathcal{C} = Sp^{\Sigma}$, the category of symmetric spectra ([**HSS, Shi**]) to get a theory of *geometric stacks in brave new algebraic geometry* (§2.4).

In §2.1 we are concerned with **classical algebraic geometry**, the base category $\mathcal{C} = \mathbb{Z} - Mod$ being endowed with the trivial model structure. We verify that if $k{-}Aff := Comm(\mathcal{C})^{\mathrm{op}} = (k{-}Alg)^{\mathrm{op}}$ is endowed with its étale Grothendieck topology and $\mathbf{P}$ is the class of smooth morphisms between (usual) commutative rings then

$$(\mathcal{C}, \mathcal{C}_0, \mathcal{A}, \tau, \mathbf{P}) := (k - Mod, k - Mod, k - Alg, \text{ét}, \mathbf{P})$$

is a HAG-context according to Def. 1.3.2.13. $n$-geometric stacks in this context will be called *Artin $n$-stacks*, and essentially coincide with geometric $n$-stacks as defined in [**S3**]; their model category will be denoted by $k - Aff^{\sim,\text{ét}}$ and its homotopy category by $\mathrm{St}(k)$.

After having established a coherent dictionary (Def. 2.1.1.4), we show in §2.1.2 how the theory of schemes, of algebraic spaces, and of Artin's algebraic stacks in groupoids ([**La-Mo**]) embeds in our theory of geometric stacks (Prop. 2.1.2.1). We also remark (Rmk. 2.1.2.2) that the general infinitesimal theory for geometric stacks developed in Part I does not apply to this context. The reason for this is that the category $\mathcal{C} = k{-}Mod$ (with its trivial model structure) is as unstable as it could possibly be: the suspension functor $S : \mathrm{Ho}(\mathcal{C}) = k - Mod \to k - Mod = \mathrm{Ho}(\mathcal{C})$ is trivial (i.e. sends each $k$-module to the zero $k$-module). The explanation for this is the following. Usual infinitesimal theory that applies to schemes, algebraic spaces or to (some classes of) Artin's algebraic stacks in groupoids is in fact (as made clear e.g. by the definition of cotangent complex of a scheme ([**Ill**, II.2]) which uses simplicial resolutions of objects $Comm(\mathcal{C})$), already conceptually part of *derived algebraic geometry* in the sense that its classical definition already requires to embed $Comm(\mathcal{C}) = k - Alg$ into the category of simplicial $k$-algebras. And in fact (as we will show in §2.2) when schemes, algebraic spaces and Artin algebraic stacks in groupoids are viewed as *derived stacks*, then their classical infinitesimal theory can be recovered (and generalized, see Cor. 2.2.4.5) and interpreted geometrically within our general formalism of Chapter 1.4 (in particular in Prop. 1.4.1.6).

Therefore we are naturally brought to §2.2 where we treat the case of **derived algebraic geometry**, i.e. the case where $\mathcal{C} := sk - Mod$, the category of simplicial modules over an arbitrary commutative ring $k$.

In §2.2.1 we describe the model categories $\mathcal{C} = sk-Mod$ and $Comm(\mathcal{C}) = sk-Alg$ whose opposite is denoted by $k-D^{-}Aff$, in particular finite cell and finitely presented objects, suspension and loop functors, Postnikov towers and stable modules. We also show that $(\mathcal{C}, \mathcal{C}_0, \mathcal{A}) := (sk - Mod, sk - Mod, sk - Alg)$ is a HA context in the sense of Def. 1.1.0.11.

In §2.2.2 we show how the general definitions of properties of modules (e.g. projective, flat, perfect) and of morphisms between commutative rings in $\mathcal{C}$ (e.g. finitely presented, flat, (formally) smooth, (formally) étale, Zariski open immersion) given in Chapter 1.2 translates concretely in the present context. The basic idea here is that of *strongness* which says that a module $M$ over a simplicial $k$-algebra $A$ (respectively, a morphism $A \longrightarrow B$ in $sk - Alg$) has the property $\mathcal{P}$, defined in the abstract setting of Chapter 1.2, if and only if $\pi_0(M)$ has the corresponding classical property as a $\pi_0(A)$-module and $\pi_0(M) \otimes_{\pi_0(A)} \pi_*(A) \simeq \pi_*(M)$ (respectively, the induced morphism $\pi_0(A) \longrightarrow \pi_0(B)$ has the corresponding classical property, and $\pi_0(B) \otimes_{\pi_0(A)} \pi_*(A) \simeq \pi_*(B)$). A straightforward extension of the étale topology to simplicial $k$-algebras, then provides us with an étale model site $(k - D^{-}Aff, \text{ét})$ satisfying assumption 1.3.2.2 (Def. 2.2.2.12 and Lemma 2.2.2.13) and with the corresponding model category $k - D^{-}Aff^{\sim,\text{ét}}$ of ét $D^{-}$-stacks (Def. 2.2.2.14). The homotopy category $\mathrm{Ho}(k - D^{-}Aff^{\sim,\text{ét}})$ of ét $D^{-}$-stacks will be simply denoted by $D^{-}\mathrm{St}(k)$. We conclude the section with two useful corollaries about topological invariance of étale and Zariski open immersions (Cor. 2.2.2.9 and 2.2.2.10), stating that ths small étale and Zariski site of a simplicial ring $A$ is equivalent to the corresponding site of $\pi_0(A)$.

In §2.2.3 we describe our HAG context (Def. 1.3.2.13) for derived algebraic geometry by choosing the class $\mathbf{P}$ to be the class of smooth morphisms in $sk - Alg$ and the model topology to be the étale topology. This HAG context $(\mathcal{C}, \mathcal{C}_0, \mathcal{A}, \tau, \mathbf{P}) := (sk-Mod, sk-Mod, sk-Alg, \text{ét}, \text{smooth})$ is shown to satisfy Artin's conditions (Def. 2.2.3.2) relative to the HA context $(\mathcal{C}, \mathcal{C}_0, \mathcal{A}) := (sk - Mod, sk - Mod, sk - Alg)$ in Prop. 1.4.3.1; as a corollary of the general theory of Part I, this gives (Cor. 2.2.3.3) an obstruction theory (respectively, a relative obstruction theory) for any $n$-geometric $D^{-}$-stack (resp., for any $n$-representable morphism between $D^{-}$-stacks), and in particular a (relative) cotangent complex for any $n$-geometric $D^{-}$-stack (resp., for any $n$-representable morphism). We finish the section showing that the properties of being flat, smooth, ètale and finitely presented can be extended to $n$-representable morphisms between $D^{-}$-stacks (Lemma 2.2.3.4), and by the definition of open and closed immersion of $D^{-}$-stacks (Def. 2.2.3.5).

In §2.2.4 we study *truncations* of derived stacks. The inclusion functor $j : k - Aff \hookrightarrow sk - Alg$, that sends a commutative $k$-algebra $R$ to the constant simplicial $k$-algebra $R$, is Quillen right adjoint to the functor $\pi_0 : sk - Alg \rightarrow k - Alg$, and this Quillen adjunction induces an adjunction

$$i := \mathbb{L}j_! : \mathrm{St}(k) \longrightarrow D^{-}\mathrm{St}(k)$$

$$\mathrm{St}(k) \longleftarrow D^{-}\mathrm{St}(k) : \mathbb{R}i^* =: t_0$$

between the (homotopy) categories of derived and un-derived stacks. The functor $i$ is fully faithful and commutes with homotopy colimits (Lemma 2.2.4.1) and embeds the theory of stacks into the theory of derived stacks, while the functor $t_0$, called the truncation functor, sends the affine stack corresponding to a simplicial $k$-algebra $A$ to the affine scheme Spec $\pi_0(A)$, and commutes with homotopy limits and colimits (Lemma 2.2.4.2). Both the inclusion and the truncation functor preserve $n$-geometric stacks and flat, smooth, étale morphisms between them (Prop. 2.2.4.4). This gives a nice compatibility between the theories in §2.1 and §2.2, and therefore between moduli spaces and their derived analogs.

In particular, we get the that for *any* Artin algebraic stack in groupoids (actually, any Artin $n$-stack) $\mathcal{X}$, $i(\mathcal{X})$ has an obstruction theory, and therefore a cotangent complex; in other words viewing Artin stacks as derived stacks simplifies and clarifies a lot their infinitesimal theory, as already remarked in this introduction. We finish the section by showing (Prop. 2.2.4.7) that for any geometric $D^-$-stack $F$, its truncation, viewed again as a derived stack (i.e. $it_0(F)$) sits inside $F$ as a closed sub-stack, and one can reasonably think of $F$ as behaving like a formal thickening of its truncation.

In §2.2.5 we give useful criteria for a $n$-representable morphism between $D^-$-stacks being smooth (respectively, étale) in terms of locally finite presentation of the induced morphism on truncations, and infinitesimal lifting properties (Prop. 2.2.5.1, resp., Prop. 2.2.5.4) or properties of the cotangent complex (Cor. 2.2.5.3, resp., Cor. 2.2.5.6).

The final §2.2.6 of Chapter 2.2 contains applications of derived algebraic geometry to the construction of various derived versions of moduli spaces as $D^-$-stacks. In 2.2.6.1 we first show (Lemma 2.2.6.1) that the stack of rank $n$ vector bundles when viewed as a $D^-$-stack using the inclusion $i : \mathrm{St}(k) \to D^-\mathrm{St}(k)$ is indeed isomorphic to the derived stack $\mathbf{Vect}_n$ of rank $n$ vector bundles defined as in §1.3.7. Then, for any simplicial set $K$, we define the derived stack $\mathbb{R}\mathbf{Loc}_n(K)$ as the derived exponentiation of $\mathbf{Vect}_n$ with respect to $K$ (Def. 2.2.6.2), show that when $K$ is finite dimensional then $\mathbb{R}\mathbf{Loc}_n(K)$ is a finitely presented 1-geometric $D^-$-stack (Lemma 2.2.6.3), and identify its truncation with the usual Artin stack of rank $n$ local systems on $K$ (Lemma 2.2.6.4). Finally we give a more concrete geometric interpretation of $\mathbb{R}\mathbf{Loc}_n(K)$ as a moduli space of derived geometric objects (*derived rank $n$ local systems*) on the topological realization $|K|$ (Prop. 2.2.6.5) and show that the tangent space of $\mathbb{R}\mathbf{Loc}_n(K)$ at a global point correponding to a rank $n$ local system $E$ on $K$ is the cohomology complex $C^*(K, E \otimes_k E^\vee)[1]$ (Prop. 2.2.6.6). The latter result shows in particular that the $D^-$- stack $\mathbb{R}\mathbf{Loc}_n(K)$ depends on strictly more than the fundamental groupoid of $K$ (because its tangent spaces can be nontrivial even if $K$ is simply connected) and therefore carries higher homotopical informations as opposed to the usual (i.e underived) Artin stack of rank $n$ local systems. In subsection 2.2.6.2 we treat the case of algebras over an operad. If $\mathcal{O}$ is an operad in the category of $k$-modules, we consider a simplicial presheaf $\mathbf{Alg}_n^{\mathcal{O}}$ on $k - D^-Aff$ which associates to any simplicial $k$-algebra $A$ the nerve of the subcategory $\mathcal{O} - Alg(A)$ of weak equivalences in the category of cofibrant algebras over the operad $\mathcal{O} \otimes_k A$ (which is an operad in simplicial $A$-modules) whose underlying $A$-module is a vector bundle of rank $n$. In Prop. 2.2.6.8 we show that $\mathbf{Alg}_n^{\mathcal{O}}$ is a 1-geometric quasi-compact $D^-$-stack, and in Prop. 2.2.6.9 we identify its tangent space in terms of derived derivations. In subsection 2.2.6.3, using a special case of J.Lurie's representability criterion (see Appendix $C$), we give sufficient conditions for the $D^-$-stack $\mathbf{Map}(\mathcal{X}, F)$ of morphisms of derived stacks $i(\mathcal{X}) \to F$ (for $\mathcal{X} \in \mathrm{St}(k)$) to be $n$-geometric (Thm. 2.2.6.11), and compute its tangent space in two particular cases (Cor. 2.2.6.14 and 2.2.6.15). The latter of these, i.e. the case $\mathbb{R}\mathcal{M}_{DR}(X) := \mathbf{Map}(i(X_{DR}), \mathbf{Vect}_n)$, where $X$ is a complex smooth projective variety, is particularly important because it is the first step in the construction of a *derived* version of *non-abelian Hodge theory* which will be investigated in a future work.

In §2.3 we treat the case of **complicial** (or **unbounded**) **algebraic geometry**, i.e. homotopical algebraic geometry over the base category $\mathcal{C} := C(k)$ of (unbounded) complexes of modules over a commutative $\mathbb{Q}$-algebra $k$ [2].

Here $Comm(\mathcal{C})$ is the model category $k - cdga$ of commutative differential graded $k$-algebras, (called shortly cdga's) whose opposite model category will be denoted by $k - DAff$. A new feature of complicial algebraic geometry with respect to derived algebraic geometry is the existence of *two* interesting HA contexts (Lemma 2.3.1.1): the *weak* one $(\mathcal{C}, \mathcal{C}_0, \mathcal{A}) := (C(k), C(k), k - cdga)$, and the *connective* one $(\mathcal{C}, \mathcal{C}_0, \mathcal{A}) := (C(k), C(k)_{\leq 0}, k - cdga_0)$ where $C(k)_{\leq 0}$ is the full subcategory of $C(k)$ consisting of complexes which are cohomologically trivial in positive degrees and $k - cdga_0$ is the full subcategory of $k - cdga$ of connected algebras (i.e. cohomologically trivial in non-zero degrees). A related phenomenon is the fact that the notion of *strongness* that describes completely the properties of modules and of morphisms between between simplicial algebras in derived algebraic geometry, is less strictly connected with standard properties of modules and of morphisms between between cdga's. For example, morphisms between cdga's which have strongly the property $\mathcal{P}$ (e.g. $\mathcal{P}=$ flat, étale etc.; see Def. 2.3.1.3) have the property $\mathcal{P}$ but the converse is true only if additional hypotheses are met (Prop. 2.3.1.4). Using the strong version of étale morphisms, we endow the category $k - DAff$ with the *strongly étale model topology* s-ét (Def. 2.3.1.6 and Lemma 2.3.1.7 ), and define the model category of *D-stacks* as $k - DAff^{\sim,\text{s-ét}}$ (Def. 2.3.1.8). The homotopy category of $k - DAff^{\sim,\text{s-ét}}$ will be simply denoted by $D\mathrm{St}(k)$.

While the notion of stack does not depend on the HA or HAG contexts chosen but only the base model site, the notion of geometric stack depends on the HA and HAG contexts. In §2.3.2 we complete the weak HA context above to the HAG context $(\mathcal{C}, \mathcal{C}_0, \mathcal{A}, \tau, \mathbf{P}) := (C(k), C(k), k - cdga, \text{s-ét}, \mathbf{P}_w)$, where $\mathbf{P}_w$ is the class of formally perfect morphisms (Def. 1.2.7.2), and call the corresponding geometric stacks *weakly geometric stacks* (Def. 2.3.2.2).

Section 2.3.3 contains some interesting examples of weakly geometric stacks. We first show (Prop. 2.3.3.1) that the stack **Perf** of perfect modules (defined in an abstract context in Def. 1.3.7.5) is weakly 1-geometric, categorically locally of finite presentation and its diagonal is $(-1)$-representable. In subsection 2.3.3.2 we first define the simplicial presheaf **Ass** sending a cdga $A$ to the nerve of the subcategory of weak equivalences in the category of associative and unital (not necessarily commutative) cofibrant $A$-dg algebras whose underlying $A$-dg module is perfect, and show that this is a weakly 1-geometric $D$-stack (Prop. 2.3.3.2 and Cor. 2.3.3.4). Then we define (using the model structure on dg-categories of [**Tab**]) a dg-categorical variation of **Ass**, denoted by $\mathbf{Cat}_*$, by sending a cdga $A$ to the nerve of the subcategory of weak equivalences of the model category of $A$-dg categories $\mathcal{D}$ which are connected and have perfect and cofibrant $A$-dg-modules of morphisms, and prove (Prop. 2.3.3.5) that the canonical classifying functor $\mathbf{Ass} \to \mathbf{Cat}_*$ is a weakly 1-representable, $fp$ epimorphism of $D$-stacks; it follows that $\mathbf{Cat}_*$ is a weakly 2-geometric $D$-stack (Cor. 2.3.3.5).

In §2.3.4 we switch to the connective HA context $(\mathcal{C}, \mathcal{C}_0, \mathcal{A}) := (C(k), C(k)_{\leq 0}, k - cdga_0)$ and complete it to the connective HAG context

$$(C(k), C(k)_{\leq 0}, k - cdga_0, \text{s-ét}, \text{fip-smooth}),$$

[2]The case of $k$ of positive characteristic can be treated as a special case of *brave new algebraic geometry* (§2.4) over the base $Hk$, $H$ being the Eilenberg-Mac Lane functor.

where fip-smooth is the class of formally perfect (Def. 1.2.7.1) and formally i-smooth morphisms (Def. 1.2.8.1). Geometric stacks in this HAG context will be simply called *geometric D-stacks* (Def. 2.3.4.2); since fip-smooth morphisms are in $\mathbf{P}_w$, any geometric $D$-stack is weakly geometric. As opposed to the weak HAG context, this connective context indeed satisfies Artin's condition of Def. 1.4.3.1 (Prop. 2.3.4.3), and therefore any geometric $D$-stack has an obstruction theory (and a cotangent complex).

In §2.3.5 we give some examples of geometric $D$-stacks. We first observe (subsection 2.3.5.1) that the normalization functor $N : sk - Alg \longrightarrow k - cdga$ induces a fully faithful functor $j := \mathbb{L}N_! : D^-\mathrm{St}(k) \hookrightarrow D\mathrm{St}(k)$. This provides us with lots of examples of (geometric) $D$-stacks. In subsection 2.3.5.2 we study the $D$-stack of CW-perfect modules. After having defined, for any cdga $A$, the notion of CW-A-dg-module of amplitude in $[a, b]$ (Def. 2.3.5.2) and proved some stability properties of this notion (Lemma 2.3.5.3), we define the sub-$D$-stack $\mathbf{Perf}^{CW}_{[a,b]} \subset \mathbf{Perf}$, consisting of all perfect modules locally equivalent to some CW-dg-modules of amplitude contained in $[a, b]$. We prove that $\mathbf{Perf}^{CW}_{[a,b]}$ is 1-geometric (Prop. 2.3.5.4), and that its tangent space at a point corresponding to a perfect CW-$A$-dg-module $E$ is given by the complex $(E^\vee \otimes^{\mathbb{L}}_A E)[1]$ (Cor. 2.3.5.6). In subsection 2.3.5.3 we define the $D$-stack of CW-dg-algebras as the homotopy pullback of $\mathbf{Ass} \longrightarrow \mathbf{Perf}$ along the inclusion $\mathbf{Perf}^{CW}_{[a,b]} \longrightarrow \mathbf{Perf}$, prove that it is 1-geometric and compute its tangent space at a global point in terms of derived derivations (Cor. 2.3.5.9). Finally subsection 2.3.5.4 is devoted to the analysis of the $D$-stack $\mathbf{Cat}^{CW}_{*,\,[n,0]}$ of CW-dg-categories of perfect amplitude in $[n, 0]$ (Def. 2.3.5.10). As opposite to the weakly geometric $D$-stack $\mathbf{Cat}_*$, that cannot have a reasonable infinitesimal theory, its full sub-$D$-stack $\mathbf{Cat}^{CW}_{*,\,[n,0]}$ is not only a 2-geometric $D$-stack but has a tangent space that can be computed in terms of Hochschild homology (Thm. 2.3.5.11). As corollaries of this important result we can prove a folklore statement (see e.g. [**Ko-So**, p. 266] in the case of $A_\infty$-categories with one object) regarding the deformation theory of certain negative dg-categories being controlled by the Hochschild complex of dg-categories (Cor. 2.3.5.12), and a result showing that if one wishes to keep the existence of the cotangent complex, the restriction to non-positively graded dg-categories is unavoidable (Cor. 2.3.5.13).

In the last, short §2.4, we establish the basics of **brave new algebraic geometry**, i.e. of homotopical algebraic geometry over the base category $\mathcal{C} = Sp^\Sigma$ of symmetric spectra ([**HSS, Shi**]). We consider $\mathcal{C}$ endowed with the positive model structure of [**Shi**], which is better behaved than the usual one when dealing with commutative monoid objects an modules over them. We denote $Comm(\mathcal{C})$ by $S - Alg$ (and call its objects commutative $S$-algebras, $S$ being the sphere spectrum, or sometimes bn-rings), and its opposite model category by $S - Aff$. Like in the case of complicial algebraic geometry, we consider two HA contexts here (Lemma 2.4.1.1): $(\mathcal{C}, \mathcal{C}_0, \mathcal{A}) := (Sp^\Sigma, Sp^\Sigma, S - Alg)$ and $(\mathcal{C}, \mathcal{C}_0, \mathcal{A}) := (Sp^\Sigma, Sp^\Sigma_c, S - Alg_0)$ where $Sp^\Sigma_c$ is the subcategory of connective symmetric spectra, and $S - Alg_0$ the subcategory of $S$-algebras with homotopy groups concentrated in degree zero. After giving some examples of formally étale and formally *thh*-étale maps between bn-rings, we define (Def. 2.4.1.3) *strong versions* of flat, (formally) étale, (formally) smooth, and Zariski open immersions, exactly like in chapters 2.2 and 2.3, and give some results relating them to the corresponding non-strong notions (Prop. 2.4.1.4). An interesting exception to this relationship occurs in the case of smooth morphism: the Eilenberg-MacLane functor $H$ from commutative rings to bn rings does not preserve (formal) smoothness in general, though it preserves (formal) strong smoothness, due to the presence of non-trivial Steenrod operations in characteristic $p > 0$ (Prop. 2.4.1.5). We endow the category $S - Aff$ with the *strong étale* model topology s-ét (Def. 2.4.1.7 and Lemma 2.4.1.8), and define the model category $S - Aff^{\sim, \text{s-ét}}$ of *S-stacks* (Def. 2.4.1.9). The homotopy category of $S - Aff^{\sim, \text{s-ét}}$ will be simply denoted by $\mathrm{St}(S)$. The two HA contexts defined above are completed (Cor. 2.4.1.11) to two different HAG contexts by choosing for both HA contexts the s-ét model topology, and the class $\mathbf{P}_{\text{s-ét}}$ of strongly étale morphsims for the first HA context (respectively, the class $\mathbf{P}$ of fip-morphisms for the second HA context). We call *geometric Deligne-Mumford S-stacks* (respectively, *geometric S-stacks*) the geometric stacks in the first (resp, second) HAG context, and observe that both contexts satisfy Artin's condition of Def. 1.4.3.1 (Prop. 2.4.1.13).

Finally in §2.4.2, we use the definition of topological modular forms of Hopkins-Miller to build a 1-geometric Deligne-Mumford stack $\overline{\mathcal{E}}_{\mathbf{S}}$ that is a "bn-derivation" of the usual stack $\overline{\mathcal{E}}$ of generalized elliptic curves [3] (i.e., the truncation of $\overline{\mathcal{E}}_{\mathbf{S}}$ is $\overline{\mathcal{E}}$), and such that the spectrum $\mathsf{tmf}$ coincide with the spectrum of *functions on* $\overline{\mathcal{E}}_{\mathbf{S}}$. We conclude by the remark that a moduli theoretic interpretation of $\overline{\mathcal{E}}_{\mathbf{S}}$ (or, most probably some variant of it), i.e. finding out which are the brave new objects that it classifies, could give not only interesting new geometry over bn rings but also new insights on classical objects of algebraic topology.

[3] See e.g. [**Del-Rap**, IV], where it is denoted by $\mathcal{M}_{(1)}$.

CHAPTER 2.1

# Geometric $n$-stacks in algebraic geometry (after C. Simpson)

All along this chapter we fix an associative commutative ring $k \in \mathbb{U}$ with unit.

## 2.1.1. The general theory

We consider $\mathcal{C} = k - Mod$, the category of $k$-modules in the universe $\mathbb{U}$. We endow the category $\mathcal{C}$ with the trivial model structure for which equivalences are isomorphisms and all morphisms are cofibrations and fibrations. The category $\mathcal{C}$ is furthermore a symmetric monoidal model category for the monoidal structure given by the tensor product of $k$-modules. The assumptions 1.1.0.1, 1.1.0.2, 1.1.0.3 and 1.1.0.4 are all trivially satisfied. The category $Comm(\mathcal{C})$ is identified with the category $k - Alg$, of commutative (associative and unital) $k$-algebras in $\mathbb{U}$, endowed with the trivial model structure. Objects in $k - Alg$ will simply be called commutative $k$-algebras, without any reference to the universe $\mathbb{U}$. For any $A \in k - Alg$, the category $A - Mod$ is the usual symmetric monoidal category of $A$-modules in $\mathbb{U}$, also endowed with its trivial model structure. Furthermore, we have $\mathbb{R}\underline{Hom}_A(M, N) \simeq \underline{Hom}_A(M, N)$ for any two objects $A \in k - Alg$, and is the usual $A$-module of morphisms from $M$ to $N$. We set $\mathcal{C}_0 := k - Mod$, and $\mathcal{A} := k - Alg$. The triplet $(k - Mod, k - Mod, k - Alg)$ is then a HA context in the sense of Def. 1.1.0.11.

The category $Aff_{\mathcal{C}}$ is identified with $k - Alg^{op}$, and therefore to the category of affine $k$-schemes in $\mathbb{U}$. It will simply be denoted by $k - Aff$ and its objects will simply be called affine $k$-schemes, without any reference to the universe $\mathbb{U}$. The model category of pre-stacks $k - Aff^{\wedge} = Aff_{\mathcal{C}}^{\wedge}$ is simply the model category of $\mathbb{V}$-simplicial presheaves on $k - Aff$, for which equivalences and fibrations are defined levelwise. The Yoneda functor

$$h : k - Aff \longrightarrow k - Aff^{\wedge}$$

is the usual one, and sends an affine $k$-scheme $X$ to the presheaf of sets it represents $h_X$ (considered as a presheaf of constant simplicial sets). Furthermore we have natural isomorphisms of functors

$$h \simeq \underline{h} \simeq \mathbb{R}\underline{h} : k - Aff \longrightarrow \mathrm{Ho}(k - Aff^{\wedge}),$$

which is nothing else than the natural composition

$$k - Aff \xrightarrow{h} k - Aff^{\wedge} \longrightarrow \mathrm{Ho}(k - Aff^{\wedge}).$$

We let $\tau =$ ét, the usual étale pre-topology on $k - Aff$ (see e.g. [**Mil**]). Recall that a family of morphisms

$$\{X_i = Spec\, A_i \longrightarrow X = Spec\, A\}_{i \in I}$$

is an ét-covering family if and only if it contains a finite sub-family $\{X_i \longrightarrow X\}_{i\in J}$, $J \subset I$, such that the corresponding morphism of commutative $k$-algebras

$$A \longrightarrow \prod_{i\in J} A_i$$

is a faithfully flat and étale morphism of commutative rings.

LEMMA 2.1.1.1. *The étale topology on $k - Aff$ satisfies assumption 1.3.2.2.*

PROOF. Points (1) and (2) of 1.3.2.2 are clear. Point (3) is induced by the faithfully flat descent of quasi-coherent modules for affine $ffqc$-hypercovers (see e.g. [**SGA4-II**, Exp. $V^{\mathrm{bis}}$])). □

The model category of stacks $k - Aff^{\sim,\text{ét}}$ is the projective model structure for simplicial presheaves on the Grothendieck site $(k - Aff, \text{ét})$, as defined for example in [**Bl**] (see also [**To1**, §1] ). Its homotopy category, denoted simply as $\mathrm{St}(k)$, is then identified with the full subcategory of $\mathrm{Ho}(k - Aff^{\wedge})$ consisting of simplicial presheaves

$$F : k - Alg = k - Aff^{op} \longrightarrow SSet_{\mathbb{V}}$$

satisfying the following two conditions

- For any two commutative $k$-algebra $A$ and $B$, the natural morphism
$$F(A \times B) \longrightarrow F(A) \times F(B)$$
is an isomorphism in $\mathrm{Ho}(SSet)$.
- For any co-augmented co-simplicial commutative $k$-algebra, $A \longrightarrow B_*$, such that the augmented simplicial object
$$Spec\, B_* \longrightarrow Spec\, A$$
is an étale hypercover, the natural morphism
$$F(A) \longrightarrow Holim_{n\in\Delta} F(B_n)$$
is an isomorphism in $\mathrm{Ho}(SSet)$.

It is well known (and also a consequence of Cor. 1.3.2.8) that the étale topology is sub-canonical; therefore there exists a fully faithful functor

$$h : k - Aff \longrightarrow \mathrm{St}(k) \subset \mathrm{Ho}(k - Aff^{\sim,\text{ét}}).$$

Furthermore, we have

$$Spec\, A \simeq \underline{Spec}\, A \simeq \mathbb{R}\underline{Spec}\, A$$

for any $A \in k - Alg$.

We set $\mathbf{P}$ to be the class of *smooth morphisms* in $k - Aff$ in the sense of [**EGAIV**, 17.3.1]. It is well known that our assumption 1.3.2.11 is satisfied (e.g. that smooth morphisms are étale-local in the source and target, see for example [**EGAIV**, 17.3.3, 17.3.4, 17.7.3]). In particular, we get that $(k - Mod, k - Mod, k - Alg, \text{ét}, \mathbf{P})$ is a HAG context in the sense of Def. 1.3.2.13. The general definition 1.3.3.1 can then be applied, and provides a notion of $n$-geometric stack in $\mathrm{St}(k)$. A first important observation (the lemma below) is that $n$-geometric stacks are $n$-stacks in the sense of [**S3**]. This is a special feature of standard algebraic geometry, and the same would be true for any theory for which the model structure on $\mathcal{C}$ is trivial: the *geometric complexity* is a bound for the *stacky complexity*.

Recall that a stack $F \in \mathrm{St}(k)$ is *$n$-truncated* if for any $X \in k - Aff$, any $s \in \pi_0(F(X))$ and any $i > n$, the sheaf $\pi_i(F, s)$ is trivial. By [**HAGI**, 3.7], this is equivalent to say that for any stack $G$ the simplicial set $\mathbb{R}_\tau \underline{Hom}(G, F)$ is $n$-truncated.

LEMMA 2.1.1.2. *Let $F$ be an $n$-geometric stack in* St($k$). *Then $F$ is $(n+1)$-truncated.*

PROOF. The proof is by induction on $n$. Representable stacks are nothing else than affine schemes, and therefore are 0-truncated. Suppose that the lemma is known for $m < n$. Let $F$ be an $n$-geometric stack, $X \in k - Aff$ and $s \in \pi_0(F(X))$. We have a natural isomorphism of sheaves on $k - Aff/X$

$$\pi_i(F, s) \simeq \pi_{i-1}(X \times_F^h X, s),$$

where $X \longrightarrow F$ is the morphisms of stacks corresponding to $s$. As the diagonal of $F$ is $(n-1)$-representable, $X \times_F^h X \simeq F \times_{F \times^h F} X \times^h X$ is $(n-1)$-geometric. By induction we find that $\pi_i(F, s) \simeq *$ for any $i > n$. □

Lemma 2.1.1.2 justifies the following terminology, closer to the usual terminology one can find in the literature.

DEFINITION 2.1.1.3. *An* Artin $n$-stack *is an $n$-truncated stack which is $m$-geometric for some integer $m$.*

The general theory of Artin $n$-stacks could then be pursued in a similar fashion as for Artin stacks in [**La-Mo**]. A part of this is done in [**S3**] and will not be reproduced here, as many of these statements will be settled down in the more general context of geometric $D^-$-stacks (see §2.2). Let us mention however, that as explained in Def. 1.3.6.2, we can define the notions of flat, smooth, étale, unramified, regular, Zariski open immersion ... morphisms between Artin $n$-stacks. These kinds of morphisms are as usual stable by homotopy pullbacks, compositions and equivalences. In particular this allows the following definition.

DEFINITION 2.1.1.4.
- *An Artin $n$-stack is a* Deligne-Mumford $n$-stack *if there exists an $n$-atlas $\{U_i\}$ for $F$ such that each morphism $U_i \longrightarrow F$ is an étale morphism.*
- *An Artin $n$-stack is an* algebraic space *if it is a Deligne-Mumford $n$-stack, and if furthermore the diagonal $F \longrightarrow F \times^h F$ is a monomorphism in the sense of Def. 1.3.6.4.*
- *An Artin $n$-stack $F$ is an* scheme *if there exists an $n$-atlas $\{U_i\}$ for $F$ such that each morphism $U_i \longrightarrow F$ is a monomorphism.*

REMARK 2.1.1.5.
(1) An algebraic space in the sense of the definition above which is automatically a 1-geometric stack, and is nothing else than an algebraic space in the usual sense. Indeed, this can be shown by induction on $n$: an algebraic space which is also $n$-geometric is by definition the quotient of a union of affine schemes $X$ by some étale equivalence relation $R \subset X \times X$ where $R$ is an algebraic space which is $(n-1)$-geometric. In particular, $R$ being a subobject in $X \times X$ we see that $R$ is a separated algebraic space, and thus is a 0-geometric stack. This implies that $X/R$ is a 1-geometric stack. In the same way, any scheme is automatically a 1-geometric stack. Moreover, algebraic spaces (resp. schemes) which are 0-geometric stacks are precisely algebraic spaces (resp. schemes) with an affine diagonal.
(2) Thought there is a small discrepancy between the notion of Artin $n$-stack and the notion of $n$-geometric stack in St($k$), our notion of Artin $n$-stack is equivalent to the notion of *slightly geometric $n$-stacks* of [**S3**].

### 2.1.2. Comparison with Artin's algebraic stacks

Artin $n$-stacks as defined in the last section are simplicial presheaves, whereas Artin stacks are usually presented in the literature using the theory of fibered categories (se e.g. [**La-Mo**]). In this section we briefly explain how the theory of fibered categories in groupoids can be embedded in the theory of simplicial presheaves, and how this can be used in order to compare the original definition of Artin stacks to our definition of Artin $n$-stacks.

In [**Hol**], it is shown that there exists a model category $Grpd/\mathcal{S}$, of cofibered categories in groupoids over a Grothendieck site $\mathcal{S}$. The fibrant objects for this model structure are precisely the stacks in groupoids in the sense of [**La-Mo**], and the equivalences in $Grpd/\mathcal{S}$ are the morphisms of cofibered categories becoming equivalences on the associated stacks (i.e. local equivalences). There exists furthermore a Quillen equivalence

$$p : P(\mathcal{S}, Grpd) \longrightarrow Grpd/\mathcal{S} \qquad P(\mathcal{S}, Grpd) \longleftarrow Grpd/\mathcal{S} : \Gamma,$$

where $P(\mathcal{S}, Grpd)$ is the local projective model category of presheaves of groupoids on $\mathcal{S}$. Finally, there exists a Quillen adjunction

$$\Pi_1 : SPr_\tau(\mathcal{S}) \longrightarrow P(\mathcal{S}, Grpd) \qquad SPr_\tau(\mathcal{S}) \longleftarrow P(\mathcal{S}, Grpd) : B,$$

where $\Pi_1$ is the natural extension of the functor sending a simplicial set to its fundamental groupoid, $B$ is the natural extension of the nerve functor, and the $SPr_\tau(\mathcal{S})$ is the local projective model structure of simplicial presheaves on the site $\mathcal{S}$ (also denoted by $\mathcal{S}^{\sim,\tau}$ in our context, at least when $\mathcal{S}$ has limits and colimits and thus can be considered as a model category with the trivial model structure). The functor $B$ preserves equivalences and the induced functor

$$B : \mathrm{Ho}(P(\mathcal{S}, Grpd)) \longrightarrow \mathrm{Ho}(SPr_\tau(\mathcal{S}))$$

is fully faithful and its image consists of all 1-truncated objects (in the sense of [**HAGI**, §3.7]). Put in another way, the model category $P(\mathcal{S}, Grpd)$ is Quillen equivalent to the $S^2$-nullification of $SPr_\tau(\mathcal{S})$ (denoted by $SPr_\tau^{\leq 1}(\mathcal{S})$ in [**HAGI**, §3.7]). In conclusion, there exists a chain of Quillen equivalences

$$Grpd/\mathcal{S} \rightleftarrows P(\mathcal{S}, Grpd) \rightleftarrows SPr_\tau^{\leq 1}(\mathcal{S}),$$

and therefore a well defined adjunction

$$t : \mathrm{Ho}(SPr_\tau(\mathcal{S})) \longrightarrow \mathrm{Ho}(Grpd/\mathcal{S}) \qquad \mathrm{Ho}(SPr_\tau(\mathcal{S})) \longleftarrow \mathrm{Ho}(Grpd/\mathcal{S}) : i,$$

such that the right adjoint

$$i : \mathrm{Ho}(Grpd/\mathcal{S}) \longrightarrow \mathrm{Ho}(SPr_\tau(\mathcal{S}))$$

is fully faithful and its image consists of all 1-truncated objects.

The category $\mathrm{Ho}(Grpd/\mathcal{S})$ can also be described as the category whose objects are stacks in groupoids in the sense of [**La-Mo**], and whose morphisms are given by 1-morphisms of stacks up to 2-isomorphisms. In other words, for two given stacks $F$ and $G$ in $Grpd/\mathcal{S}$, the set of morphisms from $F$ to $G$ in $\mathrm{Ho}(Grpd/\mathcal{S})$ is the set of isomorphism classes of the groupoid $\mathcal{H}om(F, G)$, of morphisms of stacks. This implies that the usual category of stacks in groupoids, up to 2-isomorphisms, can be identified through the functor $i$ with the full subcategory of 1-truncated objects in $\mathrm{Ho}(SPr_\tau(\mathcal{S}))$. Furthermore, the functor $i$ being defined as the composite of right derived functors and derived Quillen equivalences will commutes with homotopy limits. As homotopy limits in $Grpd/\mathcal{S}$ can also be identified with the 2-limit of stacks as defined in [**La-Mo**], the functor $i$ will send 2-limits to homotopy limits. As a particular case we obtain that $i$ sends the 2-fiber product of stacks in groupoids to the homotopy fiber product.

The 2-categorical structure of stacks in groupoids can also be recovered from the model category $SPr_{\tau}(\mathcal{S})$. Indeed, applying the simplicial localization techniques of [**D-K**1] to the Quillen adjunctions described above, we get a well defined diagram of $S$-categories

$$L(Grpd/\mathcal{S}) \longrightarrow LP(\mathcal{S}, Grpd) \longrightarrow LSPr_{\tau}(\mathcal{S}),$$

which is fully faithful in the sense of [**HAGI**, Def. 2.1.3]. In particular, the $S$-category $L(Grpd/\mathcal{S})$ is naturally equivalent to the full sub-$S$-category of $LSPr_{\tau}(\mathcal{S})$ consisting of 1-truncated objects. Using [**D-K**3], the $S$-category $L(Grpd/\mathcal{S})$ is also equivalent to the $S$-category whose objects are stacks in groupoids, cofibrant as objects in $Grpd/\mathcal{S}$, and whose morphisms simplicial sets are given by the simplicial $Hom$'s sets of $Grpd/\mathcal{S}$. These simplicial $Hom$'s sets are simply the nerves of the groupoid of functors between cofibered categories in groupoids. In other words, replacing the simplicial sets of morphisms in $L(Grpd/\mathcal{S})$ by their fundamental groupoids, we find a 2-category naturally 2-equivalent to the usual 2-category of stacks in groupoids on $\mathcal{S}$. Therefore, we see that the 2-category of stacks in groupoids can be identified, up to a natural 2-equivalence, as the 2-category obtained from the full sub-$S$-category of $LSPr_{\tau}(\mathcal{S})$ consisting of 1-truncated objects, by replacing its simplicial sets of morphisms by their fundamental groupoids.

We now come back to the case where $\mathcal{S} = (k - Aff, \text{ét})$, the Grothendieck site of affine $k$-schemes with the étale pre-topology. We have seen that there exists a fully faithful functor

$$i : \mathrm{Ho}(Grpd/k - Aff^{\sim,\text{ét}}) \longrightarrow \mathrm{St}(k),$$

from the category of stacks in groupoids up to 2-isomorphisms, to the homotopy category of stacks. The image of this functor consists of all 1-truncated objects and it is compatible with the simplicial structure (i.e. possesses a natural lifts as a morphism of $S$-categories). We also have seen that $i$ sends 2-fiber products of stacks to homotopy fiber products.

Using the functor $i$, every stack in groupoids can be seen as an object in our category of stacks $\mathrm{St}(k)$. For example, all examples of stacks presented in [**La-Mo**] give rise to stacks in our sense. The proposition below subsumes the main properties of the functor $i$, relating the usual notion of scheme, algebraic space and stack to the one of our definition Def. 2.1.1.4. Recall that a stack in groupoids $X$ is separated (resp. quasi-separated) if its diagonal is a proper (resp. separated) morphism.

PROPOSITION 2.1.2.1. (1) *For any commutative $k$-algebra $A$, there exists a natural isomorphism*

$$i(Spec\, A) \simeq Spec\, A.$$

(2) *If $X$ is a scheme (resp. algebraic space, resp. Deligne-Mumford stack, resp. Artin stack) with an affine diagonal in the sense of* [**La-Mo**], *then $i(X)$ is an Artin 0-stack which is 0-geometric (resp. an Artin 0-stack which is 0-geometric, resp. a Deligne-Mumford 1-stack which is 0-geometric, resp. an Artin 1-stack which is 0-geometric) in the sense of Def. 2.1.1.4.*

(3) *If $X$ is a scheme (resp. algebraic space, resp. Deligne-Mumford stack, resp. Artin stack) in the sense of* [**La-Mo**], *then $i(X)$ is an Artin 0-stack which is 1-geometric (resp. an Artin 0-stack which is 1-geometric, resp. a Deligne-Mumford 1-stack which is 1-geometric, resp. an Artin 1-stack which is 1-geometric) in the sense of Def. 2.1.1.4.*

(4) *Let $f : F \longrightarrow G$ be a morphism between Artin stacks in the sense of* [**La-Mo**]. *Then the morphism $f$ is flat (resp. smooth, resp. étale, resp. unramified, resp. Zariski open immersion) if and only if $i(f) : i(F) \longrightarrow i(G)$ is so.*

PROOF. This readily follows from the definition using the fact that $i$ preserves affine schemes, epimorphisms of stacks, and sends 2-fiber products to homotopy pullbacks. □

REMARK 2.1.2.2. If we try to apply the general infinitesimal and obstruction theory developed in §1.4 to the present HAG-context, we immediately see that this is impossible because the suspension functor $S : \mathrm{Ho}(\mathcal{C}) = k - Mod \to k - Mod = \mathrm{Ho}(\mathcal{C})$ is trivial. On the other hand, the reader might object that there is already a well established infinitesimal theory, at least in the case of schemes, algebraic spaces and for a certain class of algebraic stacks in groupoids, and that our theory does not seem to be able to reproduce it. The answer to this question turns out to be both conceptually and technically relevant. First of all, if we look at e.g. the definition of the cotangent complex of a scheme ([**Ill**, 2.1.2]) we realize that a basic and necessary step is to enlarge the category of rings to the category of simplicial rings in order to be able to consider free (or more generally cofibrant) resolutions of maps between rings. In our setup, this can be reformulated by saying that in order to get the correct infinitesimal theory, even for ordinary schemes, it is necessary to view them as geometric objects in *derived algebraic geometry*, i.e. on homotopical algebraic geometry over the base category $\mathcal{C} = sk - Mod$ of simplicial $k$-modules (so that $Comm(\mathcal{C})$ is exactly the category of commutative simplicial $k$-algebras). In other words, the usual infinitesimal theory of schemes is already "secretly" a part of derived algebraic geometry, that will be studied in detail in the next chapter 2.2. Moreover, as it will be shown, this approach has, even for classical objects like schemes or Artin stacks in groupoids, both conceptual advantages (like e.g. the fact that the cotangent complex of a scheme can be interpreted geometrically as a *genuine* cotangent space to the scheme when viewed as a derived stack, satisfying a natural universal property, while the cotangent complex of a scheme do not have any universal property inside the theory of schemes), and technical advantages (like the fact, proved in Cor. 2.2.4.5, that *any* Artin stack in groupoids has an obstruction theory).

CONVENTION 2.1.2.3. From now on we will omit mentioning the functor $i$, and will simply view stacks in groupoids as objects in $\mathrm{St}(k)$. In particular, we will allow ourselves to use the standard notions and vocabulary of the general theory of schemes.

CHAPTER 2.2

# Derived algebraic geometry

All along this chapter $k$ will be a fixed commutative (associative and unital) ring.

## 2.2.1. The HA context

In this section we specialize our general theory of Part I to the case where $\mathcal{C} = sk-Mod$, is the category of simplicial $k$-modules in the universe $\mathbb{U}$. The category $sk-Mod$ is endowed with its standard model category structure, for which the fibrations and equivalences are defined on the underlying simplicial sets (see for example [**Goe-Ja**]). The tensor product of $k$-modules extends naturally to a levelwise tensor product on $sk-Mod$, making it into a symmetric monoidal model category. Finally, $sk-Mod$ is known to be a $\mathbb{U}$-combinatorial proper and simplicial model category.

The model category $sk-Mod$ is known to satisfy assumptions 1.1.0.1, 1.1.0.2 and 1.1.0.3 (see [**Schw-Shi**]). Finally, it follows easily from [**Q1**, II.4, II.6] that $sk-Mod$ also satisfies assumption 1.1.0.4.

The category $Comm(sk-Mod)$ will be denoted by $sk-Alg$, and its objects will be called simplicial commutative $k$-algebras. More generally, for $A \in sk-Alg$, the category $A-Comm(\mathcal{C})$ will be denoted by $A-Alg_s$. For any $A \in sk-Alg$ we will denoted by $A-Mod_s$ the category of $A$-modules in $sk-Mod$, which is nothing else than the category of simplicial modules over the simplicial ring $A$. The model structure on $sk-Alg$, $A-Alg_s$ and $A-Mod_s$ is the usual one, for which the equivalences and fibrations are defined on the underlying simplicial sets. For an object $A \in sk-Alg$, we will denote by $\pi_i(A)$ its homotopy group (pointed at 0). The graded abelian group $\pi_*(A)$ inherits a structure of a commutative graded algebra from $A$, which defines a functor $A \mapsto \pi_*(A)$ from $sk-Alg$ to the category of commutative graded $k$-algebras. More generally, if $A$ is a simplicial commutative $k$-algebra, and $M$ is an $A$-module, the graded abelian group $\pi_*(M)$ has a natural structure of a graded $\pi_*(A)$-module.

There exists a Quillen adjunction

$$\pi_0 : sk-Alg \longrightarrow k-Alg \qquad sk-Alg \longleftarrow k-Alg : i,$$

where $i$ sends a commutative $k$-algebra to the corresponding constant simplicial commutative $k$-algebra. This Quillen adjunction induces a fully faithful functor

$$i : k-Alg \longrightarrow \mathrm{Ho}(sk-Alg).$$

From now on we will omit to mention the functor $i$, and always consider $k-Alg$ as embedded in $sk-Alg$, except if the contrary is specified. Note that when $A \in k-Alg$, also considered as an object in $sk-Alg$, we have two different notions of $A$-modules, $A-Mod$, and $A-Mod_s$. The first one is the usual category of $A$-modules, whereas the second one is the category of simplicial objects in $A-Mod$.

For any morphism of simplicial commutative $k$-algebras $A \longrightarrow B$, the $B$-module $\mathbb{L}_{B/A}$ constructed in 1.2.1.2 is naturally isomorphic in $\mathrm{Ho}(B-Mod)$ to D. Quillen's cotangent complex introduced in [**Q2**]. In particular, if $A \longrightarrow B$ is a morphism between (non-simplicial) commutative $k$-algebras, then we have $\pi_0(\mathbb{L}_{B/A}) \simeq \Omega^1_{B/A}$. More generally, we find by adjunction

$$\pi_0(\mathbb{L}_{B/A}) \simeq \Omega^1_{\pi_0(B)/\pi_0(A)}.$$

Recall also that $A \longrightarrow B$ in $k-Alg$ is étale in the sense of [**EGAIV**, 17.1.1] if and only if $\mathbb{L}_{B/A} \simeq 0$ and $B$ is finitely presented as a commutative $A$-algebra. In the same way, a morphism $A \longrightarrow B$ in $k-Alg$ is smooth in the sense of [**EGAIV**, 17.1.1] if and only if $\Omega^1_{B/A}$ is a projective $B$-module, $\pi_i(\mathbb{L}_{B/A}) \simeq 0$ for $i > 0$ and $B$ is finitely presented as a commutative $A$-algebra.. Finally, recall the existence of a natural first quadrant spectral sequence (see [**Q1**, II.6 Thm. 6(b)])

$$Tor^p_{\pi_*(A)}(\pi_*(M), \pi_*(N))_q \Rightarrow \pi_{p+q}(M \otimes^{\mathbb{L}}_A N),$$

for $A \in sk-Alg$ and any objects $M$ and $N$ in $A-Mod_s$.

We set $\mathcal{C}_0 := \mathcal{C} = sk-Mod$, and $\mathcal{A} := sk-Alg$. Then, clearly, assumption 1.1.0.6 is also satisfied. The triplet $(sk-Mod, sk-Mod, sk-Alg)$ is then a HA context in the sense of Def. 1.1.0.11. Note that for any $A \in sk-Alg$ we have $(A-Mod_s)_0 = A-Mod$, whereas $(A-Mod_s)_1$ consists of all $A$-modules $M$ such that $\pi_0(M) = 0$, also called *connected modules.*

For an integer $n \geq 0$ we define the $n$-th sphere $k$-modules by $S^n_k := S^n \otimes k \in sk-Mod$. The free commutative monoid on $S^n_k$ is an object $k[S^n] \in sk-Alg$, such that for any $A \in sk-Alg$ there are functorial isomorphisms

$$[k[S^n], A]_{sk-Alg} \simeq \pi_n(A).$$

In the same way we define $\Delta^n \otimes k \in sk-Mod$ and its associated free commutative monoid $k[\Delta^n]$. There are natural morphisms $k[S^n] \longrightarrow k[\Delta^{n+1}]$, coming from the natural inclusions $\partial\Delta^{n+1} = S^n \hookrightarrow \Delta^{n+1}$. The set of morphisms

$$\{k[S^n] \longrightarrow k[\Delta^{n+1}]\}_{n\geq 0}$$

form a generating set of cofibrations in $sk-Alg$. The model category $sk-Alg$ is then easily checked to be compactly generated in the sense of Def. 1.2.3.4. A finite cell object in $sk-Alg$ is then any object $A \in sk-Alg$ for which there exists a finite sequence in $sk-Alg$

$$A_0 = k \longrightarrow A_1 \longrightarrow A_2 \longrightarrow \cdots \longrightarrow A_m = A,$$

such that for any $i$ there exists a push-out square in $sk-Alg$

$$\begin{array}{ccc} A_i & \longrightarrow & A_{i+1} \\ \uparrow & & \uparrow \\ k[S^{n_i}] & \longrightarrow & k[\Delta^{n_i+1}] \end{array}$$

Our Prop. 1.2.3.5 implies that an object $A \in sk-Alg$ is finitely presented in the sense of Def. 1.2.3.1 if and only if it is equivalent to a retract of a finite cell object (see also [**EKMM**, III.2] or [**Kr-Ma**, Thm. III.5.7] for other proofs). More generally, for $A \in sk-Alg$, there exists a notion of finite cell object in $A-Alg_s$ using the elementary morphisms

$$A[S^n] := A \otimes_k k[S^n] \longrightarrow A[\Delta^{n+1}] := A \otimes_k k[\Delta^{n+1}].$$

In the same way, a morphism $A \longrightarrow B$ in $sk-Alg$ is finitely presented in the sense of Def. 1.2.3.1 if and only if $B$ is equivalent to a retract of a finite cell objects in $A-Alg_s$. Prop. 1.2.3.5 also implies that any morphism $A \longrightarrow B$, considered as an object in $A - Alg_s$, is equivalent to a filtered colimit of finite cell objects, so in particular to a filtered homotopy colimit of finitely presented objects.

The Quillen adjunction between $k - Alg$ and $sk - Alg$ shows that the functor

$$\pi_0 : sk - Alg \longrightarrow k - Alg$$

does preserve finitely presented morphisms. On the contrary, the inclusion functor $i : k - Alg \longrightarrow sk - Alg$ does not preserve finitely presented objects, and the finite presentation condition in $sk - Alg$ is in general stronger than in $k - Alg$.

For $A \in sk - Alg$, we also have a notion of finite cell objects in $A - Mod_s$, based the generating set for cofibrations consisting of morphisms of the form

$$S^n_A := A \otimes_k S^n_k \longrightarrow \Delta^{n+1}_A := A \otimes_k \Delta^{n+1}_A.$$

Using Prop. 1.2.3.5 we see that the finitely presented objects in $A - Mod_s$ are the objects equivalent to a retract of a finite cell objects (see also [**EKMM**, III.2] or [**Kr-Ma**, Thm. III.5.7]). Moreover, the functor

$$\pi_0 : A - Mod_s \longrightarrow \pi_0(A) - Mod$$

is left Quillen, so preserves finitely presented objects. On the contrary, for $A \in k-Alg$, the natural inclusion functor $A - Mod \longrightarrow A - Mod_s$ from $A$-modules to simplicial $A$-modules does not preserve finitely presented objects in general.

The category $sk - Mod$ is also Quillen equivalent (actually equivalent) to the model category $C^-(k)$ of non-positively graded cochain complexes of $k$-modules, through the Dold-Kan correspondence ([**We**, 8.4.1]). In particular, the suspension functor

$$S : \mathrm{Ho}(sk - Mod) \longrightarrow \mathrm{Ho}(sk - Mod)$$

corresponds to the shift functor $E \mapsto E[1]$ on the level of complexes, and is a fully faithful functor. This implies that for any $A \in sk - Alg$, the suspension functor

$$S : \mathrm{Ho}(A - Mod_s) \longrightarrow \mathrm{Ho}(A - Mod_s)$$

is also fully faithful. We have furthermore $\pi_i(S(M)) \simeq \pi_{i+1}(M)$ for all $M \in A-Mod_s$. The suspension and loop functors will be denoted respectively by

$$M[1] := S(M) \qquad M[-1] := \Omega(M).$$

For any $A \in sk - Alg$, we can construct a functorial tower in $sk - Alg$, called the *Postnikov tower*,

$$A \longrightarrow \cdots \longrightarrow A_{\leq n} \longrightarrow A_{\leq n-1} \longrightarrow \cdots \longrightarrow A_{\leq 0} = \pi_0(A)$$

in such a way that $\pi_i(A_{\leq n}) = 0$ for all $i > n$, and the morphism $A \longrightarrow A_{\leq n}$ induces isomorphisms on the $\pi_i$'s for all $i \leq n$. The morphism $A \longrightarrow A_{\leq n}$ is characterized by the fact that for any $B \in sk - Alg$ which is $n$-truncated (i.e. $\pi_i(B) = 0$ for all $i > n$), the induced morphism

$$Map_{sk-Alg}(A_{\leq n}, B) \longrightarrow Map_{sk-Alg}(A, B)$$

is an isomorphism in $\mathrm{Ho}(SSet)$. This implies in particular that the Postnikov tower is furthermore unique up to equivalence (i.e. unique as an object in the homotopy category of diagrams). There exists a natural isomorphism in $\mathrm{Ho}(sk - Alg)$

$$A \simeq Holim_n A_{\leq n}.$$

For any integer $n$, the homotopy fiber of the morphism

$$A_{\leq n} \longrightarrow A_{\leq n-1}$$

is isomorphic in $\mathrm{Ho}(sk - Mod)$ to $S^n \otimes_k \pi_n(A)$, and is also denoted by $\pi_n(A)[n]$. The $k$-module $\pi_n(A)$ has a natural structure of a $\pi_0(A)$-module, and this induces a natural structure of a simplicial $\pi_0(A)$-module on each $\pi_n(A)[i]$ for all $i$. Using the natural projection $A_{\leq n-1} \longrightarrow \pi_0(A)$, we thus see the object $\pi_n(A)[i]$ as an object in $A_{\leq n-1} - Mod_s$. Note that there is a natural isomorphism in $\mathrm{Ho}(A_{\leq n-1} - Mod_s)$

$$S(\pi_n(A)[i]) \simeq \pi_n(A)[i+1] \qquad \Omega(\pi_n(A)[i]) \simeq \pi_n(A)[i-1],$$

where $\pi_n(A)[i]$ is understood to be 0 for $i < 0$. We recall the following important and well known fact.

LEMMA 2.2.1.1. *With the above notations, there exists a unique derivation*

$$d_n \in \pi_0(\mathbb{D}er_k(A_{\leq n-1}, \pi_n(A)[n+1]))$$

*such that the natural projection*

$$A_{\leq n-1} \oplus_{d_n} \pi_n(A)[n] \longrightarrow A_{\leq n-1}$$

*is isomorphic in* $\mathrm{Ho}(sk - Alg/A_{\leq n-1})$ *to the natural morphism*

$$A_{\leq n} \longrightarrow A_{\leq n-1}.$$

SKETCH OF PROOF. (See also [**Ba**] for more details).

The uniqueness of $d_n$ follows easily from our lemma 1.4.3.7, and the fact that the natural morphism

$$\mathbb{L}QZ(\pi_n(A)[n+1]) \longrightarrow \pi_n(A)[n+1]$$

induces an isomorphism on $\pi_i$ for all $i \leq n+1$ (this follows from our lemma 2.2.2.7 below). To prove the existence of $d_n$, we consider the homotopy push-out diagram in $sk - Alg$

$$\begin{array}{ccc} A_{\leq n-1} & \longrightarrow & B \\ \uparrow & & \uparrow \\ A_{\leq n} & \longrightarrow & A_{\leq n-1}. \end{array}$$

The identity of $A_{\leq n-1}$ induces a morphism $B \longrightarrow A_{\leq n-1}$, which is a retraction of $A_{\leq n-1} \longrightarrow B$. Taking the $(n+1)$-truncation gives a commutative diagram

$$\begin{array}{ccc} A_{\leq n-1} & \longrightarrow & B_{\leq n+1} \\ \uparrow & & \uparrow {\scriptstyle s} \\ A_{\leq n} & \longrightarrow & A_{\leq n-1}. \end{array}$$

in such a way that $s$ has a retraction. This easily implies that the morphism $s$ is isomorphic, in a non-canonical way, to the zero derivation $A_{\leq n-1} \longrightarrow A_{\leq n-1} \oplus \pi_n(A)[n+1]$. The top horizontal morphism of the previous diagram then gives rise to a derivation

$$d_n : A_{\leq n-1} \longrightarrow A_{\leq n-1} \oplus \pi_n(A)[n+1].$$

The diagram

$$\begin{array}{ccc} A_{\leq n-1} & \xrightarrow{d_n} & B_{\leq n+1} \\ \uparrow & & \uparrow {\scriptstyle s} \\ A_{\leq n} & \longrightarrow & A_{\leq n-1}, \end{array}$$

is then easily checked to be homotopy cartesian, showing that $A_{\leq n} \longrightarrow A_{\leq n-1}$ is isomorphic to $A_{\leq n-1} \oplus_{d_n} \pi_n(A)[n] \longrightarrow A_{\leq n-1}$. □

Finally, the truncation construction also exists for modules. For any $A \in sk-Alg$, and $M \in A - Mod_s$, there exists a natural tower of morphisms in $A - Mod_s$

$$M \longrightarrow \cdots \longrightarrow M_{\leq n} \longrightarrow M_{\leq n-1} \longrightarrow \cdots \longrightarrow M_{\leq 0} = \pi_0(M),$$

such a way that $\pi_i(M_{\leq n}) = 0$ for all $i > n$, and the morphism $M \longrightarrow M_{\leq n}$ induces an isomorphisms on $\pi_i$ for $i \leq n$. The natural morphism

$$M \longrightarrow Holim_n M_{\leq n}$$

is an isomorphism in $\mathrm{Ho}(A - Mod_s)$. Furthermore, the $A$-module $M_{\leq n}$ is induced by a natural $A_{\leq n}$-module, still denoted by $M_{\leq n}$, through the natural morphism $A \longrightarrow A_{\leq n}$. The natural projection $M \longrightarrow M_{\leq n}$ is again characterized by the fact that for any $A$-module $N$ which is $n$-truncated, the induced morphism

$$Map_{A-Mod_s}(M_{\leq n}, N) \longrightarrow Map_{A-Mod_s}(M, N)$$

is an isomorphism in $\mathrm{Ho}(SSet)$.

For an object $A \in sk - Alg$, the homotopy category $\mathrm{Ho}(Sp(A - Mod_s))$, of stable $A$-modules can be described in the following way. By normalization, the commutative simplicial $k$-algebra $A$ can be transformed into a commutative $dg$-algebra over $k$, $N(A)$ (because $N$ is lax symmetric monoidal). We can therefore consider its model category of unbounded $N(A)$-dg-modules, and its homotopy category $\mathrm{Ho}(N(A) - Mod)$. The two categories $\mathrm{Ho}(Sp(A - Mod_s))$ and $\mathrm{Ho}(N(A) - Mod)$ are then naturally equivalent. In particular, when $A$ is a commutative $k$-algebra, then $N(A) = A$, and one finds that $\mathrm{Ho}(Sp(A - Mod_s))$ is simply the unbounded derived category of $A$, or equivalently the homotopy category of the model category $C(A)$ of unbounded complexes of $A$-modules

$$\mathrm{Ho}(Sp(A - Mod_s)) \simeq D(A) \simeq \mathrm{Ho}(C(A)).$$

Finally, using our Cor. 1.2.3.8 (see also [**EKMM**, III.7]), we see that the perfect objects in the symmetric monoidal model category $\mathrm{Ho}(Sp(A - Mod_s))$ are exactly the finitely presented objects.

We now let $k - D^{-}Aff$ be the opposite model category of $sk - Alg$. We use our general notations, $Spec\, A \in k - D^{-}Aff$ being the object corresponding to $A \in sk - Alg$. We will also sometimes use the notation

$$t_0(Spec\, A) := Spec\, \pi_0(A).$$

## 2.2.2. Flat, smooth, étale and Zariski open morphisms

According to our general definitions presented in §1.2 we have various notions of projective and perfect modules, flat, smooth, étale, unramified ... morphisms in $sk - Alg$. Our first task, before visiting our general notions of stacks, will be to give concrete descriptions of these notions.

Any object $A \in sk-Alg$ gives rise to a commutative graded $k$-algebra of homotopy $\pi_*(A)$, which is functorial in $A$. In particular, $\pi_i(A)$ is always endowed with a natural structure of a $\pi_0(A)$-module, functorially in $A$. For a morphism $A \longrightarrow B$ in $sk - Alg$ we obtain a natural morphism

$$\pi_*(A) \otimes_{\pi_0(A)} \pi_0(B) \longrightarrow \pi_*(B).$$

More generally, for $A \in sk - Alg$ and $M$ an $A$-module, one has a natural morphism of $\pi_0(A)$-modules $\pi_0(M) \longrightarrow \pi_*(M)$, giving rise to a natural morphism

$$\pi_*(A) \otimes_{\pi_0(A)} \pi_0(M) \longrightarrow \pi_*(M).$$

These two morphisms are the same when the commutative $A$-algebra $B$ is considered as an $A$-module in the usual way.

DEFINITION 2.2.2.1. *Let $A \in sk - Alg$ and $M$ be an $A$-module. The $A$-module $M$ is* strong *if the natural morphism*

$$\pi_*(A) \otimes_{\pi_0(A)} \pi_0(M) \longrightarrow \pi_*(M)$$

*is an isomorphism.*

LEMMA 2.2.2.2. *Let $A \in sk - Alg$ and $M$ be an $A$-module.*

1. *The $A$-module $M$ is projective if and only if it is strong and $\pi_0(M)$ is a projective $\pi_0(A)$-module.*
2. *The $A$-module $M$ is flat if and only if it is strong and $\pi_0(M)$ is a flat $\pi_0(A)$-module.*
3. *The $A$-module $M$ is perfect if and only if it is strong and $\pi_0(M)$ is a projective $\pi_0(A)$-module of finite type.*
4. *The $A$-module $M$ is projective and finitely presented if and only if it is perfect.*

PROOF. (1) Let us suppose that $M$ is projective.We first notice that a retract of a strong module $A$-module is again a strong $A$-module. This allows us to suppose that $M$ is free, which clearly implies that $M$ is strong and that $\pi_0(M)$ is a free $\pi_0(A)$-module (so in particular projective). Conversely, let $M$ be a strong $A$-module with $\pi_0(M)$ projective over $\pi_0(A)$. We write $\pi_0(M)$ as a retract of a free $\pi_0$-module

$$\pi_0(M) \xrightarrow{i} \pi_0(A)^{(I)} = \oplus_I \pi_0(A) \xrightarrow{r} \pi_0(M).$$

The morphism $r$ is given by a family of elements $r_i \in \pi_0(M)$ for $i \in I$, and therefore can be seen as a morphism $r' : A^{(I)} \longrightarrow M$, well defined in $\mathrm{Ho}(A - Mod_s)$. In the same way, the projector $p = i \circ r$ of $\pi_0(A)^{(I)}$, can be seen as a projector $p'$ of $A^{(I)}$ in the homotopy category $\mathrm{Ho}(A - Mod_s)$. By construction, this projector gives rise to a split fibration sequence

$$K \longrightarrow A^{(I)} \longrightarrow C,$$

and the morphism $r'$ induces a well defined morphism in $\mathrm{Ho}(A - Mod_s)$

$$r' : C \longrightarrow M.$$

By construction, this morphism induces an isomorphisms on $\pi_0$, and as $C$ and $M$ are strong modules, $r'$ is an isomorphism in $\mathrm{Ho}(A - Mod_s)$.

(2) Let $M$ be a strong $A$-module with $\pi_0(M)$ flat over $\pi_0(A)$, and $N$ be any $A$-module. Clearly, $\pi_*(M)$ is flat as a $\pi_*(A)$-module. Therefore, the Tor spectral sequence of [**Q2**]

$$Tor^*_{\pi_*(A)}(\pi_*(M), \pi_*(N)) \Rightarrow \pi_*(M \otimes^{\mathbb{L}}_A N)$$

degenerates and gives a natural isomorphism

$$\pi_*(M \otimes^{\mathbb{L}}_A N) \simeq \pi_*(M) \otimes_{\pi_*(A)} \pi_*(N) \simeq$$

$$\simeq (\pi_0(M) \otimes_{\pi_0(A)} \pi_*(A)) \otimes_{\pi_*(A)} \pi_*(N) \simeq \pi_0(M) \otimes_{\pi_0(A)} \pi_*(N).$$

As $\pi_0(M) \otimes_{\pi_0(A)} -$ is an exact functor its transform long exact sequences into long exact sequences. This easily implies that $M \otimes^{\mathbb{L}}_A -$ preserves homotopy fiber sequences, and therefore that $M$ is a flat $A$-module.

Conversely, suppose that $M$ is a flat $A$-module. Any short exact sequence $0 \rightarrow N \rightarrow P$ of $\pi_0(A)$-modules can also be seen as a homotopy fiber sequence of $A$-modules, as any morphism $N \to P$ is always a fibration. Therefore, we obtain a homotopy fiber sequence

$$0 \longrightarrow M \otimes_A^{\mathbb{L}} N \longrightarrow M \otimes_A^{\mathbb{L}} P$$

which on $\pi_0$ gives a short exact sequence

$$0 \longrightarrow \pi_0(M) \otimes_{\pi_0(A)} N \longrightarrow \pi_0(M) \otimes_{\pi_0(A)} P.$$

This shows that $\pi_0(M) \otimes_{\pi_0(A)} -$ is an exact functor, and therefore that $\pi_0(M)$ is a flat $\pi_0(A)$-module. Furthermore, taking $N = 0$ we get that for any $\pi_0(A)$-module $P$ one has $\pi_i(M \otimes_A^{\mathbb{L}} P) = 0$ for any $i > 0$. In other words, we have an isomorphism in $\mathrm{Ho}(A - Mod_s)$

$$M \otimes_A^{\mathbb{L}} P \simeq \pi_0(M) \otimes_{\pi_0(A)} P.$$

By shifting $P$ we obtain that for any $i \geq 0$ and any $\pi_0(A)$-module $P$ we have

$$M \otimes_A^{\mathbb{L}} (P[i]) \simeq (\pi_0(M) \otimes_{\pi_0(A)} P)[i].$$

Passing to Postnikov towers we see that this implies that for any $A$-module $P$ we have

$$\pi_i(M \otimes_A^{\mathbb{L}} P) \simeq \pi_0(M) \otimes_{\pi_0(A)} \pi_i(P).$$

Applying this to $P = A$ we find that $M$ is a strong $A$-module.

(3) Let us first suppose that $M$ is strong with $\pi_0(M)$ projective of finite type over $\pi_0(A)$. By point (1) we know that $M$ is a projective $A$-module. Moreover, the proof of (1) also shows that $M$ is a retract in $\mathrm{Ho}(A - Mod_s)$ of some $A^I$ with $I$ finite. By our general result Prop. 1.2.4.2 this implies that $A$ is perfect. Conversely, let $M$ be a perfect $A$-module. By (2) and Prop. 1.2.4.2 we know that $M$ is strong and that $\pi_0(M)$ is flat over $\pi_0(A)$. Furthermore, the unit $k$ of $sk - Mod$ is finitely presented, so by Prop. 1.2.4.2 $M$ is finitely presented object in $A - Mod_s$. Using the left Quillen functor $\pi_0$ from $A - Mod_s$ to $\pi_0(A)$-modules, we see that this implies that $\pi_0(M)$ is a finitely presented $\pi_0(A)$-module, and therefore is projective of finite type.

(4) Follows from (1), (3) and Prop. 1.2.4.2. □

DEFINITION 2.2.2.3. (1) *A morphism $A \longrightarrow B$ in $sk - Alg$ is* strong *if the natural morphism*

$$\pi_*(A) \otimes_{\pi_0(A)} \pi_0(B) \longrightarrow \pi_*(B)$$

*is an isomorphism (i.e. $B$ is strong as an $A$-module).*

(2) *A morphism $A \longrightarrow B$ in $sk - Alg$ is* strongly flat *(resp.* strongly smooth, *resp.* strongly étale, *resp.* a strong Zariski open immersion*) if it is strong and if the morphism of affine schemes*

$$Spec\,\pi_0(B) \longrightarrow Spec\,\pi_0(A)$$

*is flat (resp. smooth, resp. étale, resp. a Zariski open immersion).*

We start by a very useful criterion in order to recognize finitely presented morphisms. We have learned this proposition from J. Lurie (see [**Lu1**]).

PROPOSITION 2.2.2.4. *Let $f : A \longrightarrow B$ be a morphism in $sk - Alg$. Then, $f$ is finitely presented if and only if it satisfies the following two conditions.*

(1) *The morphism $\pi_0(A) \longrightarrow \pi_0(B)$ is a finitely presented morphism of commutative rings.*
(2) *The cotangent complex $\mathbb{L}_{B/A} \in \mathrm{Ho}(B - Mod_s)$ is finitely presented.*

PROOF. Let us assume first that $f$ is finitely presented. Then (1) and (2) are easily seen to be true by fact that $\pi_0$ is left adjoint and by definition of derivations.

Let us now assume that $f : A \longrightarrow B$ is a morphism in $sk - Alg$ such that (1) and (2) are satisfied. Let $k \geq 0$ be an integer, and let $P(k)$ be the following property: for any filtered diagram $C_i$ in $A - Alg_s$, such that $\pi_j(C_i) = 0$ for all $j > k$, the natural morphism

$$Hocolim_i Map_{A-Alg_s}(B, C_i) \longrightarrow Map_{A-Alg_s}(B, Hocolim_i C_i)$$

is an isomorphism in Ho$(SSet)$.

We start to prove by induction on $k$ that $P(k)$ holds for all $k$. For $k = 0$ this is hypothesis (1). Suppose $P(k-1)$ holds, and let $C_i$ be a any filtered diagram in $A - Alg_s$, such that $\pi_j(C_i) = 0$ for all $j > k$. We consider $C = Hocolim_i C_i$, as well as the $k$-th Postnikov towers

$$C \longrightarrow C_{\leq k-1} \qquad (C_i) \longrightarrow (C_i)_{\leq k-1}.$$

There is a commutative square of simplicial sets

$$\begin{array}{ccc} Hocolim_i Map_{A-Alg_s}(B, C_i) & \longrightarrow & Hocolim_i Map_{A-Alg_s}(B, (C_i)_{k-1}) \\ \downarrow & & \downarrow \\ Map_{A-Alg_s}(B, C) & \longrightarrow & Map_{A-Alg_s}(B, C_{k-1}). \end{array}$$

By induction, the morphism on the right is an equivalence. Furthermore, using Prop. 1.4.2.6, Lem. 2.2.1.1 and the fact that the cotangent complex $\mathbb{L}_{B/A}$ is finitely presented, we see that the morphism induced on the homotopy fibers of the horizontal morphisms is also an equivalence. By the five lemma this implies that the morphism

$$Hocolim_i Map_{A-Alg_s}(B, C_i) \longrightarrow Map_{A-Alg_s}(B, C)$$

is an equivalence. This shows that $P(k)$ is satisfied.

As $\mathbb{L}_{B/A}$ is finitely presented, it is a retract of a finite cell $B$-module (see Prop. 1.2.3.5). In particular, there is an integer $k_0 > 0$, such that $[\mathbb{L}_{B/A}, M]_{B-Mod_s} = 0$ for any $B$-module $M$ such that $\pi_i(M) = 0$ for all $i < k_0$ (one can chose $k_0$ strictly bigger than the dimension of the cells of a module of which $\mathbb{L}_{B/A}$ is a retract). Once again, Prop. 1.4.2.6 and Lem. 2.2.1.1 implies that for any commutative $A$-algebra $C$, the natural projection $C \longrightarrow C_{\leq k_0}$ induces a bijection

$$[B, C]_{A-Alg_s} \simeq [B, C_{\leq k_0}]_{A-Alg_s}.$$

Therefore, as the property $P(k_0)$ holds, we find that for any filtered system $C_i$ in $A - Alg_s$, the natural morphism

$$Hocolim_i Map_{A-Alg_s}(B, C_i) \longrightarrow Map_{A-Alg_s}(B, Hocolim_i C_i)$$

induces an isomorphism in $\pi_0$. As this is valid for any filtered system, this shows that the morphism

$$Hocolim_i Map_{A-Alg_s}(B, C_i) \longrightarrow Map_{A-Alg_s}(B, Hocolim_i C_i)$$

induces an isomorphism on all the $\pi_i$'s, and therefore is an equivalence. This shows that $f$ is finitely presented. □

PROPOSITION 2.2.2.5. *Let $f : A \longrightarrow B$ be a morphism in $sk - Alg$.*

(1) *The morphism $f$ is smooth if and only if it is perfect. The morphism $f$ is formally smooth if it is formally perfect. The morphism $f$ is formally i-smooth if and only if it is formally smooth.*
(2) *The morphism $f$ is (formally) unramified if and only if it is (formally) étale.*

(3) *The morphism $f$ is (formally) étale if and only if it is (formally) thh-étale.*
(4) *A morphism $A \longrightarrow B$ in $sk-Alg$ is* flat *(resp.* a Zariski open immersion*) if and only if it is* strongly flat *(resp.* a strong Zariski open immersion*).*

PROOF. (1) The first two assumptions follow from Lem. 2.2.2.2 (4). For the comparison between formally i-smooth and formally smooth morphism we notice that a $B$-module $P \in B-Mod$ is projective if and only if $[P, M[1]] = 0$ for any $M \in B-Mod$. This and Prop. 1.2.8.3 imply the statement (note that in our present context $\mathcal{A} = sk-Alg$ and $\mathcal{C}_0 = \mathcal{C}$).

(2) Follows from the fact that the suspension functor of $sk-Mod$ is fully faithful and from Prop. 1.2.6.5 (1).

(3) By Prop. 1.2.6.5 (2) (formally) thh-étale morphisms are (formally) étale. Conversely, we need to prove that a formally étale morphism $A \longrightarrow B$ is thh-étale. For this, we use the well known spectral sequence

$$\pi_*(Sym^*(\mathbb{L}_{B/A}[1])) \Rightarrow \pi_*(THH(B/A))$$

of [**Q2**, 8]. Using this spectral sequence we see that $B \simeq THH(B/A)$ if and only if $\mathbb{L}_{B/A} \simeq 0$. In particular a formally étale morphism is always formally thh-étale.

(4) For flat morphism this is Lem. 2.2.2.2 (2). Let $f : A \longrightarrow B$ be a Zariski open immersion. By definition, $f$ is a flat morphism, and therefore is strongly flat by what we have seen. Moreover, for any commutative $k$-algebra $C$, considered as an object $C \in sk-Alg$ concentrated in degree 0, we have natural isomorphisms in $\mathrm{Ho}(SSet)$

$$Map_{sk-Alg}(A, C) \simeq Hom_{k-Alg}(\pi_0(A), C)$$

$$Map_{sk-Alg}(B, C) \simeq Hom_{k-Alg}(\pi_0(B), C).$$

In particular, as $f$ is a epimorphism in $sk-Alg$ the induced morphism of affine schemes $\varphi : Spec\,\pi_0(B) \longrightarrow Spec\,\pi_0(A)$ is a monomorphism of schemes. This last morphism is therefore a flat monomorphism of affine schemes. Moreover, as $f$ is finitely presented so is $\varphi$, which is therefore a finitely presented flat monomorphism. By [**EGAIV**, 2.4.6], a finitely presented flat morphism is open, so $\varphi$ is an open flat monomorphism and thus a Zariski open immersion. This implies that $f$ is a strong Zariski open immersion.

Conversely, let $f : A \longrightarrow B$ be a strong Zariski open immersion. By (1) it is a flat morphism. It only remains to show that it is also an epimorphism in $sk-Alg$ and that it is finitely presented. For the first of these properties, we use the $Tor$ spectral sequence to see that the natural morphism $B \longrightarrow B \otimes^{\mathbb{L}}_A B$ induces an isomorphism on homotopy groups

$$\pi_*(B \otimes^{\mathbb{L}}_A B) \simeq \pi_*(B) \otimes_{\pi_*(A)} \pi_*(B) \simeq (\pi_0(B) \otimes_{\pi_0(A)} \pi_0(B)) \otimes_{\pi_0(A)} \pi_*(A) \simeq \pi_*(B).$$

In other words, the natural morphism $B \otimes^{\mathbb{L}}_A B \longrightarrow B$ is an isomorphism in $\mathrm{Ho}(sk-Alg)$, which implies that $f$ is an epinomorphism. It remain to be shown that $f$ is furthermore finitely presented, but this follows from Prop. 2.2.2.4 and Prop. 1.2.6.5. □

THEOREM 2.2.2.6. (1) *A morphism in $sk-Alg$ is étale if and only if it is strongly étale.*
(2) *A morphism in $sk-Alg$ is smooth if and only if it is strongly smooth.*

PROOF. We start by two fundamental lemmas.

Recall from §1.2.1 the Quillen adjunction

$$Q : A - Comm^{nu}(\mathcal{C}) \longrightarrow A - Mod \qquad A - Comm^{nu}(\mathcal{C}) \longleftarrow A - Mod : Z.$$

LEMMA 2.2.2.7. *Let $A \in sk - Alg$, and $M \in A - Mod_s$ be a A-module such that $\pi_i(M) = 0$ for all $i < k$, for some fixed integer $k$. Then, the adjunction morphism*

$$\mathbb{L}QZ(M) \longrightarrow M$$

*induces isomorphisms*

$$\pi_i(\mathbb{L}QZ(M)) \simeq \pi_i(M)$$

*for all $i \leq k$. In particular*

$$\pi_i(\mathbb{L}QZ(M)) \simeq 0 \; for \; i < k \qquad \pi_k(\mathbb{L}QZ(M)) \simeq \pi_k(M).$$

PROOF. The non-unital $A$-algebra $Z(M)$ being $(k-1)$-connected, we can write it, up to an equivalence, as a CW object in $A - Comm^{nu}(sk - Mod)$ with cells of dimension at least $k$ (in the sense of [**EKMM**]). In other words $Z(M)$ is equivalent to a filtered colimit

$$\dots \longrightarrow A_i \longrightarrow A_{i+1} \longrightarrow \dots$$

where at each step there is a push-out diagram in $A - Comm^{nu}(sk - Mod)$

$$\begin{array}{ccc} A_i & \longrightarrow & A_{i+1} \\ \uparrow & & \uparrow \\ \coprod k[S^{n_i}]^{nu} & \longrightarrow & \coprod k[\Delta^{n_i+1}]^{nu}, \end{array}$$

where $n_{i+1} > n_i \geq k-1$, and $k[K]^{nu}$ is the free non-unital commutative simplicial $k$-algebra generated by a simplicial set $K$. Therefore, $Q$ being left Quillen, the $A$-module $\mathbb{L}QZ(M)$ is the homotopy colimit of

$$\dots \longrightarrow Q(A_i) \longrightarrow Q(A_{i+1}) \longrightarrow \dots$$

where at each step there exists a homotopy push-out diagram in $A - Mod_s$

$$\begin{array}{ccc} Q(A_i) & \longrightarrow & Q(A_{i+1}) \\ \uparrow & & \uparrow \\ \oplus Q(k[S^{n_i}]^{nu}) & \longrightarrow & 0. \end{array}$$

Computing homotopy groups using long exact sequences, we see that the statement of the lemma can be reduced to prove that for a free non-unital commutative simplicial $k$-algebra $k[S^n]^{nu}$, we have

$$\pi_i(Q(k[S^n]^{nu})) \simeq 0 \; for \; i < n \qquad \pi_n(Q(k[S^n]^{nu})) \simeq k.$$

But, using that $Q$ is left Quillen we find

$$Q(k[S^n]^{nu}) \simeq S^n \otimes k,$$

which implies the result. □

LEMMA 2.2.2.8. *Let $A$ be any object in $sk - Alg$ and let us consider the $k$-th stage of its Postnikov tower*

$$A_{\leq k} \longrightarrow A_{\leq k-1}.$$

*There exist natural isomorphisms*

$$\pi_{k+1}(\mathbb{L}_{A_{\leq k-1}/A_{\leq k}}) \simeq \pi_k(A)$$

$$\pi_i(\mathbb{L}_{A_{\leq k-1}/A_{\leq k}}) \simeq 0 \; for \; i \leq k.$$

PROOF. This follows from lemma 2.2.1.1, lemma 1.4.3.7, and lemma 2.2.2.7. □

Let us now start the proof of theorem 2.2.2.6.

(1) Let $f : A \longrightarrow B$ be a strongly étale morphism. By definition of strongly étale the morphism $f$ is strongly flat. In particular, the square

$$\begin{array}{ccc} A & \longrightarrow & B \\ \downarrow & & \downarrow \\ \pi_0(A) & \longrightarrow & \pi_0(B) \end{array}$$

is homotopy cocartesian in $sk - Alg$. Therefore, Prop. 1.2.1.6 (2) implies

$$\mathbb{L}_{B/A} \otimes_B \pi_0(B) \simeq \mathbb{L}_{\pi_0(B)/\pi_0(A)} \simeq 0.$$

Using the Tor spectral sequence

$$Tor^*_{\pi_*(B)}(\pi_0(B), \pi_*(\mathbb{L}_{B/A})) \Rightarrow \pi_*(\mathbb{L}_{B/A} \otimes_B \pi_0(B)) = 0$$

we find by induction on $k$ that $\pi_k(\mathbb{L}_{B/A}) = 0$. This shows that the morphism $f$ is formally étale. The fact that it is also finitely presented follows then from Prop. 2.2.2.4.

Conversely, let $f : A \longrightarrow B$ be an étale morphism. As $\pi_0 : sk - Alg \longrightarrow k - Alg$ is left Quillen, we deduce immediately that $\pi_0(A) \longrightarrow \pi_0(B)$ is an étale morphism of commutative rings. We will prove by induction that all the truncations

$$f_k : A_{\leq k} \longrightarrow B_{\leq k}$$

are strongly étale and thus étale as well by what we have seen before. We thus assume that $f_{k-1}$ is strongly étale (and thus also étale by what we have seen in the first part of the proof). By adjunction, the truncations

$$f_k : A_{\leq k} \longrightarrow B_{\leq k}$$

are such that

$$(\mathbb{L}_{B_{\leq k}/A_{\leq k}})_{\leq k} \simeq (\mathbb{L}_{B/A})_{\leq k} \simeq 0.$$

Furthermore, let $M$ be any $\pi_0(B)$-module, and $d$ be any morphism in $\mathrm{Ho}(B_{\leq k} - Mod_s)$

$$d : \mathbb{L}_{B_{\leq k}/A_{\leq k}} \longrightarrow M[k+1].$$

We consider the commutative diagram in $\mathrm{Ho}(A_{\leq k}/sk - Alg/B_{\leq k})$

$$\begin{array}{ccc} A_{\leq k} & \longrightarrow & B_{\leq k} \oplus_d M[k] \\ \downarrow & & \downarrow \\ B_{\leq k} & \longrightarrow & B_{\leq k}. \end{array}$$

The obstruction of the existence of a morphism

$$u : B_{\leq k} \longrightarrow B_{\leq k} \oplus_d M[k]$$

in $\mathrm{Ho}(A_{\leq k}/sk - Alg/B_{\leq k})$ is precisely $d \in [\mathbb{L}_{B_{\leq k}/A_{\leq k}}, M[k+1]]$. On the other hand, by adjunction, the existence of such a morphism $f$ is equivalent to the existence of a morphism

$$u' : B \longrightarrow B_{\leq k} \oplus_d M[k]$$

in $\mathrm{Ho}(A/sk - Alg/B_{\leq k})$. The obstruction of the existence of $u'$ itself is $d' \in [\mathbb{L}_{B/A}, M[k+1]]$, the image of $d$ by the natural morphism, which vanishes as $A \to B$ is formally étale. This implies that $u'$ and thus $u$ exists, and therefore that for any $\pi_0(B)$-module $M$ we have

$$[\mathbb{L}_{B_{\leq k}/A_{\leq k}}, M[k+1]] = 0.$$

As the object $\mathbb{L}_{B_{\leq k}/A_{\leq k}}$ is already known to be $k$-connected, we conclude that it is furthermore is $(k+1)$-connected.

Now, there exists a morphism of homotopy cofiber sequences in $sk-Mod$

$$\begin{array}{ccccc}
\mathbb{L}_{A_{\leq k}} \otimes^{\mathbb{L}}_{A_{\leq k}} A_{\leq k-1} & \longrightarrow & \mathbb{L}_{A_{\leq k-1}} & \longrightarrow & \mathbb{L}_{A_{\leq k-1}/A_{\leq k}} \\
\downarrow & & \downarrow & & \downarrow \\
\mathbb{L}_{B_{\leq k}} \otimes^{\mathbb{L}}_{B_{\leq k}} B_{\leq k-1} & \longrightarrow & \mathbb{L}_{B_{\leq k-1}} & \longrightarrow & \mathbb{L}_{B_{\leq k-1}/B_{\leq k}}.
\end{array}$$

Base changing the first row by $A_{\leq k-1} \longrightarrow B_{\leq k-1}$ gives another morphism of homotopy cofiber sequences in $sk-Mod$

$$\begin{array}{ccccc}
\mathbb{L}_{A_{\leq k}} \otimes^{\mathbb{L}}_{A_{\leq k}} B_{\leq k-1} & \longrightarrow & \mathbb{L}_{A_{\leq k-1}} \otimes^{\mathbb{L}}_{A_{\leq k-1}} B_{\leq k-1} & \longrightarrow & \mathbb{L}_{A_{\leq k-1}/A_{\leq k}} \otimes^{\mathbb{L}}_{A_{\leq k-1}} B_{\leq k-1} \\
\downarrow & & \downarrow & & \downarrow \\
\mathbb{L}_{B_{\leq k}} \otimes^{\mathbb{L}}_{B_{\leq k}} B_{\leq k-1} & \longrightarrow & \mathbb{L}_{B_{\leq k-1}} & \longrightarrow & \mathbb{L}_{B_{\leq k-1}/B_{\leq k}}.
\end{array}$$

Passing to the long exact sequences in homotopy, and using that $f_{k-1}$ is étale, as well as the fact that $\mathbb{L}_{B_{\leq k}/A_{\leq k}} \otimes^{\mathbb{L}}_{A_{\leq k-1}} B_{\leq k-1}$ is $(k+1)$-connected, it is easy to see that the vertical morphism on the right induces an isomorphism

$$\pi_{k+1}(\mathbb{L}_{A_{\leq k-1}/A_{\leq k}} \otimes^{\mathbb{L}}_{A_{\leq k-1}} B_{\leq k-1}) \simeq \pi_{k+1}(\mathbb{L}_{B_{\leq k-1}/B_{\leq k}}).$$

Therefore, Lemma 2.2.2.8 implies that the natural morphism

$$\pi_k(A) \otimes_{\pi_0(A)} \pi_0(B) \longrightarrow \pi_k(B)$$

is an isomorphism. By induction on $k$ this shows that $f$ is strongly étale.

(2) Let $f : A \longrightarrow B$ be a strongly smooth morphism. As the morphism $f$ is strongly flat, there is a homotopy push-out square in $sk-Alg$

$$\begin{array}{ccc}
A & \longrightarrow & B \\
\downarrow & & \downarrow \\
\pi_0(A) & \longrightarrow & \pi_0(B).
\end{array}$$

Together with Prop. 1.2.1.6 (2), we thus have a natural isomorphism

$$\mathbb{L}_{B/A} \otimes^{\mathbb{L}}_{B} \pi_0(B) \simeq \mathbb{L}_{\pi_0(B)/\pi_0(A)}.$$

As the morphism $\pi_0(A) \longrightarrow \pi_0(B)$ is smooth, we have $\mathbb{L}_{\pi_0(B)/\pi_0(A)} \simeq \Omega^1_{\pi_0(B)/\pi_0(A)}[0]$. Using that $\Omega^1_{\pi_0(B)/\pi_0(A)}$ is a projective $\pi_0(B)$-module, we see that the isomorphism

$$\Omega^1_{\pi_0(B)/\pi_0(A)}[0] \simeq \mathbb{L}_{B/A} \otimes^{\mathbb{L}}_{B} \pi_0(B)$$

can be lifted to a morphism

$$P \longrightarrow \mathbb{L}_{B/A}$$

in $\mathrm{Ho}(B-Mod_s)$, where $P$ is a projective $B$-module such that $P \otimes^{\mathbb{L}}_{B} \pi_0(B) \simeq \Omega^1_{\pi_0(B)/\pi_0(A)}[0]$. We let $K$ be the homotopy cofiber of this last morphism. By construction, we have

$$K \otimes^{\mathbb{L}}_{B} \pi_0(B) \simeq 0,$$

which easily implies by induction on $k$ that $\pi_k(K) = 0$. Therefore, the morphism

$$P \longrightarrow \mathbb{L}_{B/A}$$

is in fact an isomorphism in $\mathrm{Ho}(B - Mod_s)$, showing that $\mathbb{L}_{B/A}$ is a projective $B$-module. Moreover, the homotopy cofiber sequence in $B - Mod_s$

$$\mathbb{L}_A \otimes^{\mathbb{L}}_A B \longrightarrow \mathbb{L}_B \longrightarrow \mathbb{L}_{B/A}$$

gives rise to a homotopy cofiber sequence

$$\mathbb{L}_B \longrightarrow \mathbb{L}_{B/A} \longrightarrow S(\mathbb{L}_A \otimes^{\mathbb{L}}_A B).$$

But, $\mathbb{L}_{B/A}$ being a retract of a free $B$-module, we see that $[\mathbb{L}_{B/A}, S(\mathbb{L}_A \otimes^{\mathbb{L}}_A B)]$ is a retract of a product of $\pi_0(S(\mathbb{L}_A \otimes^{\mathbb{L}}_A B)) = 0$ and thus is trivial. This implies that the morphism $\mathbb{L}_{B/A} \longrightarrow S(\mathbb{L}_A \otimes^{\mathbb{L}}_A B)$ is trivial in $\mathrm{Ho}(B - Mod_s)$, and therefore that the homotopy cofiber sequence

$$\mathbb{L}_A \otimes^{\mathbb{L}}_A B \longrightarrow \mathbb{L}_B \longrightarrow \mathbb{L}_{B/A},$$

which is also a homotopy fiber sequence, splits. In particular, the morphism $\mathbb{L}_A \otimes^{\mathbb{L}}_A B \longrightarrow \mathbb{L}_B$ has a retraction. We have thus shown that $f$ is a formally smooth morphism. The fact that $f$ is furthermore finitely presented follows from Prop. 2.2.2.4 and the fact that $\mathbb{L}_{B/A}$ if finitely presented because $\Omega^1_{B/A}$ is so (see Lem. 2.2.2.2).

Conversely, let $f : A \longrightarrow B$ be a smooth morphism in $sk - Alg$ and let us prove it is strongly smooth. First of all, using that $\pi_0 : sk - Alg \longrightarrow k - Alg$ is left Quillen, we see that $\pi_0(A) \longrightarrow \pi_0(B)$ has the required lifting property for being a formally smooth morphism. Furthermore, it is a finitely presented morphism, so is a smooth morphism of commutative rings. We then form the homotopy push-out square in $sk - Alg$

$$\begin{array}{ccc} A & \longrightarrow & B \\ \downarrow & & \downarrow \\ \pi_0(A) & \longrightarrow & C. \end{array}$$

By base change, $\pi_0(A) \longrightarrow C$ is a smooth morphism. We will start to prove that the natural morphism $C \longrightarrow \pi_0(C) \simeq \pi_0(B)$ is an isomorphism. Suppose it is not, and let $i$ be the smallest integer $i > 0$ such that $\pi_i(C) \neq 0$. Considering the homotopy cofiber sequence

$$\mathbb{L}_{C_{\leq i}/\pi_0(A)} \otimes^{\mathbb{L}}_{C_{\leq i}} \pi_0(C) \longrightarrow \mathbb{L}_{\pi_0(C)/\pi_0(A)} \simeq \Omega^1_{\pi_0(C)/\pi_0(A)}[0] \longrightarrow \mathbb{L}_{C_{\leq i}/\pi_0(C)},$$

and using lemma 2.2.2.8, we see that

$$\pi_i(\mathbb{L}_{C/\pi_0(A)} \otimes^{\mathbb{L}}_C \pi_0(C)) \simeq \pi_i(\mathbb{L}_{C_{\leq i}/\pi_0(A)} \otimes^{\mathbb{L}}_{C_{\leq i}} \pi_0(C)) \simeq \pi_{i+1}(\mathbb{L}_{C_{\leq i}/\pi_0(C)}) \simeq \pi_i(C) \neq 0.$$

But this contradicts the fact that $\mathbb{L}_{C/\pi_0(A)} \otimes^{\mathbb{L}}_C \pi_0(C)$ is a projective (and thus strong by lemma 2.2.2.2 (1)) $\pi_0(C)$-module, and thus the fact that $\pi_0(A) \longrightarrow C$ is a smooth morphism. We therefore have a homotopy push-out diagram in $sk - Alg$

$$\begin{array}{ccc} A & \longrightarrow & B \\ \downarrow & & \downarrow \\ \pi_0(A) & \longrightarrow & \pi_0(B). \end{array}$$

Using that the bottom horizontal morphism in flat and the Tor spectral sequence, we get by induction that the natural morphism

$$\pi_i(A) \otimes_{\pi_0(A)} \pi_0(B) \longrightarrow \pi_i(B)$$

is an isomorphism. We thus have seen that $f$ is a strongly smooth morphism. □

An important corollary of theorem 2.2.2.6 is the following topological invariance of étale morphisms.

COROLLARY 2.2.2.9. *Let $A \in sk - Alg$ and $t_0(X) = Spec\,(\pi_0 A) \longrightarrow X = Spec\,A$ be the natural morphism. Then, the base change functor*

$$\mathrm{Ho}(k - D^- Aff/X) \longrightarrow \mathrm{Ho}(k - D^- Aff/t_0(X))$$

*induces an equivalence from the full subcategory of étale morphism $Y \to X$ to the full subcategory of étale morphisms $Y' \to t_0(X)$. Furthermore, this equivalence preserves epimorphisms of stacks.*

PROOF. We consider the Postnikov tower

$$A \longrightarrow \cdots \longrightarrow A_{\leq k} \longrightarrow A_{\leq k-1} \longrightarrow \cdots \longrightarrow A_{\leq 0} = \pi_0(A),$$

and the associated diagram in $k - D^- Aff$

$$X_{\leq 0} = t_0(X) \longrightarrow X_{\leq 1} \longrightarrow \cdots \longrightarrow X_{\leq k-1} \longrightarrow X_{\leq k} \longrightarrow \cdots \longrightarrow X.$$

We define a model category $k - D^- Aff/X_{\leq *}$, whose objects are families of objects $Y_k \longrightarrow X_{\leq k}$ in $k - D^- Aff/X_{\leq k}$, together with transitions morphisms $Y_{k-1} \longrightarrow Y_k \times_{X_{\leq k}} X_{\leq k-1}$ in $k - D^- Aff/X_{\leq k-1}$. The morphisms in $k - D^- Aff/X_{\leq *}$ are simply the families of morphisms $Y_k \longrightarrow Y'_k$ in $k - D^- Aff/X_{\leq k}$ that commute with the transition morphisms. The model structure on $k - D^- Aff/X_{\leq *}$ is such that a morphism $f : Y_* \longrightarrow Y'_*$ in $k - D^- Aff/X_{\leq *}$ is an equivalence (resp. a fibration) if all morphisms $Y_k \longrightarrow Y'_k$ are equivalences (resp. fibrations) in $k - D^- Aff$.

There exists a Quillen adjunction

$$G : k\text{–}D^- Aff/X_{\leq *} \longrightarrow k\text{–}D^- Aff/X \qquad k\text{–}D^- Aff/X_{\leq *} \longleftarrow k\text{–}D^- Aff/X : F,$$

defined by $G(Y_*)$ to be the colimit in $k - D^- Aff/X$ of the diagram

$$Y_0 \longrightarrow Y_1 \longrightarrow \ldots Y_k \longrightarrow \cdots \longrightarrow X.$$

The right adjoint $F$ sends a morphism $Y \longrightarrow X$ to the various pullbacks

$$F(Y)_k := Y \times_X X_{\leq k} \longrightarrow X_k$$

together with the obvious transition isomorphisms. Using Thm. 2.2.2.6 (1) it is not hard to check that the derived adjunction

$$\mathbb{L}G : \mathrm{Ho}(k - D^- Aff/X_{\leq *}) \longrightarrow \mathrm{Ho}(k - D^- Aff/X)$$

$$\mathrm{Ho}(k - D^- Aff/X_{\leq *}) \longleftarrow \mathrm{Ho}(k - D^- Aff/X) : \mathbb{R}F,$$

induces an equivalence from the full subcategory of $\mathrm{Ho}(k - D^- Aff/X)$ consisting of étale morphisms $Y \longrightarrow X$, and the full subcategory of $\mathrm{Ho}(k\text{–}D^- Aff/X_{\leq *})$ consisting of objects $Y_*$ such that each $Y_k \longrightarrow X_{\leq k}$ is étale and each transition morphism

$$Y_{k-1} \longrightarrow Y_k \times^h_{X_{\leq k}} X_{\leq k-1}$$

is an isomorphism in $\mathrm{Ho}(k - D^- Aff/X_{\leq k-1})$. Let us denote these two categories respectively by $\mathrm{Ho}(k - D^- Aff/X)^{et}$ and $\mathrm{Ho}(k - D^- Aff/X_{\leq *})^{cart,et}$. Using Cor. 1.4.3.12 and Lem. 2.2.1.1, we know that each base change functor

$$\mathrm{Ho}(k - D^- Aff/X_{\leq k}) \longrightarrow \mathrm{Ho}(k - D^- Aff/X_{\leq k-1})$$

induces an equivalence on the full sub-categories of étale morphisms. This easily implies that the natural projection functor

$$\mathrm{Ho}(k-D^{-}Aff/X_{\leq *})^{cart,et} \longrightarrow \mathrm{Ho}(k-D^{-}Aff/X_{\leq 0})^{et}$$

is an equivalence of categories. Therefore, by composition, we find that the base change functor

$$\mathrm{Ho}(k-D^{-}Aff/X)^{et} \longrightarrow \mathrm{Ho}(k-D^{-}Aff/X_{\leq *})^{cart,et} \longrightarrow \mathrm{Ho}(k-D^{-}Aff/X_{\leq 0})^{et}$$

is an equivalence of categories.

Finally, the statement concerning epimorphism of stacks is obvious, as a flat morphism $Y \longrightarrow X$ in $k-D^{-}Aff$ induces an epimorphism of stacks if and only if $t_0(Y) \longrightarrow t_0(X)$ is a surjective morphism of affine schemes. □

A direct specialization of 2.2.2.9 is the following.

COROLLARY 2.2.2.10. *Let $A \in sk-Alg$ and $t_0(X) = Spec\,(\pi_0 A) \longrightarrow X = Spec\,A$ be the natural morphism. Then, the base change functor*

$$\mathrm{Ho}(k-D^{-}Aff/X) \longrightarrow \mathrm{Ho}(k-D^{-}Aff/t_0(X))$$

*induces an equivalence from full sub-categories of Zariski open immersions $Y \to X$ to the full subcategory of Zariski open immersions $Y' \to t_0(X)$. Furthermore, this equivalence preserves epimorphisms of stacks.*

PROOF. Indeed, using 2.2.2.9 it is enough to see that an étale morphism $A \longrightarrow B$ is a Zariski open immersion if and only if $\pi_0(A) \longrightarrow \pi_0(B)$ is. But this true by Thm. 2.2.2.6 and Prop. 2.2.2.5. □

From the proof of Thm. 2.2.2.6 we also extract the following more precise result.

COROLLARY 2.2.2.11. *Let $f : A \longrightarrow B$ be a morphism in $sk-Alg$. The following are equivalent.*

(1) *The morphism $f$ is smooth (resp. étale).*
(2) *The morphism $f$ is flat and $\pi_0(A) \longrightarrow \pi_0(B)$ is smooth (resp. étale).*
(3) *The morphism $f$ is formally smooth (resp. formally étale) and $\pi_0(B)$ is a finitely presented $\pi_0(A)$-algebra.*

We are now ready to define the étale model topology (Def. 1.3.1.1) on $k-D^{-}Aff$.

DEFINITION 2.2.2.12. *A family of morphisms $\{Spec\,A_i \longrightarrow Spec\,A\}_{i\in I}$ in $k-D^{-}Aff$ is an* étale covering family *(or simply* ét-covering family*) if it satisfies the following two conditions.*

(1) *Each morphism $A \longrightarrow A_i$ is étale.*
(2) *There exists a finite sub-set $J \subset I$ such that the family $\{A \longrightarrow A_i\}_{i\in J}$ is a formal covering family in the sense of 1.2.5.1.*

Using that étale morphisms are precisely the strongly étale morphisms (see Corollary 2.2.2.11) we immediately deduce that a family of morphisms $\{Spec\,A_i \longrightarrow Spec\,A\}_{i\in I}$ in $k-D^{-}Aff$ is an ét-covering family if and only if there exists a finite sub-set $J \subset I$ satisfying the following two conditions.

- For all $i \in I$, the natural morphism

$$\pi_*(A) \otimes_{\pi_0(A)} \pi_0(A_i) \longrightarrow \pi_*(A_i)$$

is an isomorphism.

- The morphism of affine schemes
$$\coprod_{i\in J} Spec\,\pi_0(A_i) \longrightarrow Spec\,\pi_0(A)$$
is étale and surjective.

LEMMA 2.2.2.13. *The ét-covering families define a model topology (Def. 1.3.1.1) on $k - D^{-}Aff$, which satisfies assumption 1.3.2.2.*

PROOF. That ét-covering families defines a model topology simply follows from the general properties of étale morphisms and formal coverings described in propositions 1.2.5.2 and 1.2.6.3. It only remain to show that the étale topology satisfies assumption 1.3.2.2.

The étale topology is quasi-compact by definition, so (1) of 1.3.2.2 is satisfied. In the same way, property (2) of 1.3.2.2 is obviously satisfied according to the explicit definition of étale coverings given above. Finally, let us check property (3) of 1.3.2.2. For this, let $A \longrightarrow B_*$ be a co-simplicial object in $sk - Alg$, corresponding to a representable étale-hypercover in $k - D^{-}Aff$ in the sense of 1.3.2.2 (3). We consider the adjunction

$$B_* \otimes^{\mathbb{L}}_A - : \mathrm{Ho}(A - Mod_s) \longrightarrow \mathrm{Ho}(csB_* - Mod)$$
$$\mathrm{Ho}(A - Mod_s) \longleftarrow \mathrm{Ho}(csB_* - Mod) : \int$$

defined in §1.2. We restrict it to the full subcategory $\mathrm{Ho}(csB_* - Mod)^{cart}$ of $\mathrm{Ho}(csB_* - Mod)$ consisting of cartesian objects

$$B_* \otimes^{\mathbb{L}}_A - : \mathrm{Ho}(A - Mod_s) \longrightarrow \mathrm{Ho}(csB_* - Mod)^{cart}$$
$$\mathrm{Ho}(A - Mod_s) \longleftarrow \mathrm{Ho}(csB_* - Mod)^{cart} : \int$$

and we need to prove that this is an equivalence. By definition of formal coverings, the base change functor

$$B_* \otimes^{\mathbb{L}}_A - : \mathrm{Ho}(A - Mod_s) \longrightarrow \mathrm{Ho}(csB_* - Mod)^{cart}$$

is clearly conservative, so it only remains to show that the adjunction map

$$Id \longrightarrow \int \circ (B_* \otimes^{\mathbb{L}}_A -)$$

is an isomorphism.

For this, let $E_* \in \mathrm{Ho}(csB_* - Mod)^{cart}$, and let us consider the adjunction morphism

$$E_1 \longrightarrow (Holim_n E_n) \otimes^{\mathbb{L}}_A B_1.$$

We need to show that this morphism is an isomorphism in $\mathrm{Ho}(B_1 - Mod_s)$. For this, we first use that $A \longrightarrow B_1$ is a strongly flat morphism, and thus

$$\pi_*((Holim_n E_n) \otimes^{\mathbb{L}}_A B_1) \simeq \pi_*((Holim_n E_n)) \otimes_{\pi_0(A)} \pi_0(B_1).$$

We then apply the spectral sequence

$$H^p(Tot(\pi_q(E_*))) \Rightarrow \pi_{q-p}((Holim_n E_n).$$

The object $\pi_q(E_*)$ is now a co-simplicial module over the co-simplicial commutative ring $\pi_0(B_*)$, which is furthermore cartesian as all the coface morphisms $B_n \longrightarrow B_{n+1}$ are flat. The morphism $\pi_0(A) \longrightarrow \pi_0(B_*)$ being an étale, and thus faithfully flat, hypercover in the usual sense, we find by the usual flat descent for quasi-coherent sheaves that $H^p(Tot(\pi_q(E_*))) \simeq 0$ for $p \neq 0$. Therefore, the above spectral sequence degenerates and gives an isomorphism

$$\pi_p((Holim_n E_n) \simeq Ker\,(\pi_p(E_0) \rightrightarrows \pi_p(E_1))\,.$$

In other words, $\pi_p((Holim_n E_n))$ is the $\pi_0(A)$-module obtained by descent from $\pi_p(E_*)$ on $\pi_p(B_*)$. In particular, the natural morphism

$$\pi_p((Holim_n E_n) \otimes_{\pi_0(A)} \pi_0(B_1) \longrightarrow \pi_p(E_1)$$

is an isomorphism. Putting all of this together we find that

$$E_1 \longrightarrow (Holim_n E_n) \otimes^{\mathbb{L}}_A B_1$$

is an isomorphism in $\mathrm{Ho}(B_1 - Mod_s)$. □

We have now the model site (Def. 1.3.1.1) $(k - D^-Aff, \text{ét})$, with the étale model topology, and we make the following definition.

DEFINITION 2.2.2.14. (1) *A* $D^-$-stack *is an object* $F \in k - D^-Aff^{\sim,\text{ét}}$ *which is a stack in the sense of Def. 1.3.2.1.*
(2) *The* model category of $D^-$-stacks *is* $k - D^-Aff^{\sim,\text{ét}}$*, and its homotopy category will be simply denoted by* $D^-\mathrm{St}(k)$.

The following result is a corollary of Proposition 2.2.2.4. It states that the property of being finitely presented is local for the étale topology defined above.

COROLLARY 2.2.2.15. *Let* $f : A \longrightarrow B$ *be a morphism in* $sk - Alg$.
(1) *If there exists an étale covering* $B \longrightarrow B'$ *such that* $A \longrightarrow B'$ *is finitely presented, then* $f$ *is finitely presented.*
(2) *If there exists an étale covering* $A \longrightarrow A'$*, such that*
$$A' \longrightarrow A' \otimes^{\mathbb{L}}_A B$$
*is finitely presented, then* $f$ *is finitely presented.*

PROOF. This follows from proposition 2.2.2.4. Indeed, it suffices to prove that both conditions of proposition 2.2.2.4 have the required local property. The first one is well known, and the second one is a consequence of corollary 1.3.7.8, as finitely presented objects in $\mathrm{Ho}(Sp(B - Mod_s))$ are precisely the perfect objects. □

## 2.2.3. The HAG context: Geometric $D^-$-stacks

We now let **P** be the class of smooth morphisms in $sk - Alg$.

LEMMA 2.2.3.1. *The class* ***P*** *of smooth morphisms and the étale model topology satisfy assumption 1.3.2.11.*

PROOF. As étale morphisms are also smooth we see that assumption 1.3.2.11 (1) is satisfied. Assumption 1.3.2.11 (2) is satisfied as smooth morphisms are stable by homotopy pullbacks, compositions and equivalences. Let us prove that smooth morphisms satisfy 1.3.2.11 (3) for the étale model topology.

Let $X \longrightarrow Y$ be a morphism in $k - D^-Aff$, and $\{U_i \longrightarrow X\}$ a finite étale covering family such that each $U_i \longrightarrow X$ and $U_i \longrightarrow Y$ is smooth. First of all, using that $\{U_i \longrightarrow X\}$ is a flat formal covering, and that each morphism $U_i \longrightarrow Y$ is flat, we see that the morphism $X \longrightarrow Y$ is flat. Using our results Prop. 2.2.2.5 and Thm. 2.2.2.6 it only remain to show that the morphism $t_0(X) \longrightarrow t_0(Y)$ is a smooth morphism between affine schemes. But, passing to $t_0$ we find a smooth covering family $\{t_0(U_i) \longrightarrow t_0(X)\}$, such that each morphism $t_0(U_i) \longrightarrow t_0(Y)$ is smooth, and we know (see e.g. [**EGAIV**]) that this implies that $t_0(X) \longrightarrow t_0(Y)$ is a smooth morphism of affine schemes.

Finally, property (4) of 1.3.2.11 is obvious. □

We have verified our assumptions 1.3.2.2 and 1.3.2.11 for the étale model topology and $\mathbf{P}$ the class of smooth morphisms. We thus have that

$$(sk - Mod, sk - Mod, sk - Alg, \text{ét}, \mathbf{P})$$

is a HAG context in the sense of Def. 1.3.2.13. We can therefore apply our general definitions to obtain a notion of $n$-geometric $D^-$-stacks in $D^-\mathrm{St}(k)$, as well as the notion of $n$-smooth morphisms. We then check that Artin's conditions of Def. 1.4.3.1 are satisfied.

PROPOSITION 2.2.3.2. *The étale model topology and the smooth morphisms satisfy Artin's conditions relative to the HA context* $(sk - Mod, sk - Mod, sk - Alg)$ *in the sense of Def. 1.4.3.1.*

PROOF. We will show that the class $\mathbf{E}$ of étale morphisms satisfies conditions (1) to (5) of Def. 1.4.3.1.

(1) is clear as $\mathbf{P}$ is exactly the class of all i-smooth morphisms.
(2) and (3) are clear by the choice of $\mathbf{E}$ and $\mathcal{A}$.

To prove (4), let $p : Y \longrightarrow X$ be a smooth morphism in $k - D^-Aff$, and let us consider the smooth morphism of affine schemes $t_0(p) : t_0(Y) \longrightarrow t_0(X)$. As $p$ is a smooth epimorphism of stacks, $t_0(p)$ is a smooth and surjective morphism of affine schemes. It is known (see e.g. [**EGAIV**]) that there exists a étale covering of affine schemes $X_0' \longrightarrow t_0(X)$ and a commutative diagram

$$\begin{array}{ccc} & & t_0(Y) \\ & \nearrow & \downarrow \\ X_0' & \longrightarrow & t_0(X). \end{array}$$

By Cor. 2.2.2.9, we know that there exists an étale covering $X' \longrightarrow X$ in $k - D^-Aff$, inducing $X_0' \longrightarrow t_0(X)$. Taking the homotopy pullback

$$Y \times^h_X X' \longrightarrow X'$$

we can replace $X$ by $X'$, and therefore assume that the morphism $t_0(p) : t_0(Y) \longrightarrow t_0(X)$ has a section. We are going to show that this section can be extended to a section of $p$, which will be enough to prove what we want.

We use the same trick as in the proof of Cor. 2.2.2.9, and consider once again the model category $k - D^-Aff/X_{\leq *}$. Using Thm. 2.2.2.6 (2) we see that the functor

$$\mathrm{Ho}(k - D^-Aff/X) \longrightarrow \mathrm{Ho}(k - D^-Aff/X_{\leq *})$$

induces an equivalences from the full subcategory $\mathrm{Ho}(k - D^-Aff/X)^{sm}$ of smooth morphisms $Z \to X$ to the full subcategory $\mathrm{Ho}(k - D^-Aff/X_{\leq *})^{cart,sm}$ consisting of objects $Z_*$ such that each $Z_k \longrightarrow X_{\leq k}$ is smooth, and each morphism

$$Z_{k-1} \longrightarrow Z_k \times^h_{Z_{\leq k}} Z_{\leq k-1}$$

is an isomorphism in $\mathrm{Ho}(k - D^-Aff)$. From this we deduce that the space of sections of $p$ can be described in the following way

$$Map_{k-D^-Aff/X}(X, Y) \simeq Holim_k Map_{k-D^-Aff/X_{\leq k}}(X_{\leq k}, Y_k),$$

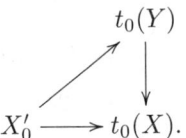

where $Y_k \simeq Y_{\leq k} \simeq Y \times^h_X X_{\leq k}$. But, as there always exists a surjection

$$\pi_0(Holim_k Map_{k-D^-Aff/X_{\leq k}}(X_{\leq k}, Y_k)) \longrightarrow Lim_k \pi_0(Map_{k-D^-Aff/X_{\leq k}}(X_{\leq k}, Y_k)),$$

we only need to check that $Lim_k \pi_0(Map_{k-D^-Aff/X_{\leq k}}(X_{\leq k}, Y_k)) \neq \emptyset$. For this, it is enough to prove that for any $k \geq 1$, the restriction map

$$\pi_0(Map_{k-D^-Aff/X_{\leq k}}(X_{\leq k}, Y_k)) \longrightarrow \pi_0(Map_{k-D^-Aff/X_{\leq k-1}}(X_{\leq k}, Y_{k-1}))$$

is surjective. In other words, we need to prove that a morphism in $\mathrm{Ho}(k\text{--}D^-Aff/X_{\leq k})$

$$\begin{array}{ccc} X_{\leq k-1} & \longrightarrow & Y_k \\ \downarrow & & \\ X_{\leq k} & & \end{array}$$

can be filled up to a commutative diagram in $\mathrm{Ho}(k - D^-Aff/X_{\leq k})$

$$\begin{array}{ccc} X_{\leq k-1} & \longrightarrow & Y_k \\ \downarrow & \nearrow & \\ X_{\leq k}. & & \end{array}$$

Using Prop. 1.4.2.6 and Lem. 2.2.1.1, we see that the obstruction for the existence of this lift lives in the group

$$[\mathbb{L}_{B_k/A_{\leq k}}, \pi_k(A)[k+1]]_{A_{\leq k}-Mod_s},$$

where $Y_k = Spec\, B_k$ and $X = Spec\, A$. But, as $Y_k \longrightarrow X_{\leq k}$ is a smooth morphism, the $A_{\leq k}$-module $\mathbb{L}_{B_k/A_{\leq k}}$ is projective, and therefore

$$[\mathbb{L}_{B_k/A_{\leq k}}, \pi_k(A)[k+1]]_{A_{\leq k}-Mod_s} \simeq 0,$$

by Lem. 2.2.2.2 (5).

It remains to prove that property (5) of Def. 1.4.3.1 is satisfied. Let $X \longrightarrow X_d[\Omega M]$ be as in the statement of Def. 1.4.3.1 (4), with $X = Spec\, A$ and some connected $A$-module $M \in A - Mod_s$, and let $U \longrightarrow X_d[\Omega M]$ be an étale morphism. We know by Lem. 1.4.3.8 (or rather its proof) that this morphism is of the form

$$X'_{d'}[\Omega M'] \longrightarrow X_d[\Omega M]$$

for some étale morphism $X' = Spec\, A' \longrightarrow X = Spec\, A$, and furthermore $X' \longrightarrow X$ is equivalent to the homotopy pullback

$$U \times_{X_d[\Omega M]} X \longrightarrow X.$$

Therefore, using Cor. 2.2.2.11, it is enough to prove that

$$t_0(X'_{d'}[\Omega M']) \longrightarrow t_0(X_d[\Omega M])$$

is a surjective morphism of affine schemes if and only if

$$t_0(X') \longrightarrow t_0(X)$$

is so. But, $M$ being connected, $t_0(X_d[\Omega M])_{red} \simeq t_0(X)$, and in the same way $t_0(X'_{d'}[\Omega M'])_{red} \simeq t_0(X')$, this gives the result, since being a surjective morphism is topologically invariant. □

COROLLARY 2.2.3.3. (1) *Any $n$-geometric $D^-$-stack has an obstruction theory. In particular, any $n$-geometric $D^-$-stack has a cotangent complex.*

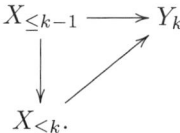

(2) *Any $n$-representable morphism of $D^-$-stacks $f : F \longrightarrow G$ has a relative obstruction theory. In particular, any $n$-representable morphism of $D^-$-stacks has a relative cotangent complex.*

PROOF. Follows from Prop. 2.2.3.2 and theorem 1.4.3.2. □

We finish by some properties of morphisms, as in Def. 1.3.6.2.

LEMMA 2.2.3.4. *Let $\mathbf{Q}$ be one the following class of morphisms in $k - D^-Aff$.*

(1) *Flat.*
(2) *Smooth.*
(3) *Etale.*
(4) *Finitely presented.*

*Then, morphisms in $\mathbf{Q}$ are compatible with the étale topology and the class $\mathbf{P}$ of smooth morphisms in the sense of Def. 1.3.6.1.*

PROOF. Using the explicit description of flat, smooth and étale morphisms given in Prop. 2.2.2.5 (1), (2), and (3) reduce to the analog well known facts for morphism between affine schemes. Finally, Cor. 2.2.2.15 implies that the class of finitely presented morphisms in $k - D^-Aff$ is compatible with the étale topology and the class **P**. □

Lemma 2.2.3.4 and definition 1.3.6.2 allows us to define the notions of flat, smooth, étale and locally finitely presented morphisms of $D^-$-stacks, which are all stable by equivalences, compositions and homotopy pullbacks. Recall that by definition, a flat (resp. smooth, resp. étale, resp. locally finitely presented) is always $n$-representable for some $n$. Using our general definition Def. 1.3.6.4 we also have notions of quasi-compact morphisms, finitely presented morphisms, and monomorphisms between $D^-$-stacks. We also make the following definition.

DEFINITION 2.2.3.5. (1) *A morphism of $D^-$-stacks is a* Zariski open immersion *if it is a locally finitely presented flat monomorphism.*

(2) *A morphism of $D^-$-stacks $F \longrightarrow G$ is a* closed immersion *if it is representable, and if for any $A \in sk-Alg$ and any morphism $X = \mathbb{R}\underline{Spec}\, A \longrightarrow G$ the induced morphism of representable $D^-$-stacks*

$$F \times^h_G X \simeq \mathbb{R}\underline{Spec}\, B \longrightarrow \mathbb{R}\underline{Spec}\, A$$

*induces an epimorphism of rings $\pi_0(A) \longrightarrow \pi_0(B)$.*

### 2.2.4. Truncations

We consider the natural inclusion functor

$$i : k - Aff \longrightarrow k - D^-Aff$$

right adjoint to the functor

$$\pi_0 : k - D^-Aff \longrightarrow k - Aff.$$

The pair $(\pi_0, i)$ is a Quillen adjunction (for the trivial model structure on $k - Aff$), and as usual we will omit to mention the inclusion functor $i$, and simply consider commutative $k$-algebras as constant simplicial objects. Furthermore, both functors preserve equivalences and thus induce a Quillen adjunction on the model category of pre-stacks (using notations of [**HAGI**, §4.8])

$$i_! : k-Aff^\wedge = SPr(k-Aff) \longrightarrow k-D^-Aff^\wedge \qquad k-Aff^\wedge \longleftarrow k-D^-Aff^\wedge : i^*.$$

The functor $i$ is furthermore continuous in the sense of [**HAGI**, §4.8], meaning that the right derived functor

$$\mathbb{R}i^* : \mathrm{Ho}(k - D^-Aff^\wedge) \longrightarrow \mathrm{Ho}(k - Aff^\wedge)$$

preserves the sub-categories of stacks. Indeed, by Lem. 1.3.2.3 (2) and adjunction this follows from the fact that $i : k - Aff \longrightarrow k - D^-Aff$ preserves co-products, equivalences and étale hypercovers. By the general properties of left Bousfield localizations we therefore get a Quillen adjunction on the model categories of stacks

$$i_! : k - Aff^{\sim,\text{ét}} \longrightarrow k - D^-Aff^{\sim,\text{ét}} \qquad k - Aff^{\sim,\text{ét}} \longleftarrow k - D^-Aff^{\sim,\text{ét}} : i^*.$$

From this we get a derived adjunction on the homotopy categories of stacks

$$\mathbb{L}i_! : \mathrm{St}(k) \longrightarrow D^-\mathrm{St}(k)$$

$$\mathrm{St}(k) \longleftarrow D^-\mathrm{St}(k) : \mathbb{R}i^*.$$

LEMMA 2.2.4.1. *The functor $\mathbb{L}i_!$ is fully faithful.*

PROOF. We need to show that for any $F \in \mathrm{St}(k)$ the adjunction morphism

$$F \longrightarrow \mathbb{R}i^* \circ \mathbb{L}i_!(F)$$

is an isomorphism. The functor $\mathbb{R}i^*$ commutes with homotopy colimits, as these are computed in the model category of simplicial presheaves and thus levelwise. Moreover, as $i_!$ is left Quillen the functor $\mathbb{L}i_!$ also commutes with homotopy colimits. Now, any stack $F \in \mathrm{St}(k)$ is a homotopy colimit of representable stacks (i.e. affine schemes), and therefore we can suppose that $F = Spec\, A$, for $A \in k - Alg$. But then $\mathbb{L}i_!(Spec\, A) \simeq \mathbb{R}\underline{Spec}\, A$. Furthermore, for any $B \in k - Alg$ there are natural isomorphisms in $\mathrm{Ho}(SSet)$

$$\mathbb{R}\underline{Spec}\, A(B) \simeq Map_{sk-Alg}(A, B) \simeq Hom_{k-Alg}(A, B) \simeq (Spec\, A)(B).$$

This shows that the adjunction morphism

$$Spec\, A \longrightarrow \mathbb{R}i^* \circ \mathbb{L}i_!(Spec\, A)$$

is an isomorphism. □

Another useful remark is the following

LEMMA 2.2.4.2. *The functor $i^* : k - D^-Aff^{\sim,\text{ét}} \longrightarrow k - Aff^{\sim,\text{ét}}$ is right and left Quillen. In particular it preserves equivalences.*

PROOF. The functor $i^*$ has a right adjoint

$$\pi_0^* : k - Aff^{\sim,\text{ét}} \longrightarrow k - D^-Aff^{\sim,\text{ét}}.$$

Using lemma 1.3.2.3 (2) we see that $\pi_0^*$ is a right Quillen functor. Therefore $i^*$ is left Quillen. □

DEFINITION 2.2.4.3. (1) *The* truncation functor *is*

$$t_0 := i^* : D^-\mathrm{St}(k) \longrightarrow \mathrm{St}(k).$$

(2) *The* extension functor *is the left adjoint to $t_0$*

$$i := \mathbb{L}i_! : \mathrm{St}(k) \longrightarrow D^-\mathrm{St}(k).$$

(3) *A $D^-$-stack $F$ is* truncated *if the adjunction morphism*

$$it_0(F) \longrightarrow F$$

*is an isomorphism in $D^-\mathrm{St}(k)$.*

By lemmas 2.2.4.1, 2.2.4.2 we know that the truncation functor $t_0$ commutes with homotopy limits and homotopy colimits. The extension functor $i$ itself commutes with homotopy colimits and is fully faithful. An important remark is that the extension $i$ does not commute with homotopy limits, as the inclusion functor $k-Alg \longrightarrow sk-Alg$ does not preserve homotopy push-outs.

Concretely, the truncation functor $t_0$ sends a functor

$$F : sk - Alg \longrightarrow SSet_{\mathbb{V}}$$

to

$$\begin{array}{rccc} t_0(F): & k - Alg & \longrightarrow & SSet_{\mathbb{V}} \\ & A & \mapsto & F(A). \end{array}$$

By adjunction we clearly have

$$t_0 \mathbb{R}\underline{Spec}\, A \simeq Spec\, \pi_0(A)$$

for any $A \in sk-Alg$, showing that the notation is compatible with the one we did use before for $t_0(Spec\, A) = Spec\, \pi_0(A)$ as objects in $k - D^-Aff$. The extension functor $i$ is characterized by

$$i(Spec\, A) \simeq \mathbb{R}\underline{Spec}\, A,$$

and the fact that it commutes with homotopy colimits.

PROPOSITION 2.2.4.4. (1) *The functor $t_0$ preserves epimorphisms of stacks.*

(2) *The functor $t_0$ sends $n$-geometric $D^-$-stacks to $n$-geometric stacks, and flat (resp. smooth, resp. étale) morphisms between $D^-$-stacks to flat (resp. smooth, resp. étale) morphisms between stacks.*

(3) *The functor $i$ preserves homotopy pullbacks of $n$-geometric stacks along a flat morphism, sends $n$-geometric stacks to $n$-geometric $D^-$-stacks, and flat (resp. smooth, resp. étale) morphisms between $n$-geometric stacks to flat (resp. smooth, resp. étale) morphisms between $n$--geometric $D^-$-stacks.*

(4) *Let $F \in \mathrm{St}(k)$ be an $n$-geometric stack, and $F' \longrightarrow i(F)$ be a flat morphism of $n$-geometric $D^-$-stacks. Then $F'$ is truncated (and therefore is the image by $i$ of an $n$-geometric stack by (2)).*

PROOF. (1) By adjunction, this follows from the fact that $i : k-Alg \longrightarrow sk-Alg$ reflects étale covering families (by Prop. 2.2.2.2).

(2) The proof is by induction on $n$. For $n = -1$, this simply follows from the formula

$$t_0 \mathbb{R}\underline{Spec}\, A \simeq Spec\, \pi_0(A),$$

and Prop. 2.2.2.5 and Thm. 2.2.2.6. Assume that the property is known for any $m < n$ and let us prove it for $n$. Let $F$ be an $n$-geometric $D^-$-stack, which by Prop. 1.3.4.2 can be written as $|X_*|$ for some $(n-1)$-smooth Segal groupoid $X_*$ in $k-D^-Aff^{\sim,\text{ét}}$. Using our property at rank $n-1$ and that $t_0$ commutes with homotopy limits shows that $t_0(X_*)$ is also a $(n-1)$-smooth Segal groupoid object in $k-Aff^{\sim,\text{ét}}$. Moreover, as $t_0$ commutes with homotopy colimits we have $t_0(F) \simeq |t_0(X_*)|$, which by Prop. 1.3.4.2 is an $n$-geometric stack. This shows that $t_0$ sends $n$-geometric $D^-$-stacks to $n$-geometric stacks and therefore preserves $n$-representable morphisms. Let $f : F \longrightarrow G$ be a flat (resp. smooth, resp. étale) morphism of $D^-$-stacks which is $n$-representable, and let us prove by induction on $n$ that $t_0(f)$ is flat (resp. smooth, resp. étale). We let $X \in \mathrm{St}(k)$ be an affine scheme, and $X \longrightarrow t_0(G)$ be a morphism of stacks. By adjunction between $i$ and $t_0$ we have

$$X \times^h_{t_0(G)} t_0(F) \simeq t_0(X \times^h_G F),$$

showing that we can assume that $G = X$ is an affine scheme and thus $F$ to be an $n$-geometric $D^-$. By definition of being flat (resp. smooth, resp. étale) there exists a smooth $n$-atlas $\{U_i\}$ of $F$ such that each composite morphism $U_i \longrightarrow X$ is $(n-1)$-representable and flat (resp. smooth, resp. étale). By induction and (1) the family $\{t_0(U_i)\}$ is a smooth $n$-atlas for $t_0(F)$, and by induction each composition $t_0(U_i) \longrightarrow t_0(X) = X$ is flat (resp. smooth, resp. étale). This implies that $t_0(F) \longrightarrow t_0(X)$ is flat (resp. smooth, resp. étale).

(3) The proof is by induction on $n$. For $n = -1$ this follows from the formula $i(Spec\, A) \simeq \mathbb{R}\underline{Spec}\, A$, the description of flat, smooth and étale morphisms (Prop. 2.2.2.5 and Thm. 2.2.2.6), and the fact that for any flat morphism of commutative $k$-algebras $A \longrightarrow B$ and any commutative $A$-algebra $C$ there is a natural isomorphism in $\mathrm{Ho}(A - Alg_s)$

$$B \otimes_A C \simeq B \otimes^{\mathbb{L}}_A C.$$

Let us now assume the property is proved for $m < n$ and let us prove it for $n$. Let $F$ be an $n$-geometric stack, and by Prop. 1.3.4.2 let us write it as $F \simeq |X_*|$ for some $(n-1)$-smooth Segal groupoid object $X_*$ in $k - Aff^{\sim,\text{ét}}$. By induction, $i(X_*)$ is again a $(n-1)$-smooth Segal groupoid objects in $k - D^-Aff^{\sim,\text{ét}}$, and as $i$ commutes with homotopy colimits we have $i(F) \simeq |i(X_*)|$. Another application of Prop. 1.3.4.2 shows that $F$ is an $n$-geometric $D^-$-stacks. We thus have seen that $i$ sends $n$-geometric stacks to $n$-geometric $D^-$-stacks.

Now, let $F \longrightarrow G$ be a flat morphism between $n$-geometric stacks, and $H \longrightarrow G$ any morphism between $n$-geometric stacks. We want to show that the natural morphism

$$i(F \times^h_G H) \longrightarrow i(F) \times^h_{i(G)} i(H)$$

is an isomorphism in $\mathrm{St}(k)$. For this, we write $G \simeq |X_*|$ for some $(n-1)$-smooth Segal groupoid object in $k - Aff^{\sim,\text{ét}}$, and we consider the Segal groupoid objects

$$F_* := F \times^h_G X_* \qquad H_* := H \times^h_G X_*,$$

where $X_* \longrightarrow |X_*| = G$ is the natural augmentation in $\mathrm{St}(k)$. The Segal groupoid objects $F_*$ and $H_*$ are again $(n-1)$-smooth Segal groupoid objects in $k - Aff^{\sim,\text{ét}}$ as $G$ is an $n$-geometric stack. The natural morphisms of Segal groupoid objects

$$F_* \longrightarrow X_* \longleftarrow H_*$$

gives rise to another $(n-1)$-smooth Segal groupoid object $F_* \times^h_{X_*} H_*$. Clearly, we have

$$F \times^h_G H \simeq |F_* \times^h_{X_*} H_*|.$$

As $i$ commutes with homotopy colimits, and by induction on $n$ we have

$$i(F \times^h_G H) \simeq |i(F_*) \times^h_{i(X_*)} i(H_*)| \simeq |i(F_*)| \times^h_{|i(X_*)|} |i(H_*)| \simeq i(F) \times^h_{i(G)} i(H).$$

It remains to prove that if $f : F \longrightarrow G$ is a flat (resp. smooth, resp. étale) morphism between $n$-geometric stacks then $i(f) : i(F) \longrightarrow i(G)$ is a flat (resp. smooth, resp. étale) morphism between $n$-geometric $D^-$-stacks. For this, let $\{U_i\}$ be a smooth $n$-atlas for $G$. We have seen before that $\{i(U_i)\}$ is a smooth $n$-atlas for $i(G)$. As we have seen that $i$ commutes with homotopy pullbacks along flat morphisms, and because of the local properties of flat (resp. smooth, resp. étale) morphisms (see Prop. 1.3.6.3 and Lem. 2.2.3.4), we can suppose that $G$ is one of the $U_i$'s and thus is an affine scheme. Now, let $\{V_i\}$ be a smooth $n$-atlas for $F$. The family $\{i(V_i)\}$ is a smooth $n$-atlas for $F$, and furthermore each morphism $i(V_i) \longrightarrow i(G)$ is the image by $i$ of a flat (resp. smooth, resp. étale) morphism between affine schemes and therefore is a flat (resp. smooth, resp. étale morphism) of $D^-$-stacks. By definition this implies that $i(F) \longrightarrow i(G)$ is flat (resp. smooth, resp. étale).

(4) The proof is by induction on $n$. For $n = -1$ this is simply the description of flat morphisms of Prop. 2.2.2.5. Let us assume the property is proved for $m < n$ and let us prove it for $n$. Let $F' \longrightarrow i(F)$ be a flat morphism, with $F$ an $n$-geometric stack and $F'$ an $n$-geometric $D^-$-stack. Let $\{U_i\}$ be a smooth $n$-atlas for $F$. Then, $\{i(U_i)\}$ is a smooth $n$-atlas for $i(F)$. We consider the commutative diagram of $D^-$-stacks

$$\begin{array}{ccc} it_0(F') & \xrightarrow{f} & F' \\ & \searrow & \downarrow \\ & & i(F). \end{array}$$

We need to prove that $f$ is an isomorphism in $D^-\mathrm{St}(k)$. As this is a local property on $i(F)$, we can take the homotopy pullback over the atlas $\{i(U_i)\}$, and thus suppose that $F$ is an affine scheme. Let now $\{V_i\}$ be a smooth $n$-atlas for $F'$, and we consider the homotopy nerve of the morphism

$$X_0 := \coprod_i V_i \longrightarrow F'.$$

This is a $(n-1)$-smooth Segal groupoid objects in $D^-Aff^{\sim,\text{ét}}$, which is such that each morphism $X_i \longrightarrow i(G)$ is flat. Therefore, by induction on $n$, the natural morphism of Segal groupoid objects

$$it_0(X_*) \longrightarrow X_*$$

is an equivalence. Therefore, as $i$ and $t_0$ commutes with homotopy colimits we find that the adjunction morphism

$$it_0(F') \simeq |it_0(X_*)| \longrightarrow |X_*| \simeq F'$$

is an isomorphism in $D^-\mathrm{St}(k)$. □

An important corollary of Prop. 2.2.2.5 is the following fact.

COROLLARY 2.2.4.5. *For any Artin $n$-stack, the $D^-$-stack $i(F)$ has an obstruction theory.*

PROOF. Follows from 2.2.2.5 (3) and Cor. 2.2.3.3. □

One also deduces from Prop. 2.2.2.5 and Lem. 2.1.1.2 the following corollary.

COROLLARY 2.2.4.6. *Let $F$ be an $n$-geometric $D^-$-stack. Then, for any $A \in sk-Alg$, such that $\pi_i(A) = 0$ for $i > k$, the simplicial set $\mathbb{R}F(A)$ is $(n+k+1)$-truncated.*

PROOF. This is by induction on $k$. For $k = 0$ this is Lem. 2.1.1.2 and the fact that $t_0$ preserves $n$-geometric stacks. To pass from $k$ to $k+1$, we consider for any $A \in sk-Alg$ with $\pi_i(A) = 0$ for $i > k+1$, the natural morphisms

$$\mathbb{R}F(A) \longrightarrow \mathbb{R}F(A_{\leq k}),$$

whose homotopy fibers can be described using Prop. 1.4.2.5 and Lem. 2.2.1.1. We find that this homotopy fiber is either empty, or equivalent to

$$Map_{A_{\leq k}-Mod}(\mathbb{L}_{F,x}, \pi_{k+1}(A)[k+1]),$$

which is $(k+1)$-truncated. By induction, $\mathbb{R}F(A_{\leq k})$ is $(k+1+n)$-truncated and the homotopy fibers of

$$\mathbb{R}F(A) \longrightarrow \mathbb{R}F(A_{\leq k}),$$

are $(k+1)$-truncated, and therefore $\mathbb{R}F(A)$ is $(k+n+2)$-truncated. □

Another important property of the truncation functor is the following local description of the truncation $t_0(F)$ sitting inside the $D^-$-stack $F$ itself.

PROPOSITION 2.2.4.7. *Let $F$ be an $n$-geometric $D^-$-stack. The adjunction morphism $it_0(F) \longrightarrow F$ is a representable morphism. Moreover, for any $A \in sk-Alg$, and any flat morphism $\mathbb{R}\underline{Spec}\,A \longrightarrow F$, the square*

$$\begin{array}{ccc} it_0(F) & \longrightarrow & F \\ \uparrow & & \uparrow \\ \mathbb{R}\underline{Spec}\,\pi_0(A) & \longrightarrow & \mathbb{R}\underline{Spec}\,A \end{array}$$

*is homotopy cartesian. In particular, the morphism $it_0(F) \longrightarrow F$ is a closed immersion in the sense of Def. 2.2.3.5.*

PROOF. By the local character of representable morphisms it is enough to prove that for any flat morphism $X = \mathbb{R}\underline{Spec}\,A \longrightarrow F$, the square

$$\begin{array}{ccc} it_0(F) & \longrightarrow & F \\ \uparrow & & \uparrow \\ it_0(X) = \mathbb{R}\underline{Spec}\,\pi_0(A) & \longrightarrow & X \end{array}$$

is homotopy cartesian. The morphism

$$it_0(F) \times^h_F X \longrightarrow it_0(F)$$

is flat, and by Prop. 2.2.2.5 (4) this implies that the $D^-$-stack $it_0(F) \times^h_F X$ is truncated. In other words the natural morphism

$$it_0(it_0(F) \times^h_F X) \simeq it_0(F) \times^h_{it_0(F)} it_0(X) \simeq it_0(X) \longrightarrow it_0(F) \times^h_F X$$

is an isomorphism. □

Using our embedding

$$i : \mathrm{St}(k) \longrightarrow D^-\mathrm{St}(k)$$

we will see stacks in $\mathrm{St}(k)$, and in particular Artin $n$-stacks, as $D^-$-stacks. However, as the functor $i$ does not commute with homotopy pullbacks we will still mention it in order to avoid confusions.

### 2.2.5. Infinitesimal criteria for smooth and étale morphisms

Recall that $sk-Mod_1$ denotes the full subcategory of $sk-Mod$ of connected simplicial $k$-modules. It consists of all $M \in sk-Mod$ for which the adjunction $S(\Omega M) \longrightarrow M$ is an isomorphism in $\mathrm{Ho}(sk-Mod)$, or equivalently for which $\pi_0(M) = 0$.

PROPOSITION 2.2.5.1. *Let $f : F \longrightarrow G$ be an $n$-representable morphism between $D^-$-stacks. The morphism $f$ is smooth if and only if it satisfies the following two conditions*

(1) *The morphism $t_0(f) : t_0(F) \longrightarrow t_0(G)$ is a locally finitely presented morphism in $k-Aff^{\sim,\text{ét}}$.*

(2) *For any $A \in sk-Alg$, any connected $M \in A-Mod_s$, and any derivation $d \in \pi_0(\mathbb{D}er(A,M))$, the natural projection $A \oplus_d \Omega M \longrightarrow A$ induces a surjective morphism*

$$\pi_0(\mathbb{R}F(A \oplus_d \Omega M)) \longrightarrow \pi_0\left(\mathbb{R}G(A \oplus_d \Omega M) \times^h_{\mathbb{R}F(A)} \mathbb{R}G(A)\right).$$

PROOF. First of all we can suppose that $F$ and $G$ are fibrant objects in $k - D^-Aff^{\sim,\text{ét}}$.

Suppose first that the morphism $f$ is smooth and let us prove that is satisfies the two conditions of the proposition. We know by 2.2.4.4 (3) that $t_0(f)$ is then a smooth morphism in $k - Aff^{\sim,\text{ét}}$, so condition (1) is satisfied. The proof that (2) is also satisfied goes by induction on $n$. Let us start with the case $n = -1$, and in other words when $f$ is a smooth and representable morphism.

We fix a point $x$ in $\pi_0\left(G(A \oplus_d \Omega M) \times^h_{F(A)} G(A)\right)$, and we need to show that the homotopy fiber taken at $x$ of the morphism

$$F(A \oplus_d \Omega M) \longrightarrow G(A \oplus_d \Omega M) \times^h_{F(A)} G(A)$$

is non empty. The point $x$ corresponds via Yoneda to a commutative diagram in $\mathrm{Ho}(k - D^-Aff^{\sim,\text{ét}}/G)$

$$\begin{array}{ccc} X & \longrightarrow & F \\ \downarrow & & \downarrow \\ X_d[\Omega M] & \longrightarrow & G, \end{array}$$

where $X := \mathbb{R}\underline{Spec}\,A$, and $X_d[\Omega M] := \mathbb{R}\underline{Spec}\,(A \oplus_d \Omega M)$. Making a homotopy base change

$$\begin{array}{ccc} X & \longrightarrow & F \times^h_G X_d[\Omega M] \\ \downarrow & & \downarrow \\ X_d[\Omega M] & \longrightarrow & X_d[\Omega M], \end{array}$$

we see that we can replace $G$ be $X_d[\Omega M]$ and $f$ by the projection $F \times^h_G X_d[\Omega M] \longrightarrow X_d[\Omega M]$. In particular, we can assume that $G$ is a representable $D^-$-stack. The morphism $f$ can then be written as

$$f : F \simeq \mathbb{R}\underline{Spec}\,C \longrightarrow \mathbb{R}\underline{Spec}\,B \simeq G,$$

and corresponds to a morphism of commutative simplicial $k$-algebras $B \longrightarrow C$. Then, using Prop. 1.4.2.6 and Cor. 2.2.3.3 we see that the obstruction for the point $x$ to lifts to a point in $\pi_0(F(A \oplus_d \Omega M))$ lives in the abelian group $[\mathbb{L}_{C/B} \otimes^{\mathbb{L}}_C A, M]$. But, as $B \longrightarrow C$ is assumed to be smooth, the $A$-module $\mathbb{L}_{C/B} \otimes^{\mathbb{L}}_C A$ is a retract of a free $A$-module. This implies that $[\mathbb{L}_{C/B} \otimes^{\mathbb{L}}_C A, M]$ is a retract of a product of $\pi_0(M)$, and therefore is 0 by hypothesis on $M$. This implies that condition (2) of the proposition is satisfied when $n = -1$.

Let us now assume that condition (2) is satisfied for all smooth $m$-representable morphisms for $m < n$, and let us prove it for a smooth $n$-representable morphism $f : F \longrightarrow G$. Using the same trick as above, se see that we can always assume that $G$ is a representable $D^-$-stack, and therefore that $F$ is an $n$-geometric $D^-$-stack. Then, let us chose a point $x$ in $\pi_0\left(G(A \oplus_d \Omega M) \times^h_{F(A)} G(A)\right)$, and we need to show that $x$ lifts to a point in $\pi_0(F(A \oplus_d \Omega M))$. For this, we use Cor. 2.2.3.3 for $F$, and consider its cotangent complex $\mathbb{L}_{F,y} \in \mathrm{Ho}(Sp(A - Mod_s))$, where $y \in F(A)$ is the image of $x$. There exists a natural functoriality morphism

$$\mathbb{L}_{G,f(y)} \longrightarrow \mathbb{L}_{F,y}$$

whose homotopy cofiber is $\mathbb{L}_{F/G,y} \in \mathrm{Ho}(Sp(A - Mod_s))$. Then, Prop. 1.4.2.6 tell us that the obstruction for the existence of this lift lives in $[\mathbb{L}_{F/G,y}, M]$. It is therefore enough to show that $[\mathbb{L}_{F/G,y}, M] \simeq 0$ for any $M \in A - Mod_s$ such that $\pi_0(M) = 0$.

LEMMA 2.2.5.2. *Let $F \longrightarrow G$ be a smooth morphism between $n$-geometric $D^-$-stacks with $G$ a representable stack. Let $A \in sk - Alg$ and $y : Y = \mathbb{R}\underline{Spec}\, A \longrightarrow F$ be a point. Then the object*

$$\mathbb{L}_{F/G,y} \in \mathrm{Ho}(Sp(A - Mod_s))$$

*is perfect, and its dual $\mathbb{L}^{\vee}_{F/G,y} \in \mathrm{Ho}(Sp(A - Mod_s))$ is 0-connective (i.e. belongs to the image of $\mathrm{Ho}(A - Mod_s) \hookrightarrow \mathrm{Ho}(Sp(A - Mod_s))$.*

PROOF. Recall first that an object in $\mathrm{Ho}(Sp(A - Mod_s))$ is perfect if and only if it is finitely presented, and if and only if it is a retract of a finite cell stable $A$-module (see Cor. 1.2.3.8, and also [**EKMM**, III.2] or [**Kr-Ma**, Thm. III.5.7]).

The proof is then by induction on $n$. When $F$ is representable, this is by definition of smooth morphisms, as then $\mathbb{L}_{F/G,y}$ is a projective $A$-module of finite presentation, and so is its dual. Let us suppose the lemma proved for all $m < n$, and lets prove it for $n$. First of all, the conditions on $\mathbb{L}_{F/G,y}$ we need to prove are local for the étale topology on $A$, because of Cor. 1.3.7.8. Therefore, one can assume that the point $y$ lifts to a point of an $n$-atlas for $F$. One can thus suppose that there exists a representable $D^-$-stack $U$, a smooth morphism $U \longrightarrow F$, such that $y \in \pi_0(\mathbb{R}F(A))$ is the image of a point $z \in \mathbb{R}U(A)$. There exists an fibration sequence of stable $A$-modules

$$\mathbb{L}_{F/G,y} \longrightarrow \mathbb{L}_{U/G,z} \longrightarrow \mathbb{L}_{U/F,z}.$$

As $U \longrightarrow G$ is smooth, $\mathbb{L}_{U/G,z}$ is a projective $A$-module. Furthermore, the stable $A$-module $\mathbb{L}_{U/F,y}$ can be identified with $\mathbb{L}_{U\times^h_F Y/Y,s}$, where $s$ is the natural section $Y \longrightarrow U \times^h_F Y$ induced by the point $z : Y \longrightarrow U$. The morphism $U \times^h_F Y \longrightarrow Y$ being a smooth and $(n-1)$-representable morphism, induction tells us that the stable $A$-module $\mathbb{L}_{U/F,y}$ satisfies the conditions of the lemma. Therefore, $\mathbb{L}_{F/G,y}$ is the homotopy fiber of a morphism between stable $A$-module satisfying the conditions of the lemma, and is easily seen to satisfies itself these conditions. □

By the above lemma, we have

$$[\mathbb{L}_{F/G,y}, M] \simeq \pi_0(\mathbb{L}^{\vee}_{F/G,y} \otimes^{\mathbb{L}}_A M) \simeq \pi_0(\mathbb{L}^{\vee}_{F/G,y}) \otimes_{\pi_0(A)} \pi_0(M) \simeq 0$$

for any $A$-module $M$ such that $\pi_0(M) = 0$. This finishes the proof of the fact that $f$ satisfies the conditions of Prop. 2.2.5.1 when it is smooth.

Conversely, let us assume that $f : F \longrightarrow G$ is a morphism satisfying the lifting property of 2.2.5.1, and let us show that $f$ is smooth. Clearly, one can suppose that $G$ is a representable stack, and thus that $F$ is an $n$-geometric $D^-$-stack. We need to show that for any representable $D^-$-stack $U$ and any smooth morphism $U \longrightarrow F$ the composite morphism $U \dashrightarrow G$ is smooth. By what we have seen in the first part of the proof, we known that $U \longrightarrow F$ also satisfies the lifting properties, and thus so does the composition $U \longrightarrow G$. We are therefore reduced to the case where $f$ is a morphism between representable $D^-$-stacks, and thus corresponds to a morphism of commutative simplicial $k$-algebras $A \longrightarrow B$. By hypothesis on $f$, $\pi_0(A) \longrightarrow \pi_0(B)$ is a finitely presented morphism of commutative rings. Furthermore, Prop. 1.4.2.6 and

Cor. 2.2.3.3 show that for any $B$-module $M$ with $\pi_0(M) = 0$, we have $[\mathbb{L}_{B/A}, M] = 0$. Let $B^{(I)} \longrightarrow \mathbb{L}_{B/A}$ be a morphism of $B$-modules, with $B^{(I)}$ free over some set $I$, and such that the induced morphism $\pi_0(B)^{(I)} \longrightarrow \pi_0(\mathbb{L}_{B/A})$ is surjective. Let $K$ be the homotopy fiber of the morphism $B^{(I)} \longrightarrow \mathbb{L}_{B/A}$, that, according to our choice, induces a homotopy fiber sequence of $A$-modules

$$B^{(I)} \longrightarrow \mathbb{L}_{B/A} \longrightarrow K[1].$$

The short exact sequence

$$[\mathbb{L}_{B/A}, B^{(I)}] \longrightarrow [\mathbb{L}_{B/A}, \mathbb{L}_{B/A}] \longrightarrow [\mathbb{L}_{B/A}, K[1]] = 0,$$

shows that $\mathbb{L}_{B/A}$ is a retract of $B^{(I)}$, and thus is a projective $B$-module. Furthermore, the homotopy cofiber sequence

$$\mathbb{L}_A \otimes^{\mathbb{L}}_A B \longrightarrow \mathbb{L}_B \longrightarrow \mathbb{L}_{B/A},$$

induces a short exact sequence

$$[\mathbb{L}_B, \mathbb{L}_A \otimes^{\mathbb{L}}_A B] \longrightarrow [\mathbb{L}_A \otimes^{\mathbb{L}}_A B, \mathbb{L}_A \otimes^{\mathbb{L}}_A B] \longrightarrow [\mathbb{L}_{B/A}, \mathbb{L}_A \otimes^{\mathbb{L}}_A B[1]] = 0,$$

shows that the morphism $\mathbb{L}_A \otimes^{\mathbb{L}}_A B \longrightarrow \mathbb{L}_B$ has a retraction. We conclude that $f$ is a formally smooth morphism such that $\pi_0(A) \longrightarrow \pi_0(B)$ is finitely presented, and by Cor. 2.2.2.11 that $f$ is a smooth morphism. □

From the proof of Prop. 2.2.5.1 we extract the following corollary.

COROLLARY 2.2.5.3. *Let $F \longrightarrow G$ be an $n$-representable morphism of $D^-$-stacks such that the morphism $t_0(F) \longrightarrow t_0(G)$ is a locally finitely presented morphism of stacks. The following three conditions are equivalent.*

1. *The morphism $f$ is smooth.*
2. *For any $A \in sk - Alg$ and any morphism of stacks $x : X = \mathbb{R}\underline{Spec}\, A \longrightarrow F$, the object*

$$\mathbb{L}_{F/G,x} \in \mathrm{Ho}(Sp(A - Mod_s))$$

*is perfect, and its dual $\mathbb{L}^{\vee}_{F/G,y} \in \mathrm{Ho}(Sp(A - Mod_s))$ is 0-connective (i.e. belongs to the image of $\mathrm{Ho}(A - Mod_s) \hookrightarrow \mathrm{Ho}(Sp(A - Mod_s))$).*

3. *For any $A \in sk - Alg$, any morphism of stacks $x : X = \mathbb{R}\underline{Spec}\, A \longrightarrow F$, and any $A$-module $M$ in $sk - Mod_1$, we have*

$$[\mathbb{L}_{F/G,x}, M] = 0.$$

PROOF. That (1) implies (2) follows from lemma 2.2.5.2. Conversely, if $\mathbb{L}_{F/G,x}$ satisfies the conditions of the corollary, then for any $A$-module $M$ such that $\pi_0(M)$ one has $[\mathbb{L}_{F/G,x}, M] = 0$. Therefore, Prop. 1.4.2.6 shows that the lifting property of Prop. 2.2.5.1 holds, and thus that (2) implies (1). Furthermore, clearly (2) implies (3), and conversely (3) together with Prop. 1.4.2.6 implies the lifting property of Prop. 2.2.5.1. □

PROPOSITION 2.2.5.4. *Let $f : F \longrightarrow G$ be an $n$-representable morphism between $D^-$-stacks. The morphism $f$ is étale if and only if it satisfies the following two conditions*

1. *The morphism $t_0(f) : t_0(F) \longrightarrow t_0(G)$ is locally finitely presented as a morphism in $k - Aff^{\sim,\tau}$.*

(2) *For any $A \in sk - Alg$, any $M \in A - Mod_s$ whose underlying $k$-module is in $sk - Mod_1$, and any derivation $d \in \pi_0(\mathbb{D}er(A, M))$, the natural projection $A \oplus_d \Omega M \longrightarrow A$ induces an isomorphism in* $\mathrm{Ho}(SSet)$

$$\mathbb{R}F(A \oplus_d \Omega M) \longrightarrow \mathbb{R}G(A \oplus_d \Omega M) \times^h_{\mathbb{R}F(A)} \mathbb{R}G(A).$$

PROOF. First of all one can suppose that $F$ and $G$ are fibrant objects in $k - D^- Aff^{\sim, \text{ét}}$.

Suppose first that the morphism $f$ is étale and let us prove that is satisfies the two conditions of the proposition. We know by 2.2.4.4 (3) that $t_0(f)$ is then a étale morphism in $k - Aff^{\sim, \text{ét}}$, so condition (1) is satisfied. The proof that (2) is also satisfied goes by induction on $n$. Let us start with the case $n = -1$, and in other words when $f$ is an étale and representable morphism. In this case the result follows from Cor. 1.2.8.4 (2). We now assume the result for all $m < n$ and prove it for $n$. Let $A \in sk - Alg$, $M$ an $A$-module with $\pi_0(M) = 0$, and $d \in \pi_0(\mathbb{D}er(A, M))$ be a derivation. Let $x$ be a point in $\pi_0(\mathbb{R}G(A \oplus_d \Omega M) \times^h_{\mathbb{R}F(A)} \mathbb{R}G(A))$, with image $y$ in $\pi_0(\mathbb{R}F(A))$. By Prop. 1.4.2.6 the homotopy fiber of the morphism

$$\mathbb{R}F(A \oplus_d \Omega M) \longrightarrow \mathbb{R}G(A \oplus_d \Omega M) \times^h_{\mathbb{R}F(A)} \mathbb{R}G(A)$$

at the point $x$ is non empty if and only if a certain obstruction in $[\mathbb{L}_{F/G,y}, M]$ vanishes. Furthermore, if nonempty this homotopy fiber is then equivalent to

$$Map_{A-Mod}(\mathbb{L}_{F/G,y}, \Omega(M)).$$

It is then enough to prove that $\mathbb{L}_{F/G,y} = 0$, and this is contained in the following lemma.

LEMMA 2.2.5.5. *Let $F \longrightarrow G$ be a étale morphism between $n$-geometric $D^-$-stacks. Let $A \in sk - Alg$ and $y : Y = \mathbb{R}\underline{Spec}\, A \longrightarrow F$ be a point. Then $\mathbb{L}_{F/G,y} \simeq 0$.*

PROOF. We easily reduce to the case when $G$ is a representable $D^-$-stack. The proof is then by induction on $n$. When $F$ is representable, this is by definition of étale morphisms. Let us suppose the lemma proved for all $m < n$, and lets prove it for $n$. First of all, the vanishing of $\mathbb{L}_{F/G,y}$ is clearly a local condition for the étale topology on $A$. Therefore, we can assume that the point $y$ lifts to a point of an $n$-atlas for $F$. We can thus suppose that there exists a representable $D^-$-stack $U$, a smooth morphism $U \longrightarrow F$, such that $y \in \pi_0(\mathbb{R}F(A))$ is the image of a point $z \in \mathbb{R}U(A)$. There exists an fibration sequence of stable $A$-modules

$$\mathbb{L}_{F/G,y} \longrightarrow \mathbb{L}_{U/G,z} \longrightarrow \mathbb{L}_{U/F,z}.$$

As $U \longrightarrow G$ is étale, $\mathbb{L}_{U/G,z} \simeq 0$. Furthermore, the stable $A$-module $\mathbb{L}_{U/F,y}$ can be identified with $\mathbb{L}_{U \times^h_F Y/Y,s}$, where $s$ is the natural section $Y \longrightarrow U \times^h_F Y$ induced by the point $z : Y \longrightarrow U$. The morphism $U \times^h_F Y \longrightarrow Y$ being an étale and $(n-1)$-representable morphism, induction tells us that the stable $A$-module $\mathbb{L}_{U/F,y}$ vanishes. We conclude that $\mathbb{L}_{F/G,y} \simeq 0$. □

The lemma finishes the proof that if $f$ is étale then it satisfies the two conditions of the proposition.

Conversely, let us assume that $f$ satisfies the properties (1) and (2) of the proposition. To prove that $f$ is étale we can suppose that $G$ is a representable $D^-$-stack. We then need to show that for any representable $D^-$-stack $U$ and any smooth morphism $u : U \longrightarrow F$, the induced morphism $v : U \longrightarrow G$ is étale. But Prop. 1.4.2.6 and our assumption (2) easily implies $\mathbb{L}_{F/G,v} \simeq 0$. Furthermore, the obstruction for the homotopy cofiber sequence

$$\mathbb{L}_{U/G,v} \longrightarrow \mathbb{L}_{F/G,u} \longrightarrow \mathbb{L}_{U/F,u},$$

to splits lives in $[\mathbb{L}_{U/F,u}, S(\mathbb{L}_{U/G,v})]$, which is zero by Cor. 2.2.5.3. Therefore, $\mathbb{L}_{U/G,v}$ is a retract of $\mathbb{L}_{F/G,u}$ and thus vanishes. This implies that $U \longrightarrow G$ is a formally étale morphism of representable $D^-$-stacks. Finally, our assumption (1) and Cor. 2.2.2.11 implies that $U \longrightarrow G$ is an étale morphism as required. □

COROLLARY 2.2.5.6. *Let $F \longrightarrow G$ be an $n$-representable morphism of $D^-$-stacks such that the morphism $t_0(F) \longrightarrow t_0(G)$ is a locally finitely presented morphism of stacks. The following two conditions are equivalent.*

(1) *The morphism $f$ is étale.*
(2) *For any $A \in sk - Alg$ and any morphism of stacks $x : X = \mathbb{R}\underline{Spec}\,A \longrightarrow F$, one has $\mathbb{L}_{F/G,x} \simeq 0$.*

PROOF. This follows from Prop. 2.2.5.4 and Prop. 1.4.2.6. □

### 2.2.6. Some examples of geometric $D^-$-stacks

We present here some basic examples of geometric $D^-$-stacks. Of course we do not claim to be exhaustive, and many other interesting examples will not be discussed here and will appear in future works (see e.g. [**Go, To-Va1**]).

**2.2.6.1. Local systems.** Recall from Def. 1.3.7.5 the existence of the $D^-$-stack $\mathbf{Vect}_n$, of rank $n$ vector bundles. Recall by 2.2.2.2 that for $A \in sk - Alg$, an $A$-module $M \in A - Alg_s$ is a rank $n$ vector bundle, if and only if $M$ is a strong $A$-module and $\pi_0(M)$ is a projective $\pi_0(A)$-module of rank $n$. Recall also from Lem. 2.2.2.2 that vector bundles are precisely the locally perfect modules. The conditions of 1.3.7.12 are all satisfied in the present context and therefore we know that $\mathbf{Vect}_n$ is a smooth 1-geometric $D^-$-stack. As a consequence of Prop. 2.2.4.4 (4) we deduce that the $D^-$-stack $\mathbf{Vect}_n$ is truncated in the sense of 2.2.4.3. We also have a stack of rank $n$ vector bundles $\mathbf{Vect}_n$ in $\mathrm{St}(k)$. Using the same notations for these two different objects is justified by the following lemma.

LEMMA 2.2.6.1. *There exists a natural isomorphism in $D^-\mathrm{St}(k)$*

$$i(\mathbf{Vect}_n) \simeq \mathbf{Vect}_n.$$

PROOF. As we know that the $D^-$-stack $\mathbf{Vect}_n$ is truncated, it is equivalent to show that there exists a natural isomorphism

$$\mathbf{Vect}_n \simeq t_0(\mathbf{Vect}_n)$$

in $\mathrm{Ho}(k - Aff^{\sim,\tau})$.

We start by defining a morphism of stacks

$$\mathbf{Vect}_n \longrightarrow t_0(\mathbf{Vect}_n).$$

For this, we construct for a commutative $k$-algebra $A$, a natural functor

$$\phi_A : A - QCoh^c_W \longrightarrow i(A) - QCoh^c_W,$$

where $i(A) \in sk - Alg$ is the constant simplicial commutative $k$-algebra associated to $A$, and where $A - QCoh^c_W$ and $i(A) - QCoh^c_W$ are defined in §1.3.7. Recall that $A - QCoh^c_W$ is the category whose objects are the data of a $B$-module $M_B$ for any morphism $A \longrightarrow B$ in $k - Alg$, together with isomorphisms $M_B \otimes_B B' \simeq M_{B'}$ for any

$A \longrightarrow B \longrightarrow B'$ in $k - Alg$, satisfying the usual cocycle conditions. The morphisms $M \to M'$ in $A - QCoh^c_W$ are simply the families of isomorphisms $M_B \simeq M'_B$ of $B$-modules which commute with the transitions isomorphisms. In the same way, the objects in $i(A) - QCoh^c_W$ are the data a cofibrant $B$-module $M_B \in B - Mod_s$ for any morphism $i(A) \longrightarrow B$ in $sk - Alg$, together with equivalences $M_B \otimes_B B' \longrightarrow M_{B'}$ for any $i(A) \longrightarrow B \longrightarrow B'$ in $sk - Alg$, satisfying the usual cocycle conditions. The morphisms $M \to M'$ in $i(A) - QCoh^c_W$ are simply the equivalences $M_B \longrightarrow M'_B$ which commutes with the transition equivalences. The functor

$$\phi_A : A - QCoh^c_W \longrightarrow i(A) - QCoh^c_W,$$

sends an object $M$ to the object $\phi_A(M)$, where $\phi_A(M)_B$ is the simplicial $B$-module defined by

$$(\phi_A(M)_B)_n := M_{B_n}$$

(note that $B$ can be seen as a simplicial commutative $A$-algebra). The functor $\phi_A$ is clearly functorial in $A$ and thus defines a morphism of simplicial presheaves

$$\mathbf{QCoh} \longrightarrow t_0(\mathbf{QCoh}).$$

We check easily that the sub-stack $\mathbf{Vect}_n$ of $\mathbf{QCoh}$ is sent to the sub-stack $t_0(\mathbf{Vect}_n)$ of $t_0(\mathbf{QCoh})$, and therefore we get a morphism of stacks

$$\mathbf{Vect}_n \longrightarrow t_0(\mathbf{Vect}_n).$$

To see that this morphism is an isomorphism of stacks we construct a morphism in the other direction by sending a simplicial $i(A)$-module $M$ to the $\pi_0(A)$-module $\pi_0(M)$. By 2.2.2.2 we easily see that this defines an inverse of the above morphism. □

For a simplicial set $K \in SSet_{\mathbb{U}}$, and an object $F \in k - D^-Aff^{\sim,\text{ét}}$, we can use the simplicial structure of the category $k - D^-Aff^{\sim,\text{ét}}$ in order to define the exponential $F^K \in k - D^-Aff^{\sim,\text{ét}}$. The model category $k - D^-Aff^{\sim,\text{ét}}$ being a simplicial model category the functor

$$(-)^K : k - D^-Aff^{\sim,\text{ét}} \longrightarrow k - D^-Aff^{\sim,\text{ét}}$$

is right Quillen, and therefore can be derived on the right. Its right derived functor will be denoted by

$$\begin{array}{ccc} D^-\mathrm{St}(k) & \longrightarrow & D^-\mathrm{St}(k) \\ F & \mapsto & F^{\mathbb{R}K}. \end{array}$$

Explicitly, we have

$$F^{\mathbb{R}K} \simeq (RF)^K$$

where $RF$ is a fibrant replacement of $F$ in $k - D^-Aff^{\sim,\text{ét}}$.

DEFINITION 2.2.6.2. *Let $K$ be a $\mathbb{U}$-small simplicial set. The* derived moduli stack of rank $n$ local systems on $K$ *is defined to be*

$$\mathbb{R}\mathbf{Loc}_n(K) := \mathbf{Vect}_n^{\mathbb{R}K}.$$

We start by the following easy observation.

LEMMA 2.2.6.3. *Assume that $K$ is a finite dimensional simplicial set. Then, the $D^-$-stack $\mathbb{R}\mathbf{Loc}_n(K)$ is a finitely presented 1-geometric $D^-$-stack.*

PROOF. We consider the following homotopy co-cartesian square of simplicial sets

$$\begin{array}{ccc} Sk_iK & \longrightarrow & Sk_{i+1}K \\ \uparrow & & \uparrow \\ \coprod_{Hom(\partial\Delta^{i+1},K)} \partial\Delta^{i+1} & \longrightarrow & \coprod_{Hom(\Delta^{i+1},K)} \Delta^{i+1}, \end{array}$$

where $Sk_iK$ is the skeleton of dimension of $i$ of $K$. This gives a homotopy pullback square of $D^-$-stacks

$$\begin{array}{ccc} \mathbb{R}\mathbf{Loc}_n(Sk_{i+1}K) & \longrightarrow & \mathbb{R}\mathbf{Loc}_n(Sk_iK) \\ \downarrow & & \downarrow \\ \prod^h_{Hom(\Delta^{i+1},K)} \mathbb{R}\mathbf{Loc}_n(\Delta^{i+1}) & \longrightarrow & \prod^h_{Hom(\partial\Delta^{i+1},K)} \mathbb{R}\mathbf{Loc}_n(\partial\Delta^{i+1}). \end{array}$$

As finitely presented 1-geometric $D^-$-stacks are stable by homotopy pullbacks, we see by induction on the skeleton that it only remains to show that $\mathbb{R}\mathbf{Loc}_n(\partial\Delta^{i+1})$ is a finitely presented 1-geometric $D^-$-stack. But there is an isomorphism in $\mathrm{Ho}(SSet)$,

$$\partial\Delta^{i+1} \simeq * \coprod^{\mathbb{L}}_{\partial\Delta^i} *,$$

giving rise to an isomorphism of $D^-$-stacks

$$\mathbb{R}\mathbf{Loc}_n(\partial\Delta^{i+1}) \simeq \mathbb{R}\mathbf{Loc}_n \times^h_{\mathbb{R}\mathbf{Loc}_n(\partial\Delta^i)} \mathbb{R}\mathbf{Loc}_n.$$

By induction on $i$, we see that it is enough to show that $\mathbf{Vect}_n$ is a finitely presented 1-geometric $D^-$-stack which is known from Cor. 1.3.7.12. □

Another easy observation is the description of the truncation $t_0\mathbb{R}\mathbf{Loc}_n(K)$. For this, recall that the simplicial set $K$ has a fundamental groupoid $\Pi_1(K)$. The usual Artin stack of rank $n$ local systems on $K$ is the stack in groupoids defined by

$$\begin{array}{cccc} \mathbf{Loc}_n(K): & k-Alg & \longrightarrow & \{Groupoids\} \\ & A & \mapsto & \underline{Hom}(\Pi_1(K), \mathbf{Vect}_n(A)). \end{array}$$

In other words, $\mathbf{Loc}_n(K)(A)$ is the groupoid of functors from $\Pi_1(K)$ to the groupoid of rank $n$ projective $A$-modules. As usual, this Artin stack is considered as an object in $\mathrm{St}(k)$.

LEMMA 2.2.6.4. *There exists a natural isomorphism in* $\mathrm{St}(k)$

$$\mathbf{Loc}_n(K) \simeq t_0\mathbb{R}\mathbf{Loc}_n(K).$$

PROOF. The truncation $t_0$ being the right derived functor of a right Quillen functor commutes with derived exponentials. Therefore, we have

$$t_0\mathbb{R}\mathbf{Loc}_n(K) \simeq (t_0\mathbb{R}\mathbf{Loc}_n)^{\mathbb{R}K} \simeq (\mathbf{Vect}_n)^{\mathbb{R}K} \in \mathrm{St}(k).$$

The stack $\mathbf{Vect}_n$ is 1-truncated, and therefore we also have natural isomorphisms

$$(\mathbf{Vect}_n)^{\mathbb{R}K}(A) \simeq Map_{SSet}(K, \mathbf{Vect}_n(A)) \simeq \underline{Hom}(\Pi_1(K), \Pi_1(\mathbf{Vect}_n(A))).$$

This equivalence is natural in $A$ and provides the isomorphism of the lemma. □

Our next step is to give a more geometrical interpretation of the $D^-$-stack $\mathbb{R}\mathbf{Loc}_n(K)$, in terms of certain local systems of objects on the topological realization of $K$.

Let now $X$ be a $\mathbb{U}$-small topological space. Let $A \in sk-Mod$, and let us define a model category $A-Mod_s(X)$, of $A$-modules over $X$. The category $A-Mod_s(X)$ is simply the category of presheaves on $X$ with values in $A-Mod_s$. The model structure on $A - Mod_s(X)$ is of the same type as the local projective model structure on simplicial presheaves. We first define an intermediate model structure on $A-Mod_s(X)$ for which equivalences (resp. fibrations) are morphism $\mathcal{E} \longrightarrow \mathcal{F}$ in $A - Mod_s(X)$ such that for any open subset $U \subset X$ the induced morphism $\mathcal{E}(U) \longrightarrow \mathcal{F}(U)$ is an equivalence (resp. a fibration). This model structure exists as $A - Mod_s$ is a $\mathbb{U}$-cofibrantly generated model category, and let us call it the *strong* model structure. The final model structure on $A - Mod_s(X)$ is the one for which cofibrations are the same cofibrations as for the strong model structure, and equivalences are the morphisms $\mathcal{E} \longrightarrow \mathcal{F}$ such that for any point $x \in X$ the induced morphism on the stalks $\mathcal{E}_x \longrightarrow \mathcal{F}_x$ is an equivalence in $A-Mod_s$. The existence of this model structure is proved the same way as for the case of simplicial presheaves (we can also use the forgetful functor $A-Mod_s(X) \longrightarrow SPr(X)$ to lift the local projective model structure on $SPr(X)$ in a standard way).

For a commutative simplicial $k$-algebra $A$, we consider $A - Mod_s(X)^c_W$, the sub-category of $A - Mod_s(X)$ consisting of cofibrant objects and equivalences between them. For a morphism of commutative simplicial $k$-algebras $A \longrightarrow B$, we have a base change functor

$$\begin{array}{ccc} A - Mod_s(X) & \longrightarrow & B - Mod_s(X) \\ \mathcal{E} & \mapsto & \mathcal{E} \otimes_A B \end{array}$$

which is a left Quillen functor, and therefore induces a well defined functor

$$- \otimes_A B : A - Mod_s(X)^c_W \longrightarrow B - Mod_s(X)^c_W.$$

This defines a lax functor $A \mapsto A - Mod_s(X)^c_W$, from $sk - Alg$ to $Cat$, which can be strictified in the usual way. We will omit to mention explicitly this strictification here and will do as if $A \mapsto A-Mod_s(X)^c_W$ does define a genuine functor $sk-Alg \longrightarrow Cat$.

We then define a sub-functor of $A \mapsto A - Mod_s(X)^c_W$ in the following way. For $A \in sk-Alg$, let $A-Loc_n(X)$ be the full subcategory of $A-Mod_s(X)^c_W$ consisting of objects $\mathcal{E}$, such that there exists an open covering $\{U_i\}$ on $X$, such that each restriction $\mathcal{E}_{|U_i}$ is isomorphic in $\mathrm{Ho}(A-Mod_s(U_i))$ to a constant presheaf with fibers a projective $A$-module of rank $n$ (i.e. projective $A$-module $E$ such that $\pi_0(E)$ is a projective $\pi_0(A)$-module of rank $n$). This defines a sub-functor of $A \mapsto A - Mod_s(X)^c_W$, and thus a functor from $sk - Alg$ to $Cat$. Applying the nerve functor we obtain a simplicial presheaf $\mathbb{R}\mathbf{Loc}_n(X) \in k - D^-Aff^{\sim,\text{ét}}$, defined by

$$\mathbb{R}\mathbf{Loc}_n(X)(A) := N(A - Loc_n(X)).$$

PROPOSITION 2.2.6.5. *Let $K$ be a simplicial set in $\mathbb{U}$ and $|K|$ be its topological realization. The simplicial presheaf $\mathbb{R}\mathbf{Loc}_n(|K|)$ is a $D^-$-stack, and there exists an isomorphism in $D^-\mathrm{St}(k)$*

$$\mathbb{R}\mathbf{Loc}_n(|K|) \simeq \mathbb{R}\mathbf{Loc}_n(K).$$

PROOF. We first remark that if $f : X \longrightarrow X'$ is a homotopy equivalence of topological spaces, then the induced morphism

$$f^* : \mathbb{R}\mathbf{Loc}_n(X') \longrightarrow \mathbb{R}\mathbf{Loc}_n(X')$$

is an equivalence of simplicial presheaves. Indeed, a standard argument reduces to the case where $f$ is the projection $X \times [0, 1] \longrightarrow X$, and then to $[0, 1] \longrightarrow *$, for which one can use the same argument as in [**To3**, Lem. 2.16].

Let us first prove that $\mathbb{R}\mathbf{Loc}_n(|K|)$ is a $D^-$-stack, and for this we will prove that the simplicial presheaf $\mathbb{R}\mathbf{Loc}_n(|K|)$ can be written, in $SPr(k - D^-Aff)$, as a certain homotopy limit of $D^-$-stacks. As $D^-$-stacks in $SPr(k - D^-Aff)$ are stable by homotopy limits this will prove what we want.

Let $U_*$ an open hypercovering of $|K|$ such that each $U_i$ is a coproduct of contractible open subsets in $|K|$ (such a hypercover exists as $|K|$ is a locally contractible space). It is not hard to show using Cor. B.0.8 and standard cohomological descent, that for any $A \in sk - Alg$, the natural morphism

$$N(A - Mod_s(|K|)^c_W) \longrightarrow Holim_{m\in\Delta} N(A - Mod_s(U_m)^c_W)$$

is an isomorphism in $\mathrm{Ho}(SSet)$. From this, we easily deduce that the natural morphism

$$\mathbb{R}\mathbf{Loc}_n(|K|) \longrightarrow Holim_{m\in\Delta}\mathbb{R}\mathbf{Loc}_n(U_m)$$

is an isomorphism in $\mathrm{Ho}(SPr(k - D^-Aff))$. Therefore, we are reduced to show that $\mathbb{R}\mathbf{Loc}_n(U_m)$ is a $D^-$-stack. But, $U_m$ being a coproduct of contractible topological spaces, $\mathbb{R}\mathbf{Loc}_n(U_m)$ is a product of some $\mathbb{R}\mathbf{Loc}_n(U)$ for some contractible space $U$. Moreover $\mathbb{R}\mathbf{Loc}_n(U)$ is naturally isomorphic in $\mathrm{Ho}(SPr(k-D^-Aff))$ to $\mathbb{R}\mathbf{Loc}_n(*) = \mathbf{Vect}_n$. As we know that $\mathbf{Vect}_n$ is a $D^-$-stack, this shows that the simplicial presheaf $\mathbb{R}\mathbf{Loc}_n(|K|)$ is a homotopy limit of $D^-$-stacks and thus is itself a $D^-$-stack.

We are left to prove that $\mathbb{R}\mathbf{Loc}_n(|K|)$ and $\mathbb{R}\mathbf{Loc}_n(K)$ are isomorphic. But, we have seen that

$$\mathbb{R}\mathbf{Loc}_n(|K|) \simeq Holim_{m\in\Delta}\mathbb{R}\mathbf{Loc}_n(U_m),$$

for an open hypercover $U_*$ of $|K|$, such that each $U_m$ is a coproduct of contractible open subsets. We let $K' := \pi_0(U_*)$ be the simplicial set of connected components of $U_*$, and thus

$$Holim_{m\in\Delta}\mathbb{R}\mathbf{Loc}_n(U_m) \simeq Holim_{m\in\Delta}\mathbb{R}\mathbf{Loc}_n(K'_m),$$

where $K'_m$ is considered as a discrete topological space. We have thus proved that

$$\mathbb{R}\mathbf{Loc}_n(|K|) \simeq Holim_{m\in\Delta}\prod^h_{K'_m}\mathbf{Vect}_n \simeq (\mathbf{Vect}_n)^{\mathbb{R}K'}.$$

But, by [**To3**, Lem. 2.10], we know that $|K|$ is homotopically equivalent to $|K'|$, and thus that $K$ is equivalent to $K'$. This implies that

$$\mathbb{R}\mathbf{Loc}_n(|K|) \simeq (\mathbf{Vect}_n)^{\mathbb{R}K'} \simeq (\mathbf{Vect}_n)^{\mathbb{R}K} \simeq \mathbb{R}\mathbf{Loc}_n(|K|).$$

□

We will now describe the cotangent complex of $\mathbb{R}\mathbf{Loc}_n(K)$. For this, we fix a global point

$$E : * \longrightarrow \mathbb{R}\mathbf{Loc}_n(K),$$

which by Lem. 2.2.6.4 corresponds to a functor

$$E : \Pi_1(K) \longrightarrow \Pi_1(\mathbf{Vect}_n(k)),$$

where $\Pi_1(\mathbf{Vect}_n(k))$ can be identified with the groupoid of rank $n$ projective $k$-modules. The object $E$ is thus a local system of rank $n$ projective $k$-modules on $K$ in the usual sense. We will compute the cotangent complex $\mathbb{L}_{\mathbb{R}\mathbf{Loc}_n(K),E} \in \mathrm{Ho}(Sp(sk - Mod))$ (recall that $\mathrm{Ho}(Sp(sk - Mod))$ can be naturally identified with the unbounded derived category of $k$). For this, we let $E \otimes_k E^\vee$ be the local system on $K$ of endomorphisms of $E$, and $C_*(K, E \otimes_k E^\vee)$ will be the complex of homology of $K$ with coefficients in the local system $E \otimes_k E^\vee$. We consider $C_*(K, E \otimes_k E^\vee)$ as an unbounded complex of $k$-modules, and therefore as an object in $\mathrm{Ho}(Sp(sk - Mod))$.

PROPOSITION 2.2.6.6. *There exists an isomorphism in* $\mathrm{Ho}(Sp(sk - Mod)) \simeq \mathrm{Ho}(C(k))$

$$\mathbb{L}_{\mathbb{R}\mathbf{Loc}_n(K),E} \simeq C_*(K, E \otimes_k E^\vee)[-1].$$

PROOF. Let $M \in sk - Mod$, and let us consider the simplicial set

$$\mathbb{D}er_E(\mathbb{R}\mathbf{Loc}_n(K), M),$$

of derivations of $\mathbb{R}\mathbf{Loc}_n(K)$ at the point $E$ and with coefficients in $M$. By definition of $\mathbb{R}\mathbf{Loc}_n(K)$ we have

$$\mathbb{D}er_E(\mathbb{R}\mathbf{Loc}_n(K), M) \simeq Map_{SSet/\mathbf{Vect}_n(k)}(K, \mathbf{Vect}_n(k \oplus M)),$$

where $K \longrightarrow \mathbf{Vect}_n(k)$ is given by the object $E$, and $\mathbf{Vect}_n(k \oplus M) \longrightarrow \mathbf{Vect}_n(k)$ is the natural projection. At this point we use Prop. A.0.6 in order to describe, functorially in $M$, the morphism $\mathbf{Vect}_n(k \oplus M) \longrightarrow \mathbf{Vect}_n(k)$. For this, we let $\mathcal{G}(k)$ to be the groupoid of projective $k$-modules of rank $n$. We also define an $S$-category $\mathcal{G}(k \oplus M)$ in the following way. Its objects are projective $k$-modules of rank $n$. The simplicial set of morphisms in $\mathcal{G}(k \oplus M)$ between two such $k$-modules $E$ and $E'$ is defined to be

$$\mathcal{G}(k \oplus M)_{(E,E')} := \underline{Hom}^{Eq}_{(k \oplus M)-Mod_s}(E \oplus (E \otimes_k M), E' \oplus (E' \otimes_k M)),$$

the simplicial set of equivalences from $E \oplus (E \otimes_k M)$ to $E' \oplus (E' \otimes_k M)$, in the model category $(k \oplus M) - Mod_s$. It is important to note that $E \oplus (E \otimes_k M)$ is isomorphic to $E \otimes_k (k \oplus M)$, and therefore is a cofibrant object in $(k \oplus M) - Mod_s$ (as the base change of a cofibrant object $E$ in $sk - Mod$). There exists a natural morphism of $S$-categories

$$\mathcal{G}(k \oplus M) \longrightarrow \mathcal{G}(k)$$

being the identity on the set of objects, and the composition of natural morphisms

$$\underline{Hom}^{Eq}_{(k \oplus M)-Mod_s}(E \oplus (E \otimes_k M), E' \oplus (E' \otimes_k M)) \longrightarrow \underline{Hom}^{Eq}_{k-Mod_s}(E, E') \longrightarrow$$

$$\pi_0(\underline{Hom}^{Eq}_{k-Mod_s}(E, E')) \simeq \mathcal{G}(k)_{(E,E')}.$$

on the simplicial sets of morphisms. Clearly, Prop. A.0.6 and its functorial properties, show that the morphism

$$\mathbf{Vect}_n(k \oplus M) \longrightarrow \mathbf{Vect}_n(k)$$

is equivalent to

$$N(\mathcal{G}(k \oplus M)) \longrightarrow N(\mathcal{G}(k)),$$

and in a functorial way in $M$.

For any $E \in \mathcal{G}(k)$, we can consider the classifying simplicial set $K(E \otimes_k E^\vee \otimes_k M, 1)$ of the simplicial abelian group $E \otimes_k E^\vee \otimes_k M$, and for any isomorphism of projective $k$-modules of rank $n$, $E \simeq E'$, the corresponding isomorphism of simplicial set

$$K(E \otimes_k E^\vee \otimes_k M, 1) \simeq K(E' \otimes_k (E')^\vee \otimes_k M, 1).$$

This defines a local system $L$ of simplicial sets on the groupoid $\mathcal{G}(k)$, for which the total space

$$Hocolim_{\mathcal{G}(k)} L \longrightarrow N(\mathcal{G}(k))$$

is easily seen to be isomorphic to the projection

$$N(\mathcal{G}(k \oplus M)) \longrightarrow N(\mathcal{G}(k)).$$

The conclusion is that the natural projection

$$\mathbf{Vect}_n(k \oplus M) \longrightarrow \mathbf{Vect}_n(k)$$

is equivalent, functorially in $M$, to the total space of the local system $E \mapsto K(E \otimes_k E^\vee \otimes_k M, 1)$ on $\mathbf{Vect}_n(k)$. Therefore, one finds a natural equivalence of simplicial sets

$$\mathbb{D}er_E(\mathbb{R}\mathbf{Loc}_n(K), M) \simeq Map_{SSet/\mathbf{Vect}_n(k)}(K, \mathbf{Vect}_n(k \oplus M)) \simeq$$

$$C^*(K, E \otimes_k E^\vee \otimes_k M[1]) \simeq Map_{C(k)}(C_*(K, E \otimes_k E^\vee)[-1], M).$$

As this equivalence is functorial in $M$, this shows that

$$\mathbb{L}_{\mathbb{R}\mathbf{Loc}_n(K),E} \simeq C_*(K, E \otimes_k E^\vee)[-1],$$

as required. □

REMARK 2.2.6.7. An important consequence of Prop. 2.2.6.6 is that the $D^-$-stack depends on strictly more than the fundamental groupoid of $K$. Indeed, the tangent space of $\mathbb{R}\mathbf{Loc}_n(K)$ at a global point corresponding to a local system $E$ on $K$ is $C^*(K, E \otimes_k E^\vee)[1]$, which can be non-trivial even when $K$ is simply connected. In general, there is a closed immersion of $D^-$-stacks (Prop. 2.2.4.7)

$$it_0(\mathbb{R}\mathbf{Loc}_n(K)) \longrightarrow \mathbb{R}\mathbf{Loc}_n(K),$$

which on the level of tangent spaces induces the natural morphism

$$\tau_{\leq 0}(C^*(K, E \otimes_k E^\vee)[1]) \longrightarrow C^*(K, E \otimes_k E^\vee)[1]$$

giving an isomorphism on $H^0$ and $H^1$. This shows that the $D^-$-stack $\mathbb{R}\mathbf{Loc}_n(K)$ contains strictly more information than the usual Artin stack of local systems on $K$, and does encode some higher homotopical invariants of $K$.

**2.2.6.2. Algebras over an operad.** Recall (e.g. from [**Re, Sp**]) the notions of operads and algebras over them, as well as their model structures. Let $\mathcal{O}$ be an operad in $k - Mod$, the category of $k$-modules. We assume that for any $n$, the $k$-module $\mathcal{O}(n)$ is projective. Then, for any $A \in sk - Alg$, we can consider $\mathcal{O} \otimes_k A$, which is an operad in the symmetric monoidal category $A - Mod_s$ of $A$-modules. We can therefore consider $\mathcal{O} \otimes_k A - Alg_s$, the category of $\mathcal{O}$-algebras in $A - Mod_s$. According to [**Hin1**] the category $\mathcal{O} \otimes_k A - Alg_s$ can be endowed with a natural structure of a $\mathbb{U}$-cofibrantly generated model category for which equivalences and fibrations are defined on the underlying objects in $sk - Mod$. For a morphism $A \longrightarrow B$ in $sk - Alg$, there is a Quillen adjunction

$$B \otimes_A - : \mathcal{O} \otimes_k A - Alg_s \longrightarrow \mathcal{O} \otimes_k B - Alg_s \qquad \mathcal{O} \otimes_k A - Alg_s \longleftarrow \mathcal{O} \otimes_k B - Alg_s : F,$$

where $F$ is the forgetful functor. The rule $A \mapsto \mathcal{O} \otimes_k A - Alg_s$ together with base change functors $B \otimes_A -$ is almost a left Quillen presheaf in the sense of Appendix B, except that the associativity of composition of base change is only valid up to a natural isomorphism. However, the standard strictification techniques can be applied in order to replace, up to a natural equivalence, this by a genuine left Quillen presheaf. We will omit to mention this replacement and will proceed as if $A \mapsto \mathcal{O} \otimes_k A - Alg_s$ actually defines a left Quillen presheaf on $k - D^-Aff$.

For any $A \in sk - Alg$, we consider $\mathcal{O} - Alg(A)$, the category of cofibrant objects in $\mathcal{O} \otimes_k A - Alg_s$ and equivalences between them. Finally, we let $\mathcal{O} - Alg_n(A)$ be the full subcategory consisting of objects $B \in \mathcal{O} - Alg(A)$, such that the underlying $A$-module of $B$ is a vector bundle of rank $n$ (i.e. that $B$ is a strong $A$-module, and $\pi_0(B)$ is a projective $\pi_0(A)$-module of rank $n$). The base change functors clearly preserves the sub-categories $\mathcal{O} - Alg_n(A)$, and we get this way a well defined presheaf of $\mathbb{V}$-small categories

$$\begin{array}{ccc} k - D^-Aff & \longrightarrow & Cat_{\mathbb{V}} \\ A & \mapsto & \mathcal{O} - Alg_n(A). \end{array}$$

Applying the nerve functor we obtain a simplicial presheaf

$$\begin{array}{ccc} k - D^-Aff & \longrightarrow & SSet_{\mathbb{V}} \\ A & \mapsto & N(\mathcal{O} - Alg_n(A)). \end{array}$$

This simplicial presheaf will be denoted by $\mathbf{Alg}_n^{\mathcal{O}}$, and is considered as an object in $k - D^-Aff^{\sim,ét}$.

PROPOSITION 2.2.6.8. (1) *The object* $\mathbf{Alg}_n^{\mathcal{O}} \in k - D^-Aff^{\sim,ét}$ *is a* $D^-$-*stack.*

(2) *The* $D^-$*-stack* $\mathbf{Alg}_n^{\mathcal{O}}$ *is* 1*-geometric and quasi-compact.*

PROOF. (1) The proof relies on the standard argument based on Cor. B.0.8, and is left to the reader.

(2) We consider the natural morphism of $D^-$-stacks

$$p : \mathbf{Alg}_n^{\mathcal{O}} \longrightarrow \mathbf{Vect}_n,$$

defined by forgetting the $\mathcal{O}$-algebra structure. Precisely, it sends an object $B \in \mathcal{O} - Alg_n(A)$ to its underlying $A$-modules, which is a rank $n$-vector bundle by definition. We are going to prove that the morphism $p$ is a representable morphism, and this will imply the result as $\mathbf{Vect}_n$ is already known to be 1-geometric and quasi-compact. For this, we consider the natural morphism $* \longrightarrow \mathbf{Vect}_n$, which is a smooth 1-atlas for $\mathbf{Vect}_n$, and consider the homotopy pullback

$$\widetilde{\mathbf{Alg}_n^{\mathcal{O}}} := \mathbf{Alg}_n^{\mathcal{O}} \times^h_{\mathbf{Vect}_n} *.$$

It is enough by Prop. 1.3.3.4 to show that $\widetilde{\mathbf{Alg}_n^{\mathcal{O}}}$ is a representable stack. For this, we use [**Re**, Thm. 1.1.5] in order to show that the $D^-$-stack $\widetilde{\mathbf{Alg}_n^{\mathcal{O}}}$ is isomorphic in $D^-\mathrm{St}(k)$ to $Map(\mathcal{O}, \underline{End}(k^n))$, defined as follows. For any $A \in sk - Alg$, we consider $\underline{End}(A^n)$, the operad in $A - Mod_s$ of endomorphisms of the object $A^n$ (recall that for $M \in A - Mod_s$, the operad $\underline{End}(M)$ is defined by $\underline{End}(M)(n) := \underline{Hom}_A(M^{\otimes_k n}, M)$). We let $Q\mathcal{O}$ be a cofibrant replacement of the operad $\mathcal{O}$ in the model category $sk - Mod$. Finally, $Map(\mathcal{O}, \underline{End}(k^n))$ is defined as

$$\begin{array}{ccc} sk - Alg & \longrightarrow & SSet \\ A & \mapsto & Map(\mathcal{O}, \underline{End}(k^n))(A) := \underline{Hom}_{Op}(Q\mathcal{O}, \underline{End}(A^n)), \end{array}$$

where $\underline{Hom}_{Op}$ denotes the simplicial set of morphism in the simplicial category of operads in $sk - Mod$. As we said, [**Re**, Thm. 1.1.5] implies that $\widetilde{\mathbf{Alg}_n^{\mathcal{O}}}$ is isomorphic to $Map(\mathcal{O}, \underline{End}(k^n))$. Therefore, it remain to show that $Map(\mathcal{O}, \underline{End}(k^n))$ is a representable $D^-$-stack.

For this, we can write $\mathcal{O}$, up to an equivalence, as the homotopy colimit of free operads

$$\mathcal{O} \simeq Hocolim_i \mathcal{O}_i.$$

Then, we have

$$Map(\mathcal{O}, \underline{End}(k^n)) \simeq Holim_i Map(\mathcal{O}_i, \underline{End}(k^n))$$

and as representable $D^-$-stacks are stable by homotopy limits, we are reduced to the case where $\mathcal{O}$ is a free operad. This means that there exists an integer $m \geq 0$, such that for any other operad $\mathcal{O}'$ in $sk - Mod$, we have a natural isomorphism of sets

$$Hom_{Op}(\mathcal{O}, \mathcal{O}') \simeq Hom(k, \mathcal{O}'(m)).$$

In particular, we find that the $D^-$-stack $Map(\mathcal{O}, \underline{End}(k^n))$ is isomorphic to the $D^-$-stack sending $A \in sk - Alg$ to the simplicial set $\underline{Hom}_{A-Mod_s}((A^n)^{\otimes_k m}, A)$, of morphisms from the $A$-module $(A^n)^{\otimes_k m}$ to $A$. But this last $D^-$-stack is clearly isomorphic to $\mathbb{R}\underline{Spec}\, B$, where $B$ is a the free commutative simplicial $k$-algebra on $k^{n^m}$, or in other words a polynomial algebra over $k$ with $n^m$ variables. □

We still fix an operad $\mathcal{O}$ in $k - Mod$ (again requiring $\mathcal{O}(m)$ to be a projective $k$-module for any $m$), and we let $B$ be an $\mathcal{O}$-algebra in $k - Mod$, such that $B$ is projective of rank $n$ as a $k$-module. This defines a well defined morphism of stacks

$$B : * \longrightarrow \mathbf{Alg}_n^{\mathcal{O}}.$$

We are going to describe the cotangent complex of $\mathbf{Alg}_n^{\mathcal{O}}$ at $B$ using the notion of (derived) derivations for $\mathcal{O}$-algebras.

For this, recall from [**Goe-Hop**] the notion of $\mathcal{O}$-derivations from $B$ and with coefficients in a $B$-module. For any $B$-module $M$, one can define the square zero extension $B \oplus M$ of $B$ by $M$, which is another $\mathcal{O}$-algebra together with a natural projection $B \oplus M \longrightarrow M$. The $k$-module of derivations from $B$ to $M$ is defined by

$$Der_k^{\mathcal{O}}(B, M) := Hom_{\mathcal{O}-Alg/B}(B, B \oplus M).$$

In the same way, for any $\mathcal{O}$-algebra $B'$ with a morphism $B' \longrightarrow B$ we set

$$Der_k^{\mathcal{O}}(B', M) := Hom_{\mathcal{O}-Alg/B}(B', B \oplus M).$$

The functor $B' \mapsto Der_k^{\mathcal{O}}(B', M)$ can be derived on the left, to give a functor

$$\mathbb{R}Der_k^{\mathcal{O}}(-, M) : \mathrm{Ho}(\mathcal{O} - Alg_s/B)^{op} \longrightarrow \mathrm{Ho}(SSet_{\mathbb{V}}).$$

As shown in [**Goe-Hop**], the functor $M \mapsto \mathbb{R}Der_k^{\mathcal{O}}(B, M)$ is co-represented by an object $\mathbb{L}_B^{\mathcal{O}} \in \mathrm{Ho}(sB - Mod)$, thus there is a natural isomorphism in $\mathrm{Ho}(SSet_{\mathbb{V}})$

$$\mathbb{R}Der_k^{\mathcal{O}}(B, M) \simeq Map_{sB-Mod}(\mathbb{L}_B^{\mathcal{O}}, M).$$

The category $sB - Mod$ of simplicial $B$-modules is a closed model category for which equivalences and fibrations are detected in $sk - Mod$. Furthermore, the category $sB - Mod$ is tensored and co-tensored over $sk - Mod$ making it into a $sk - Mod$-model category in the sense of [**Ho1**]. Passing to model categories of spectra, we obtain a model category $Sp(sB - Mod)$ which is a $Sp(sk - Mod)$-model category. We will then set

$$\mathbb{R}\underline{Der}_k^{\mathcal{O}}(B, M) := \mathbb{R}\underline{Hom}_{Sp(sB-Mod)}(\mathbb{L}_B^{\mathcal{O}}, M) \in \mathrm{Ho}(Sp(sk - Mod)),$$

where $\mathbb{R}\underline{Hom}_{Sp(sB-Mod)}$ denotes the $\mathrm{Ho}(Sp(sk - Mod))$-enriched derived $Hom$ of $Sp(sB - Mod)$. Note that $\mathrm{Ho}(sB - Mod)$ is equivalent to the unbounded derived category of $B$-modules, and as well that $\mathrm{Ho}(Sp(sk - Mod))$ is equivalent to the unbounded derived category of $k$-modules.

PROPOSITION 2.2.6.9. *With the notations as above, the tangent complex of the $D^-$-stack $\mathbf{Alg}_n^{\mathcal{O}}$ at the point $B$ is given by*

$$\mathbb{T}_{\mathbf{Alg}_n^{\mathcal{O}}, B} \simeq \mathbb{R}\underline{Der}_k^{\mathcal{O}}(B, B)[1] \in \mathrm{Ho}(Sp(sk - Mod)).$$

PROOF. We have an isomorphism in $\mathrm{Ho}(Sp(sk - Mod))$

$$\mathbb{T}_{\mathbf{Alg}_n^{\mathcal{O}}, B} \simeq \mathbb{T}_{\Omega_B \mathbf{Alg}_n^{\mathcal{O}}, B}[1].$$

Using Prop. A.0.6, we see that the $D^-$-stack $\Omega_B \mathbf{Alg}_n^{\mathcal{O}}$ can be described as

$$\begin{array}{ccc} sk - Alg & \longrightarrow & SSet \\ A & \mapsto & Map_{\mathcal{O}-Alg_s}^{eq}(B, B \otimes_k A) \end{array}$$

where $Map_{\mathcal{O}-Alg_s}^{eq}$ denotes the mapping space of equivalences in the category of $\mathcal{O}$-algebras in $sA - Mod$. In particular, we see that for $M \in sk - Mod$, the simplicial set $\mathbb{D}er_{\Omega_B \mathbf{Alg}_n^{\mathcal{O}}}(B, M)$ is naturally equivalent to the homotopy fiber, taken at the identity, of the morphism

$$Map_{\mathcal{O}-Alg_s}(B, B \otimes_k (k \oplus M)) \longrightarrow Map_{\mathcal{O}-Alg_s}(B, B).$$

The $\mathcal{O}$-algebra $B \otimes_k (k \oplus M)$ can be identified with $B \oplus (M \otimes_k B)$, the square zero extension of $B$ by $M \otimes_k B$, as defined in [**Goe-Hop**]. Therefore, by definition of derived derivations this homotopy fiber is naturally equivalent to $\mathbb{R}Der_k^{\mathcal{O}}(B, M)$. This shows that

$$\mathbb{D}er_{\Omega_B \mathbf{Alg}_n^{\mathcal{O}}}(B, M) \simeq \mathbb{R}Der_k^{\mathcal{O}}(B, M \otimes_k B) \simeq Map_{Sp(sB-Mod)}(\mathbb{L}_B^{\mathcal{O}}, M \otimes_k B).$$

This implies that for any $N \in sk - Mod$, we have

$$Map_{Sp(sk-Mod)}(N, \mathbb{T}_{\Omega_B \mathbf{Alg}_n^{\mathcal{O}}, B}) \simeq \mathbb{D}er_{\Omega_B \mathbf{Alg}_n^{\mathcal{O}}}(B, N^{\vee}) \simeq Map_{Sp(sB-Mod)}(\mathbb{L}_B^{\mathcal{O}}, N^{\vee} \otimes_k B) \simeq$$

$$Map_{Sp(sk-Mod)}(N, \mathbb{R}\underline{Hom}_{Sp(sB-Mod)}(\mathbb{L}_B^{\mathcal{O}}, B)) \simeq Map_{Sp(sk-Mod)}(N, \mathbb{R}\underline{Der}_k(B, B)).$$

The Yoneda lemma implies the existence of a natural isomorphism in $\mathrm{Ho}(Sp(sk - Mod))$

$$\mathbb{T}_{\Omega_B \mathbf{Alg}_n^{\mathcal{O}}, B} \simeq \mathbb{R}\underline{Der}_k(B, B).$$

We thus we have

$$\mathbb{T}_{\mathbf{Alg}_n^{\mathcal{O}}, B} \simeq \mathbb{R}\underline{Der}_k(B, B)[1].$$

□

REMARK 2.2.6.10. The proof of Prop. 2.2.6.9 actually shows that for any $M \in sk - Mod$ we have

$$\mathbb{R}\underline{Hom}_k^{Sp}(\mathbb{L}_{\mathbf{Alg}_n^{\mathcal{O}}, B}, M) \simeq \mathbb{R}Der_k(B, B \otimes_k M)[1].$$

**2.2.6.3. Mapping $D^-$-stacks.** We let $X$ be a stack in $\mathrm{St}(k)$, and $F$ be an $n$-geometric $D^-$-stack. We are going to investigate the geometricity of the $D^-$-stack of morphisms from $i(X)$ to $F$.

$$\mathbf{Map}(X, F) := \mathbb{R}_{\acute{e}t}\underline{\mathcal{H}om}(i(X), F) \in D^-\mathrm{St}(k).$$

Recall that $\mathbb{R}_{\acute{e}t}\underline{\mathcal{H}om}$ denotes the internal Hom of the category $D^-\mathrm{St}(k)$.

THEOREM 2.2.6.11. *With the notations above, we assume the following three conditions are satisfied.*

(1) *The stack*

$$t_0(\mathbf{Map}(X, F)) \simeq \mathbb{R}_{\acute{e}t}\underline{\mathcal{H}om}(X, t_0(F)) \in \mathrm{St}(k)$$

*is $n$-geometric.*

(2) *The $D^-$-stack* $\mathbf{Map}(X, F)$ *has a cotangent complex.*

(3) *The stack $X$ can be written in* $\mathrm{St}(k)$ *has a homotopy colimit $Hocolim_i U_i$, where $U_i$ is a affine scheme, flat over $Spec\, k$.*

*Then the $D^-$-stack* $\mathbf{Map}(X, F)$ *is $n$-geometric.*

PROOF. The only if part is clear. Let us suppose that $\mathbf{Map}(X, F)$ satisfies the three conditions. To prove that it is an $n$-geometric $D^-$-stack we are going to lift an $n$-atlas of $t_0(\mathbf{Map}(X, F))$ to an $n$-atlas of $\mathbf{Map}(X, F)$. For this we use the following special case of J.Lurie's representability criterion, which can be proved using the material of this paper (see Appendix C).

THEOREM 2.2.6.12. *(J. Lurie, see* [**Lu1**] *and Appendix C) Let $F$ be a $D^-$-stack. The following conditions are equivalent.*

(1) *$F$ is an $n$-geometric $D^-$-stack.*
(2) *$F$ satisfies the following three conditions.*
    (a) *The truncation $t_0(F)$ is an Artin $(n+1)$-stack.*
    (b) *$F$ has an obstruction theory.*

(c) *For any $A \in sk - Alg$, the natural morphism*

$$\mathbb{R}F(A) \longrightarrow Holim_k \mathbb{R}F(A_{\leq k})$$

*is an isomorphism in* $\mathrm{Ho}(SSet)$.

We need to prove that $\mathbf{Map}(X,F)$ satisfies the conditions $(a) - (c)$ of theorem 2.2.6.12. Condition $(a)$ is clear by assumption. The existence of a cotangent complex for $\mathbf{Map}(X,F)$ is guaranteed by assumption. The fact that $\mathbf{Map}(X,F)$ is moreover inf-cartesian follows from the general fact.

LEMMA 2.2.6.13. *Let $F$ be a $D^-$-stack which is inf-cartesian. Then, for any $D^-$-stack $F'$, the $D^-$-stack $\mathbb{R}_{\acute{e}t}\underline{\mathcal{H}om}(F',F)$ is inf-cartesian.*

PROOF. Writing $F'$ has a homotopy colimit of representable $D^-$-stacks

$$F' \simeq Hocolim_i U_i,$$

we find

$$\mathbb{R}_{\acute{e}t}\underline{\mathcal{H}om}(F',F) \simeq Holim_i \mathbb{R}_{\acute{e}t}\underline{\mathcal{H}om}(U_i,F).$$

As being inf-cartesian is clearly stable by homotopy limits, we reduce to the case where $F' = \mathbb{R}\underline{Spec\,B}$ is a representable $D^-$-stack. Let $A \in sk - Alg$, $M$ be an $A$-module with $\pi_0(M) = 0$, and $d \in \pi_0(\mathbb{D}er(A,M))$ be a derivation. Then, the commutative square

$$\begin{array}{ccc} \mathbb{R}_{\acute{e}t}\underline{\mathcal{H}om}(F',F)(A \oplus_d \Omega M) & \longrightarrow & \mathbb{R}_{\acute{e}t}\underline{\mathcal{H}om}(F',F)(A) \\ \downarrow & & \downarrow \\ \mathbb{R}_{\acute{e}t}\underline{\mathcal{H}om}(F',F)(A) & \longrightarrow & \mathbb{R}_{\acute{e}t}\underline{\mathcal{H}om}(F',F)(A \oplus M) \end{array}$$

is equivalent to the commutative square

$$\begin{array}{ccc} \mathbb{R}F((A \oplus_d \Omega M) \otimes^{\mathbb{L}}_k B) & \longrightarrow & \mathbb{R}F(A \otimes^{\mathbb{L}}_k B) \\ \downarrow & & \downarrow \\ \mathbb{R}F(A \otimes^{\mathbb{L}}_k B) & \longrightarrow & \mathbb{R}F((A \oplus M) \otimes^{\mathbb{L}}_k B). \end{array}$$

Using the $F$ is inf-cartesian with respect to $A \otimes^{\mathbb{L}}_k B \in sk - Alg$, and the derivation $d \otimes_k B \in \pi_0(\mathbb{D}er(A \otimes^{\mathbb{L}}_k B, M \otimes^{\mathbb{L}}_k B))$ implies that this last square is homotopy cartesian. This shows that $\mathbb{R}_{\acute{e}t}\underline{\mathcal{H}om}(F',F)$ is inf-cartesian. $\square$

In order to finish the proof of Thm. 2.2.6.11 it only remain to show that $\mathbf{Map}(X,F)$ also satisfies the condition $(c)$ of Thm. 2.2.6.12. For this, we write $X$ as $Hocolim_i U_i$ with $U_i$ affine and flat over $k$, and therefore $\mathbf{Map}(X,F)$ can be written as the homotopy limit $Holim_i \mathbf{Map}(U_i,F)$. In order to check condition $(c)$ we can therefore assume that $X$ is an affine scheme, flat over $k$. Let us write $X = Spec\,B$, with $B$ a commutative flat $k$-algebra. Then, for any $A \in sk - Mod$, the morphism

$$\mathbb{R}\mathbf{Map}(X,F)(A) \longrightarrow Holim_k \mathbb{R}\mathbf{Map}(X,F)(A_{\leq k})$$

is equivalent to

$$\mathbb{R}F(A \otimes_k B) \longrightarrow Holim_k \mathbb{R}F((A_{\leq k}) \otimes_k B).$$

But, as $B$ is flat over $k$, we have $(A_{\leq k}) \otimes_k B \simeq (A \otimes_k B)_{\leq k}$, and therefore the above morphism is equivalent to

$$\mathbb{R}F(A \otimes_k B) \longrightarrow Holim_k \mathbb{R}F((A \otimes_k B)_{\leq k})$$

and is therefore an equivalence by $(1) \Rightarrow (2)$ of Thm. 2.2.6.12 applied to $F$. $\square$

The following corollaries are direct consequences of Thm. 2.2.6.11. The only non trivial part consists of proving the existence of a cotangent complex, which we will assume in the present version of this work.

COROLLARY 2.2.6.14. *Let $X$ be a projective and flat scheme over $Spec\, k$, and $Y$ a projective and smooth scheme over $Spec\, k$. Then, the $D^-$-stack $\mathbf{Map}(i(X), i(Y))$ is a 1-geometric $D^-$-scheme. Furthermore, for any morphism of schemes $f : X \longrightarrow Y$, the cotangent complex of $\mathbf{Map}(i(X), i(Y))$ at the point $f$ is given by*

$$\mathbb{L}_{\mathbf{Map}(i(X),i(Y)),f} \simeq (C^*(X, f^*(T_Y)))^\vee,$$

*where $T_Y$ is the tangent sheaf of $Y \longrightarrow Spec\, k$.*

Let us now suppose that $k = \mathbb{C}$ is the field of complex numbers, and let $X$ be a smooth and projective variety. We will be interested in the sheaf $X_{DR}$ of [**S1**], defined by $X_{DR}(A) := A_{red}$, for a commutative $\mathbb{C}$-algebra $A$. Recall that the stack $\mathcal{M}_{DR}(X)$ is defined as the stack of morphisms from $X_{DR}$ to $\mathbf{Vect}_n$, and is identified with the stack of flat bundles on $X$ (see [**S1**]). It is known that $\mathcal{M}_{DR}(X)$ is an Artin 1-stack.

COROLLARY 2.2.6.15. *The $D^-$-stack $\mathbb{R}\mathcal{M}_{DR}(X) := \mathbf{Map}(i(X_{DR}), \mathbf{Vect}_n)$ is 1-geometric. For a point $E : * \longrightarrow \mathbb{R}\mathcal{M}_{DR}(X)$, corresponding to a flat vector bundle $E$ on $X$, the cotangent complex of $\mathbb{R}\mathcal{M}_{DR}(X)$ at $E$ is given by*

$$\mathbb{L}_{\mathbb{R}\mathcal{M}_{DR}(X),E} \simeq (C^*_{DR}(X, E \otimes E^*))^\vee[-1],$$

*where $C^*_{DR}(X, E \otimes E^*)$ is the algebraic de Rham cohomology of $X$ with coefficients in the flat bundle $E \otimes E^*$.*

Corollary 2.2.6.15 is only the beginning of the story; in fact we can also produce, in a similar way, $\mathbb{R}\mathcal{M}_{Dol}(X)$ and $\mathbb{R}\mathcal{M}_{Hod}(X)$, which are derived versions of the moduli stacks of Higgs bundles and $\lambda$-connections of [**S1**], and this would lead us to a *derived version of non-abelian Hodge theory.* We think this is very interesting research direction because these derived moduli also encode higher homotopical data in their tangent complexes. We hope to come back to this topic in a future work.

CHAPTER 2.3

# Complicial algebraic geometry

In this chapter we present a second context of application of the general formalism of Part I, in which the base model category is $C(k)$, the category of unbounded complexes over some ring $k$ of characteristic zero. Contrary to the previously considered applications, the general notions presented in §1.2 does not produce here notions which are very close to the usual ones for commutative rings. As a consequence the geometric intuition is here only a very loose guide.

We will present two different HAG contexts over $C(k)$. The first one is very weak in the sense that it is very easy for a stack to be geometric in this context (these geometric stacks will be called *weakly geometric*). The price to pay for this abundance of geometric stacks is that this context does not satisfy Artin's conditions and thus there is no good infinitesimal theory.

The second HAG context we consider is a bit closer to the geometric intuition, and satisfies the Artin's conditions so it behaves well infinitesimally.

Both of these "unbounded" contexts seem interesting as we are able to produce examples of geometric stacks which cannot be represented by geometric $D^-$-stacks, i.e. by the kind of geometric stacks studied in the previous chapter.

In this chapter $k$ will be a commutative $\mathbb{Q}$-algebra.

## 2.3.1. Two HA contexts

We let $\mathcal{C} := C(k)$, the model category of unbounded complexes of $k$-modules in $\mathbb{U}$. The model structure on $C(k)$ is the projective one, for which fibrations are epimorphisms and equivalences are quasi-isomorphisms. The model category $C(k)$ is a symmetric monoidal model category for the tensor product of complexes. Furthermore, it is well known that our assumptions 1.1.0.1, 1.1.0.3, 1.1.0.2 and 1.1.0.4 are satisfied.

The category $Comm(C(k))$ is the usual model category of unbounded commutative differential graded algebras over $k$, for which fibrations are epimorphisms and equivalences are quasi-isomorphisms. The category $Comm(C(k))$ will be denoted by $k - cdga$, and its objects will simply be called cdga's. For $A \in k - cdga$, the category $A - Mod$ is the category of unbounded $A$-dg-modules, again with its natural model structure. In order to avoid confusion of notations we will denote the category $A - Mod$ by $A - Mod_{dg}$. Note that for a usual commutative $k$-algebra $k'$ we have $k' - Mod_{dg} = C(k')$, whereas $k' - Mod$ will denote the usual category of $k'$-modules. Objects in $A - Mod_{dg}$ will be called *$A$-dg-modules.*

As for the case of simplicial algebras, we will set for any $E \in C(k)$

$$\pi_i(E) := H^{-i}(E).$$

When $A$ is a $k$-cdga, the $\mathbb{Z}$-graded $k$-module $\pi_*(A)$ has a natural structure of a commutative graded $k$-algebra. In the same way for $M$ an $A$-dg-module, $\pi_*(M)$ becomes a graded $\pi_*(A)$-module. Objects $A \in k - cdga$ such that $\pi_i(A) = 0$ for any

$i<0$ will be called $(-1)$-connected. Any $A \in k-cdga$ possesses a $(-1)$-connected cover $A' \longrightarrow A$, which is such that $\pi_i(A') \simeq \pi_i(A)$ for any $i \geq 0$.

As in the context of derived algebraic geometry we will denote by $M \mapsto M[1]$ the suspension functor. In the same way, $M \mapsto M[n]$ is the $n$-times iterated suspension functor (when $n$ is negative this means the $n$-times iterated loop functor).

The first new feature of complicial algebraic geometry is the existence of two interesting choices for the subcategory $\mathcal{C}_0$, both of them of particular interest depending on the context. We let $C(k)_{\leq 0}$ be the full subcategory of $C(k)$ consisting of complexes $E$ such that $\pi_i(E)=0$ for any $i<0$ (or equivalently, such that $H^i(E)=0$ for all $i>0$, which explains better the notation). We also set $k-cdga_0$ to be the full subcategory of $k-cdga$ consisting of $A \in k-cdga$ such that $\pi_i(A)=0$ for any $i \neq 0$. In the same way we denote by $k-cdga_{\leq 0}$ for the full subcategory of $k-cdga$ consisting of $A$ such that $\pi_i(A) \equiv H^{-i}(A)=0$ for any $i<0$.

LEMMA 2.3.1.1. (1) *The triplet* $(C(k), C(k), k-cdga)$ *is a HA context.*
(2) *The triplet* $(C(k), C(k)_{\leq 0}, k-cdga_0)$ *is a HA context.*

PROOF. The only non-trivial point is to show that any $A \in k-cdga_0$ is $C(k)_{\leq 0}$-good in the sense of Def. 1.1.0.10. By definition $A$ is equivalent to some usual commutative $k$-algebra, so we can replace $A$ by $k$ itself. We are then left to prove that the natural functor

$$\mathrm{Ho}(C(k))^{op} \longrightarrow \mathrm{Ho}((C(k)_{\leq 0}^{op})^{\wedge})$$

is fully faithful.

To prove this, we consider the restriction functor

$$i^*: \mathrm{Ho}((C(k)^{op})^{\wedge}) \longrightarrow \mathrm{Ho}((C(k)_{\leq 0}^{op})^{\wedge})$$

induced by the inclusion $i: C(k)_{\leq 0} \subset C(k)$. We restrict the functor $i^*$ to the full sub-categories of corepresentable objects

$$i^*: \mathrm{Ho}((C(k)^{op})^{\wedge})^{\mathrm{corep}} \longrightarrow \mathrm{Ho}((C(k)_{\leq 0}^{op})^{\wedge})^{\mathrm{corep}}.$$

A precision here: we say that a functor $F: C(k)_{\leq 0} \longrightarrow SSet$ is *corepresentable* if it is of the form

$$D \mapsto Map_{C(k)}(E, i(D))$$

for some object $E \in C(k)$ (i.e. it belongs the essential image of $\mathrm{Ho}(C(k))^{op} \longrightarrow \mathrm{Ho}((C(k)_{\leq 0}^{op})^{\wedge})$. We claim that this restricted functor $i^*$ is an equivalence of categories. Indeed, an inverse $f$ can be constructed by sending a functor $F: C(k)_{\leq 0} \longrightarrow SSet$ to the functor

$$\begin{array}{cccc} f(F): & C(k) & \longrightarrow & SSet \\ & D & \mapsto & Holim_{n \geq 0} \Omega^n F(D(\leq n)[n]), \end{array}$$

where $D(\leq n)$ is the naive truncation of $D$ defined by $D(\leq n)_m = D_m$ if $m \leq n$ and $D(\leq n)_m = 0$ if $m>n$, and $\Omega^n F(D(\leq n)[n])$ is the $n$-fold loop space of the simplicial set $F(D(\leq n)[n])$, based at the natural point $* \simeq F(0) \rightarrow F(D(\leq n)[n])$. As any $D \in C(k)$ is functorially equivalent to the homotopy limit $Holim_n D(\leq n)$, it is easy to check that the functor $f$ and $i^*$ are inverse to each other.

To finish the proof, it is enough to notice that there exists a commutative diagram (up to a natural isomorphism)

$$\begin{array}{ccc} \mathrm{Ho}(C(k))^{op} & \xrightarrow{\mathbb{R}\underline{h}} & \mathrm{Ho}((C(k))^{\wedge})^{corep} \\ & \searrow & \downarrow i^* \\ & & \mathrm{Ho}((C(k)^{op}_{\leq 0})^{\wedge})^{corep}. \end{array}$$

The Yoneda embedding $\mathbb{R}\underline{h}$ being fully faithful, this implies that the functor

$$\mathrm{Ho}(C(k))^{op} \longrightarrow \mathrm{Ho}((C(k)^{op}_{\leq 0})^{\wedge})$$

is also fully faithful. □

REMARK 2.3.1.2. There are objects $A \in k - cdga$ which are *not* $C(k)_{\leq 0}$-good. For example, we can take $A$ with $\pi_*(A) \simeq k[T, T^{-1}]$ where $T$ is in degree 2. Then, there is no nontrivial $A$-dg-module $M$ such that $M \in C(k)_{\leq 0}$: this clearly implies that $A$ can not be $C(k)_{\leq 0}$-good.

Another example is given by $A = k[T]$ where $T$ is in degree $-2$, and $M$ be the $A$-module $A[T^{-1}]$. As there are no morphisms from $M$ to $A$-modules in $C(k)_{\leq 0}$, then $M$ is sent to zero by the functor

$$\mathrm{Ho}(A - Mod)^{op} \longrightarrow \mathrm{Ho}((A - Mod^{op}_{\leq 0})^{\wedge}).$$

DEFINITION 2.3.1.3. (1) *Let $A \in k - cdga$, and $M$ be an $A$-dg-module. The $A$-dg-module $M$ is* strong *if the natural morphism*

$$\pi_*(A) \otimes_{\pi_0(A)} \pi_0(M) \longrightarrow \pi_*(M)$$

*is an isomorphism.*

(2) *A morphism $A \longrightarrow B$ in $k - cdga$ is* strongly flat *(resp.* strongly smooth, *resp.* strongly étale, *resp.* a strong Zariski open immersion*) if $B$ is strong as an $A$-dg-module, and if the morphism of affine schemes*

$$Spec\, \pi_0(B) \longrightarrow Spec\, \pi_0(A)$$

*is flat (resp. smooth, resp. étale, resp. a Zariski open immersion).*

One of the main difference between derived algebraic geometry and complicial algebraic geometry lies in the fact that the strong notions of flat, smooth, étale and Zariski open immersion are not as easily related to the general notions presented in §1.2. We have the following partial comparison result.

PROPOSITION 2.3.1.4. *Let $f : A \longrightarrow B$ be a morphism in $k - cdga$.*

(1) *If $A$ and $B$ are $(-1)$-connected, the morphism $f$ is smooth (resp. i-smooth, resp. resp. étale, resp. a Zariski open immersion) in the sense of Def. 1.2.6.1, 1.2.7.1, if and only if $f$ is strongly smooth (resp. strongly étale, resp. a strong Zariski open immersion).*

(2) *If the morphism $f$ is strongly flat (resp. strongly smooth, resp. strongly étale, resp. a strong Zariski open immersion), then it is flat (resp. smooth, resp. étale, resp. a Zariski open immersion) in the sense of Def. 1.2.6.1, 1.2.7.1.*

PROOF. (1) The proof is precisely the same as for Thm. 2.2.2.6, as the homotopy theory of $(-1)$-connected $k - cdga$ is equivalent to the one of commutative simplicial $k$-algebras, and as this equivalence preserves cotangent complexes.

(2) For flat morphisms there is nothing to prove, as all morphisms are flat in the sense of Def. 1.2.6.1 since the model category $C(k)$ is stable. Let us suppose that $f : A \longrightarrow B$ is strongly smooth (resp. strongly étale, resp. a strong Zariski open immersion). we can consider the morphism induced on the $(-1)$-connected covers

$$f' : A' \longrightarrow B',$$

where we recall that the $(-1)$-connected cover $A' \longrightarrow A$ induces isomorphims on $\pi_i$ for all $i \geq 0$, and is such that $\pi_i(A') = 0$ for all $i < 0$. As the morphism $f$ is strongly flat that the square

$$\begin{array}{ccc} A' & \xrightarrow{f'} & B' \\ \downarrow & & \downarrow \\ A & \xrightarrow[f]{} & B \end{array}$$

is homotopy co-cartesian. Therefore, as our notions of smooth, étale and Zariski open immersion are stable by homotopy push-out, it is enough to show that $f'$ is smooth (resp. étale, resp. a Zariski open immersion). But this follows from (1). □

EXAMPLE 2.3.1.5. Before going further into complicial algebraic geometry we would like to present an example illustrating the difference between the strong notions of flat, smooth, étale and Zariski open immersion and the general notions presented in §1.2, showing in particular that proposition 2.3.1.4 (2) does not have a converse.

Let $A$ be any commutative $k$-algebra, $X = Spec\, A$ the corresponding affine scheme, and $U \subset X$ be a quasi-compact open subscheme. It is easy to see that there exists a perfect complex of $A$-modules $K$, such that $U$ is the open subscheme of $X$ on which $K$ is acyclic. By Prop. 1.2.10.1 there exists then a morphism $A \longrightarrow A_K$ in $k - cdga$ which is a Zariski open immersion. Moreover, the universal property of $A \longrightarrow A_K$ shows that if $A_K$ is cohomologically concentrated in degree 0 then $U$ is affine and we have $U \simeq Spec\, \pi_0(A_K)$. Therefore, as soon as $U$ is not affine, $A_K$ cannot be concentrated in degree 0. The morphism $A \longrightarrow A_K$ is thus a Zariski open immersion, and therefore étale, but is not a strong morphism.

This example also shows that if the scheme $U$ is considered as a scheme over $C(k)$ (see §2.3.5.1), then $U$ is equivalent to $\mathbb{R}\underline{Spec}\, A_K$, and thus is affine as a stack over $C(k)$, even though $U$ is not necessarily an affine subscheme of $X$ over $k$.

The opposite model category $k - cdga^{op}$ will be denoted by $k - DAff$. We will endow it with the strong étale model topology.

DEFINITION 2.3.1.6. *A family of morphisms* $\{Spec\, A_i \longrightarrow Spec\, A\}_{i \in I}$ *in* $k - DAff$ *is a* strong étale covering family *(or simply* s-ét covering family*) if it satisfies the following two conditions.*

1. *Each morphism* $A \longrightarrow A_i$ *is strongly étale.*
2. *There exists a finite subset* $J \subset I$ *such that the family* $\{A \longrightarrow A_i\}_{i \in J}$ *is a formal covering family in the sense of 1.2.5.1.*

Using the definition of strong étale morphisms, we immediately check that a family of morphisms $\{Spec\, A_i \longrightarrow Spec\, A\}_{i \in I}$ in $k - DAff$ is a s-ét covering family if and only if there exists a finite sub-set $J \subset I$ satisfying the following two conditions.

- For all $i \in I$, the natural morphism

$$\pi_*(A) \otimes_{\pi_0(A)} \pi_0(A_i) \longrightarrow \pi_*(A_i)$$

is an isomorphism.

- The morphism of affine schemes
$$\coprod_{i\in J} Spec\,\pi_0(A_i) \longrightarrow Spec\,\pi_0(A)$$
is étale and surjective.

LEMMA 2.3.1.7. *The s-ét covering families define a model topology on $k-DAff$, which satisfies assumption 1.3.2.2.*

PROOF. The same as for Lem. 2.2.2.13. □

The model topology s-ét gives rise to a model category of stacks $k-DAff^{\sim,\text{s-ét}}$.

DEFINITION 2.3.1.8. (1) *A* $D$-stack *is an object $F \in k-DAff^{\sim,\text{s-ét}}$ which is a stack in the sense of Def. 1.3.2.1.*
(2) *The* model category of $D$-stacks *is $k-DAff^{\sim,\text{s-ét}}$, and its homotopy category is simply denoted by* $\mathrm{DSt}(k)$.

From Prop. 2.3.1.4 we get the following generalization of Cor. 2.2.2.9.

COROLLARY 2.3.1.9. *Let $A \in k-cdga$ and $A' \longrightarrow A$ be its $(-1)$-connected cover. Let us consider the natural morphisms*
$$t_0(X) := Spec\,(\pi_0 A) \longrightarrow X' = Spec\,A' \qquad X = Spec\,A \longrightarrow X' := SpecA'.$$
*Then, the homotopy base change functors*
$$\mathrm{Ho}(k-DAff/X') \longrightarrow \mathrm{Ho}(k-DAff/t_0(X))$$
$$\mathrm{Ho}(k-DAff/X') \longrightarrow \mathrm{Ho}(k-DAff/X)$$
*induces equivalences between the full sub-categories of strongly étale morphisms. Furthermore, these equivalences preserve epimorphisms of stacks.*

PROOF. Using Cor. 2.2.2.9 we see that it is enough to show that the base change along the connective cover $A' \longrightarrow A$
$$\mathrm{Ho}(k-DAff/X') \longrightarrow \mathrm{Ho}(k-DAff/X)$$
induces an equivalences from the full sub-categories of strongly étale morphism $Y' \to X'$ to the full subcategory of étale morphisms $Y \to X$. But an inverse to this functor is given by sending a strongly étale morphism $A \longrightarrow B$ to its connective cover $A' \longrightarrow B'$. □

## 2.3.2. Weakly geometric $D$-stacks

We now let $\mathbf{P}_w$ be the class of formally perfect morphisms in $k-DAff$, also called in the present context *weakly smooth morphisms*.

LEMMA 2.3.2.1. *The class $\mathbf{P}_w$ of fp morphisms and the s-ét model topology satisfy assumptions 1.3.2.11.*

PROOF. The only non-trivial thing to show is that the notion of fp morphism is local for the s-ét topology on the source. As strongly étale morphisms are also formally étale, this easily reduces to showing that being a perfect module is local for the s-ét model topology. But this follows from corollary 1.3.7.4. □

We can now state that $(C(k), C(k), k-cdga, \text{s-ét}, \mathbf{P}_w)$ is a HAG context in the sense of Def. 1.3.2.13. From our general definitions we obtain a first notion of geometric $D$-stacks.

DEFINITION 2.3.2.2. (1) *A* weakly $n$-geometric $D$-stack *is a a $D$-stack $F \in \mathrm{DSt}(k)$ which is $n$-geometric for the HAG context $(C(k), C(k), k-cdga, s\text{-ét}, \mathbf{P}_w)$.*

(2) $A$ weakly $n$-representable morphism of $D$-stacks *is an* $n$*-representable morphism for the HAG context* $(C(k), C(k), k-cdga, s\text{-}ét, \mathbf{P}_w)$.

In the context $(C(k), C(k), k-cdga, \text{s-ét}, \mathbf{P}_w)$ the i-smooth morphisms are the formally etale morphisms, and therefore Artin's conditions of Def. 1.4.3.1 can not be satisfied. There are actually many interesting weakly geometric $D$-stacks which do not have cotangent complexes, as we will see.

### 2.3.3. Examples of weakly geometric $D$-stacks

**2.3.3.1. Perfect modules.** We consider the $D$-stack **Perf**, as defined in Def. 1.3.7.5. As we have seen during the proof of Lem. 2.3.2.1 the notion of being perfect is local for the s-ét topology. In particular, for any $A \in k-cdga$, the simplicial set $\mathbf{Perf}(A)$ is naturally equivalent to the nerve of the category of equivalences between perfect $A$-dg-modules.

PROPOSITION 2.3.3.1. (1) *The* $D$*-stack* **Perf** *is categorically locally of finite presentation in the sense of Def. 1.3.6.4.*
(2) *The* $D$*-stack* **Perf** *is weakly* 1*-geometric. Furthermore its diagonal is* $(-1)$*-representable.*

PROOF. (1) Let $A = Colim_{i\in I}A_i$ be a filtered colimit of objects in $k-cdga$. We need to prove that the morphism

$$Colim_i\mathbf{Perf}(A_i) \longrightarrow \mathbf{Perf}(A)$$

is an equivalence. Let $j \in I$, and two perfect $A_j$-dg-modules $P$ and $Q$. Then, we have

$$Map_{A-Mod}(A\otimes^{\mathbb{L}}_{A_j} P, A\otimes^{\mathbb{L}}_{A_j} Q) \simeq Map_{A_j-Mod}(P, A\otimes^{\mathbb{L}}_{A_j} Q) \simeq$$

$$\simeq Map_{A_j-Mod}(P, Colim_{i\in j/J}A_i\otimes^{\mathbb{L}}_{A_j} Q).$$

As $P$ is perfect it is finitely presented in the sense of Def. 1.2.3.1, and thus we get

$$Map_{A-Mod}(A\otimes^{\mathbb{L}}_{A_j} P, A\otimes^{\mathbb{L}}_{A_j} Q) \simeq Colim_{i\in j/I}Map_{A_i-Mod}(A_i\otimes^{\mathbb{L}}_{A_j} P, A_i\otimes^{\mathbb{L}}_{A_j} Q).$$

Furthermore, as during the proof of Prop. 1.3.7.14 we can show that the same formula holds when $Map$ is replaced by the sub-simplicial set $Map^{eq}$ of equivalences. Invoking the relations between mapping spaces and loop spaces of nerves of model categories (see Appendix B), this clearly implies that the morphism

$$Colim_i\mathbf{Perf}(A_i) \longrightarrow \mathbf{Perf}(A)$$

induces isomorphisms on $\pi_i$ for $i > 0$ and an injective morphism on $\pi_0$.

It only remains to show that for any perfect $A$-dg-module $P$, there exists $j \in I$ and a perfect $A_j$-dg-module $Q$ such that $P \simeq A\otimes^{\mathbb{L}}_{A_j} Q$. For this, we use that perfect modules are precisely the retract of finite cell modules. By construction, it is clear that a finite cell $A$-dg-module is defined over some $A_j$. We can therefore write $P$ as a direct factor of $A\otimes^{\mathbb{L}}_{A_j} Q$ for some $j \in I$ and $Q$ a perfect $A_j$-dg-module. The direct factor $P$ is then determined by a projector

$$p \in [A\otimes^{\mathbb{L}}_{A_j} Q, A\otimes^{\mathbb{L}}_{A_j} Q] \simeq Colim_{i\in j/I}[A_i\otimes^{\mathbb{L}}_{A_j} Q, A_i\otimes^{\mathbb{L}}_{A_j} Q].$$

Thus, $p$ defines a projector $p_i$ in $[A_i\otimes^{\mathbb{L}}_{A_j} Q, A_i\otimes^{\mathbb{L}}_{A_j} Q]$ for some $i$, corresponding to a direct factor $P_i$ of $A_i\otimes^{\mathbb{L}}_{A_j} Q$. Clearly, we have

$$P \simeq A\otimes^{\mathbb{L}}_{A_i} P_i.$$

(2) We start by showing that the diagonal of **Perf** is $(-1)$-representable. In other words, we need to prove that for any $A \in k-cdga$, and any two perfect $A$-dg-modules $M$ and $N$, the $D$-stack

$$\begin{array}{cccc} Eq(M,N): & A-cdga & \longrightarrow & SSet_{\mathbb{V}} \\ & (A \to B) & \mapsto & Map^{eq}_{B-Mod}(M \otimes^{\mathbb{L}}_{A} B, N \otimes^{\mathbb{L}}_{A} B), \end{array}$$

(where $Map^{eq}$ denotes the sub-simplicial set of $Map$ consisting of equivalences) is a representable $D$-stack. This is true, and the proof is exactly the same as the proof of Prop. 1.3.7.14.

To construct a 1-atlas, let us chose a $\mathbb{U}$-small set $\mathcal{F}$ of representative for the isomorphism classes of finitely presented objects in $\mathrm{Ho}(k-cdga)$. Then, for any $A \in \mathcal{F}$, let $\mathcal{M}_A$ be a $\mathbb{U}$-small set of representative for the isomorphism classes of perfect objects in $\mathrm{Ho}(A-Mod_{dg})$. The fact that these $\mathbb{U}$-small sets $\mathcal{F}$ and $\mathcal{M}$ exists follows from the fact finitely presented objects are retracts of finite cell objects (see Cor. 1.2.3.8).

We consider the natural morphism

$$p: U := \coprod_{A \in \mathcal{F}} \coprod_{P \in \mathcal{M}_A} U_{A,P} := \mathbb{R}\underline{Spec}\, A \longrightarrow \mathbf{Perf}.$$

By construction the morphism $p$ is an epimorphism of stacks. Furthermore, as the diagonal of **Perf** is $(-1)$-representable each morphism $U_{A,P} = \mathbb{R}\underline{Spec}\, A \longrightarrow \mathbf{Perf}$ is $(-1)$-representable. Finally, as $U_{A,P} = \mathbb{R}\underline{Spec}\, A$ and **Perf** are both categorically locally of finite presentation, we see that for any $B \in k-cdga$ and any $Y := \mathbb{R}\underline{Spec}\, B \longrightarrow \mathbf{Perf}$, the morphism

$$U_{A,P} \times^{h}_{\mathbf{Perf}} Y \longrightarrow Y$$

is a finitely presented morphism between representable $D$-stacks. In particular, it is a perfect morphism. We therefore conclude that

$$p: U := \coprod_{A \in \mathcal{F}} \coprod_{P \in \mathcal{M}_A} U_{A,P} := \mathbb{R}\underline{Spec}\, A \longrightarrow \mathbf{Perf}$$

is a 1-atlas for **Perf**. This finishes the proof that **Perf** is weakly 1-geometric. □

**2.3.3.2. The $D$-stacks of dg-algebras and dg-categories.** We consider for any $A \in k-cdga$ the model category $A-dga$, of associative and unital $A$-algebras (in $\mathbb{U}$). The model structure on $A-dga$ is the usual one for which equivalences are quasi-isomorphisms and fibrations are epimorphisms. We consider $Ass(A)$ to be the subcategory of $A-dga$ consisting of equivalences between objects $B \in A-dga$ satisfying the following two conditions.

- The object $B$ is cofibrant in $A-dga$.
- The underlying $A$-dg-module of $B$ is perfect.

For a morphism $A \longrightarrow A'$ in $k-cdga$, we have a base change functor

$$A' \otimes_A - : Ass(A) \longrightarrow Ass(A'),$$

making $A \mapsto Ass(A)$ into a pseudo-functor on $k-cdga$. Using the usual strictification procedure, and passing to the nerve we obtain a simplicial presheaf

$$\begin{array}{cccc} \mathbf{Ass}: & k-DAff & \longrightarrow & SSet_{\mathbb{V}} \\ & A & \mapsto & N(Ass(A)). \end{array}$$

PROPOSITION 2.3.3.2. (1) *The simplicial presheaf* **Ass** *is a $D$-stack.*
(2) *The natural projection* **Ass** $\longrightarrow$ **Perf**, *which forget the algebra structure, is $(-1)$-representable.*

PROOF. (1) This is exactly the same proof as Thm. 1.3.7.2.

(2) Let $A \in k-cdga$ and $\mathbb{R}\underline{Spec}\, A \longrightarrow \mathbf{Perf}$ be a point corresponding to a perfect $A$-dg-module $E$. We denote by $\widetilde{\mathbf{Ass}_E}$ the homotopy fiber taken at $E$ of the morphism $\mathbf{Ass} \longrightarrow \mathbf{Perf}$.

LEMMA 2.3.3.3. *The D-stack* $\widetilde{\mathbf{Ass}_E}$ *is representable.*

PROOF. This is the same argument as for Prop. 2.2.6.8. We see using [**Re**, Thm. 1.1.5] that $\widetilde{\mathbf{Ass}_E}$ is the $D$-stack (over $Spec\, A$) $Map(\mathcal{A}ss, \mathbb{R}\underline{End}(E))$, of morphisms from the associative operad to the the (derived) endomorphism operad of $E$, defined the same way as in the proof of 2.2.6.8. Again the same argument as for 2.2.6.8, consisting of writing $\mathcal{A}ss$ as a homotopy colimit of free operads, reduces the statement of the lemma to prove that for a perfect $A$-dg-module $K$ of, the $D$-stack

$$\begin{array}{ccc} A-cdga & \longrightarrow & SSet \\ A' & \mapsto & Map_{(C(k))}(K, A') \end{array}$$

is representable. But this is true as it is equivalent to $\mathbb{R}\underline{Spec}\, B$, where $B$ is the (derived) free $A$-cdga over $K$. □

The previous lemma shows that $\mathbf{Ass} \longrightarrow \mathbf{Perf}$ is a $(-1)$-representable morphism, and finishes the proof of Prop. 2.3.3.2. □

COROLLARY 2.3.3.4. *The D-stack* $\mathbf{Ass}$ *is weakly* 1*-geometric.*

PROOF. Follows from Prop. 2.3.3.1 and Prop. 2.3.3.2. □

We now consider a slight modification of **Ass**, by considering dg-algebras as dg-categories with only one objects. For this, let $A \in k - cdga$. Recall that a $A$-dg-category is by definition a category enriched over the symmetric monoidal category $A - Mod_{dg}$, of $A$-dg-modules. More precisely, a $A$-dg-category $\mathcal{D}$ consists of the following data

- A set of objects $Ob(\mathcal{D})$.
- For any pair of objects $(x, y)$ in $Ob(\mathcal{D})$ an $A$-dg-module $\mathcal{D}(x, y)$.
- For any triple of objects $(x, y, z)$ in $Ob(\mathcal{D})$ a composition morphism

$$\mathcal{D}(x, y) \otimes_A \mathcal{D}(y, z) \longrightarrow \mathcal{D}(x, z),$$

  which satisfies the usual unital and associativity conditions.

The $A$-dg-categories (in the universe $\mathbb{U}$) form a category $A - dgCat$, with the obvious notion of morphisms. For an $A - dg$-category $\mathcal{D}$, we can form a category $\pi_0(\mathcal{D})$, sometimes called the *homotopy category of* $\mathcal{D}$, whose objects are the same as $\mathcal{D}$ and for which morphisms from $x$ to $y$ is the set $\pi_0(\mathcal{D}(x, y))$ (with the obvious induced compositions). The construction $\mathcal{D} \mapsto \pi_0(\mathcal{D})$ defines a functor from $A-dgCat$ to the category of $\mathbb{U}$-small categories. Recall that a morphism $f : \mathcal{D} \longrightarrow \mathcal{E}$ is then called a *quasi-equivalence* (or simply an *equivalence*) if it satisfies the following two conditions.

- For any pair of objects $(x, y)$ in $\mathcal{D}$ the induced morphism

$$f_{x,y} : \mathcal{D}(x, y) \longrightarrow \mathcal{E}(f(x), f(y))$$

  is an equivalence in $A - Mod_{dg}$.
- The induced functor

$$\pi_0(f) : \pi_0(\mathcal{D}) \longrightarrow \pi_0(\mathcal{E})$$

  is an equivalence of categories.

We also define a notion of fibration, as the morphisms $f : \mathcal{D} \longrightarrow \mathcal{E}$ satisfying the following two conditions.

- For any pair of objects $(x, y)$ in $\mathcal{D}$ the induced morphism
$$f_{x,y} : \mathcal{D}(x, y) \longrightarrow \mathcal{E}(f(x), f(y))$$
is a fibration in $A - Mod_{dg}$.
- For any object $x$ in $\mathcal{D}$, and any isomorphism $v : f(x) \to z$ in $\pi_0(\mathcal{E})$, there exists an isomorphism $u : x \to y$ in $\pi_0(\mathcal{D})$ such that $\pi_0(f)(u) = v$.

With these notions of fibrations and equivalences, the category $A - dgCat$ is a model category. This is proved in [**Tab**] when $A$ is a commutative ring. The general case of categories enriched in a well behaved monoidal model category has been worked out recently by J. Tapia (private communication).

For $A \in k - cdga$, we denote by $Cat_*(A)$ the category of equivalences between objects $\mathcal{D} \in A - dgCat$ satisfying the following two conditions.

- For any two objects $x$ and $y$ in $\mathcal{D}$ the $A$-dg-module $\mathcal{D}(x, y)$ is perfect and cofibrant in $A - Mod_{dg}$.
- The category $\pi_0(\mathcal{D})$ possesses a unique object up to isomorphism.

For a morphism $A \longrightarrow A'$ in $k - cdga$, we have a base change functor
$$- \otimes_A A' : Cat_*(A) \longrightarrow Cat_*(A')$$
obtained by the formula
$$Ob(\mathcal{D} \otimes_A A') := Ob(\mathcal{D}) \qquad (\mathcal{D} \otimes_A A')(x, y) := \mathcal{D}(x, y) \otimes_A A'.$$
This makes $A \mapsto Cat_*(A)$ into a pseudo-functor on $k - cdga$. Strictifying and applying the nerve construction we get a well defined simplicial presheaf
$$\begin{array}{cccc} \mathbf{Cat}_* : & k - cdga & \longrightarrow & SSet_{\mathbb{V}} \\ & A & \mapsto & N(Cat_*(A)). \end{array}$$

It is worth mentioning that $\mathbf{Cat}_*$ is not a stack, since there are non trivial twisted forms of objects in $Cat_*(k)$ with respect to the étale topology on $k$. These twisted forms can be interpreted as certain *stacks in dg-categories* having *locally* a unique object up to equivalences, but they might have either no global objects or several nonequivalent global objects. We will not explicitly describe the stack associated to $\mathbf{Cat}_*$, as this will be irrelevant for the sequel, and will simply consider $\mathbf{Cat}_*$ as an object in $D\mathrm{St}(k)$.

There exists a morphism of simplicial presheaves
$$B : \mathbf{Ass} \longrightarrow \mathbf{Cat}_*,$$
which sends an associative $A$-algebra $C$ to the $A$-dg-category $BC$, having a unique object $*$ and $C$ as the endomorphism $A$-algebra of $*$. The morphism $B$ will be considered as a morphism in $D\mathrm{St}(k)$.

PROPOSITION 2.3.3.5. *The morphism $B : \mathbf{Ass} \longrightarrow \mathbf{Cat}_*$ is weakly 1-representable, fp and an epimorphism of $D$-stacks.*

PROOF. We will prove a more precise result, giving explicit description of the homotopy fibers of $B$. For this, we start by some model category considerations relating associative dga to dg-categories. Recall that for any $A \in k - cdga$, we have two model categories, $A - dga$ and $A - dgCat$, of associative $A$-algebras and $A$-dg-categories. We consider $\mathbf{1}/A - dgCat$, the model category of dg-categories together with a distinguised object. More precisely, $\mathbf{1}$ is the $A$-dg-category with a unique object and $A$ as its endomorphism dg-algebra, and $\mathbf{1}/A - dgCat$ is the comma model category. The functor $B : A - dga \longrightarrow \mathbf{1}/A - dgCat$ is a left Quillen functor. Indeed, its right adjoint $\Omega_*$, sends a pointed $A$-dg-category $\mathcal{D}$ to the $A$-algebra $\mathcal{D}(*,*)$ of endomorphisms of the distinguished point $*$. Clearly, the adjunction morphism $\Omega_* B \Rightarrow Id$ is an equivalence, and thus the functor

$$B : \mathrm{Ho}(A - dga) \longrightarrow \mathrm{Ho}(\mathbf{1}/A - dgCat)$$

is fully faithful. As a consequence, we get that for any $C, C' \in A - dga$ there exists a natural homotopy fiber sequence of simplicial sets

$$Map_{A-dgCat}(\mathbf{1}, BC') \longrightarrow Map_{A-dga}(B, B') \longrightarrow Map_{A-dgCat}(BC, BC').$$

Now, let $A \in k - cgda$ and $B$ be an associative $A$-algebra, corresponding to a morphism

$$x : X := \mathbb{R}\underline{Spec}\, A \longrightarrow \mathbf{Ass}.$$

We consider the $D$-stack

$$F := \mathbf{Ass} \times^h_{\mathbf{Cat}_*} X.$$

Using the relations between mapping spaces in $A - dga$ and $A - dgCat$ above we easily see that the $D$-stack $F$ is connected (i.e. $\pi_0(F) \simeq *$). In particular, in order to show that $F$ is weakly 1-geometric, it is enough to prove that $\Omega_B F$ is a representable $D$-stack. Using again the homotopy fiber sequence of mapping spaces above we see that the $D$-stack $\Omega_B F$ can be described as

$$\begin{array}{cccc} \Omega_B F : & A - cdga & \longrightarrow & SSet_{\mathbb{V}} \\ & (A \to A') & \mapsto & Map_{\mathbf{1}/A'-dgCat}(S^1 \otimes^{\mathbb{L}} \mathbf{1}, B(B \otimes^{\mathbb{L}}_A A')) \end{array}$$

where $S^1 \otimes^{\mathbb{L}} \mathbf{1}$ is computed in the model category $A - dgCat$. One can easily check that one has an isomorphism in $\mathbf{1}/A - dgCat$

$$S^1 \otimes^{\mathbb{L}} \mathbf{1} \simeq B(A[T, T^{-1}]),$$

where $A[T, T^{-1}] := A \otimes_k k[T, T^{-1}]$. Therefore, there is a natural equivalence

$$Map_{\mathbf{1}/A'-dgCat}(S^1 \otimes^{\mathbb{L}} \mathbf{1}, B(B \otimes^{\mathbb{L}}_A A')) \simeq Map_{A-dga}(A[T, T^{-1}], B \otimes^{\mathbb{L}}_A A'),$$

and thus the $D$-stack $\Omega_B F$ can also be described by

$$\begin{array}{cccc} \Omega_B F : & A - cdga & \longrightarrow & SSet_{\mathbb{V}} \\ & (A \to A') & \mapsto & Map_{A-dga}(A[T, T^{-1}], B \otimes^{\mathbb{L}}_A A'). \end{array}$$

The morphism of $k$-algebras $k[T] \longrightarrow k[T, T^{-1}]$ induces natural morphisms (here $B^\vee$ is the dual of the perfect $A$-dg-module $B$)

$$Map_{A-dga}(A[T, T^{-1}], B \otimes^{\mathbb{L}}_A A') \longrightarrow Map_{A-dga}(A[T], B \otimes^{\mathbb{L}}_A A') \simeq Map_{A-Mod}(B^\vee, A').$$

It is not difficult to check that this gives a morphism of $D$-stacks

$$\Omega_B F \longrightarrow \mathbb{R}\underline{Spec}\, \mathbb{L}F(B^\vee),$$

where $\mathbb{L}F(B^\vee)$ is the derived free $A$-cdga over the $A$-dg-module $B^\vee$. Furthermore, applying Prop. 1.2.10.1 this morphism is easily seen to be representable by an open Zariski immersion, showing that $\Omega_B F$ is thus a representable $D$-stack. □

COROLLARY 2.3.3.6. *The $D$-stack $\mathbf{Cat}_*$ is weakly 2-geometric.*

PROOF. This follows immediately from Cor. 2.3.3.4, Prop. 2.3.3.5 and the general criterion of Cor. 1.3.4.5. □

Let $A \in k-cdga$ and $B$ be an associative $A$-algebra corresponding to a morphism of $D$-stacks

$$B : X := \mathbb{R}\underline{Spec}\,A \longrightarrow \mathbf{Ass}.$$

We define a $D$-stack $B^*$ on $A - cdga$ in the following way

$$\begin{array}{rccc} B^* : & A - cdga & \longrightarrow & SSet_{\mathbb{V}} \\ & (A \to A') & \mapsto & Map_{A-dga}(A[T,T^{-1}], B \otimes^{\mathbb{L}}_A A'). \end{array}$$

The $D$-stack $B^*$ is called the *$D$-stack of invertible elements in $B$*. The $D$-stack $B^*$ possesses in fact a natural loop stack structure (i.e. the above functor factors naturally, up to equivalence, through a functor from $A - cdga$ to the category of loop spaces), induced by the Hopf algebra structure on $A[T,T^{-1}]$. This loop stack structure is also the one induced by the natural equivalences

$$B^*(A') \simeq \Omega_* Map_{A-dgCat}(*, B \otimes^{\mathbb{L}}_A A').$$

Delooping gives another $D$-stack

$$\begin{array}{rccc} K(B^*,1) : & A - cdga & \longrightarrow & SSet_{\mathbb{V}} \\ & (A \to A') & \mapsto & K(B^*(A'),1). \end{array}$$

The following corollary is a reformulation of Prop. 2.3.3.5 and of its proof.

COROLLARY 2.3.3.7. *Let $A \in k - cdga$ and $B \in \mathbf{Ass}(A)$ corresponding to an associative $A$-algebra $B$. Then, there is a natural homotopy cartesian square of $D$-stacks*

$$\begin{array}{ccc} \mathbf{Ass} & \longrightarrow & \mathbf{Cat}_* \\ \uparrow & & \uparrow{\scriptstyle B} \\ K(B^*,1) & \longrightarrow & \mathbb{R}\underline{Spec}\,A. \end{array}$$

## 2.3.4. Geometric $D$-stacks

We now switch to the HA context $(C(k), C(k)_{\leq 0}, k - cdga_0)$. Recall that $C(k)_{\leq 0}$ is the subcategory of $C(k)$ consisting of $(-1)$-connected object, and $k - cdga_0$ is the subcategory of $k - cdga$ of objects cohomologically concentrated in degree 0. Within this HA context we let $\mathbf{P}$ to be the class of formally perfect and formally i-smooth morphisms in $k - DAff$. Morphisms in $\mathbf{P}$ will simply be called *fip-smooth morphisms*. There are no easy description of them, but Prop. 1.2.8.3 implies that a morphism $f : A \longrightarrow B$ be a morphism is fip-smooth if it satisfies the following two conditions.

- The cotangent complex $\mathbb{L}_{B/A} \in \mathrm{Ho}(A - Mod_{dg})$ is perfect.
- For any $R \in k - cdga_0$, any connected module $M \in R - Mod_{dg}$ and any morphism $B \longrightarrow R$, one has
$$\pi_0(\mathbb{L}^{\vee}_{B/A} \otimes^{\mathbb{L}}_B M) = 0.$$

The converse is easily shown to be true if $A$ and $B$ are both $(-1)$-connected.

LEMMA 2.3.4.1. *The class $\boldsymbol{P}$ of fip-smooth morphisms and the s-ét model topology satisfy assumptions 1.3.2.11.*

PROOF. We see that the only non trivial part is to show that the notion of fip-smooth morphism is local for the s-ét topology. For fp-morphisms this is Cor. 1.3.7.4. For fi-smooth morphisms this is an easy consequence of the definitions. □

From Lem. 2.3.4.1 we get a HAG context $(C(k), C(k)_{\leq 0}, k - cdga_0, s - et, \mathbf{P})$.

DEFINITION 2.3.4.2. (1) *An $n$-*geometric $D$-stack *is a $D$-stack $F \in D\mathrm{St}(k)$ which is $n$-geometric for the HAG context $(C(k), C(k)_{\leq 0}, k - cdga_0, s - et, \mathbf{P})$.*

(2) *A* strongly $n$-representable morphism of $D$-stacks *is an $n$-representable morphism of $D$-stacks for the HAG context* $(C(k), C(k)_{\leq 0}, k - cdga_0, s - et, \mathbf{P})$.

Note that $\mathbf{P} \subset \mathbf{P}_w$ and therefore that any $n$-geometric $D$-stack is weakly $n$-geometric.

LEMMA 2.3.4.3. *The s-ét topology and the class* $\mathbf{P}$ *of fip-smooth morphisms satisfy Artin's conditions of Def. 1.4.3.1.*

PROOF. This is essentially the same proof as Prop. 2.2.3.2, and is even more simple as one uses here the strongly étale topology. □

COROLLARY 2.3.4.4. *Any $n$-geometric $D$-stack possesses an obstruction theory (relative to the HA context* $(C(k), C(k)_{\leq 0}, k - cdga_0)$*).*

PROOF. This follows from Lem. 2.3.4.3 and Thm. 1.4.3.2. □

### 2.3.5. Examples of geometric $D$-stacks

**2.3.5.1. $D^-$-stacks and $D$-stacks.** We consider the normalization functor $N : sk - Alg \longrightarrow k - cdga$, sending a simplicial commutative $k$-algebra $A$ to its normalization $N(A)$, with its induced structure of commutative differential graded algebra. The pullback functor gives a Quillen adjunction

$$N_! : k\text{-}D^-Aff^{\sim,et} \longrightarrow k\text{-}DAff^{\sim,s-et} \qquad k\text{-}D^-Aff^{\sim,et} \longleftarrow k\text{-}DAff^{\sim,s-et} : N^*.$$

As the functor $N$ is known to be homotopically fully faithful, the left derived functor

$$j := \mathbb{L}N_! : D^-\mathrm{St}(k) \longrightarrow D\mathrm{St}(k)$$

is fully faithful. We can actually characterize the essential image of the functor $j$ as consisting of all $D$-stacks $F$ for which for any $A \in k - cdga$ with $(-1)$-connected cover $A' \longrightarrow A$, the morphism $\mathbb{R}F(A') \longrightarrow \mathbb{R}F(A)$ is an equivalence. In other words, for any $F \in k - D^-Aff^{\sim,et}$, and any $A \in k - cdga$, we have

$$\mathbb{R}j(F)(A) \simeq \mathbb{R}F(D(A')),$$

where $D(A') \in sk - Alg$ is a denormalization of $A'$, the $(-1)$-connected cover of $A$.

PROPOSITION 2.3.5.1. (1) *For any $A \in sk - Alg$, we have*

$$j(\mathbb{R}\underline{Spec}\,A) \simeq \mathbb{R}\underline{Spec}\,N(A).$$

(2) *The functor $j$ commutes with homotopy limits and homotopy colimits.*
(3) *The functor $j$ sends $n$-geometric $D^-$-stacks to $n$-geometric $D$-stacks.*

PROOF. This is clear. □

The previous proposition shows that any $n$-geometric $D^-$-stack gives rise to an $n$-geometric $D$-stack, and thus provides us with a lot of examples of those.

**2.3.5.2. CW-perfect modules.** Let $A \in k - cdga$. We define by induction on $n = b - a$ the notion of a perfect CW-$A$-dg-module of amplitude contained in $[a, b]$.

DEFINITION 2.3.5.2. (1) *A* perfect CW-$A$-dg-module of amplitude contained in $[a, a]$ *is an $A$-dg-module $M$ isomorphic in* $\mathrm{Ho}(A - Mod_{dg})$ *to* $P[-a]$, *with $P$ a projective and finitely presented $A$-dg-module (as usual in the sense of definitions 1.2.3.1 and 1.2.4.1).*

(2) *Assume that the notion of perfect CW-A-dg-module of amplitude contained in* $[a, b]$ *has been defined for any* $a \leq b$ *such that* $b - a = n - 1$. *A* perfect CW-$A$-dg-module of amplitude contained in $[a, b]$, *with* $b - a = n$, *is an* $A$*-dg-module* $M$ *isomorphic in* $\mathrm{Ho}(A - Mod_{dg})$ *to the homotopy cofiber of a morphism*

$$P[-a-1] \longrightarrow N,$$

*where* $P$ *is projective and finitely presented, and* $N$ *is a perfect CW-A-dg-module of amplitude contained in* $[a+1, b]$.

The perfect CW-$A$-dg-modules satisfy the following stability conditions.

LEMMA 2.3.5.3. (1) *If* $M$ *is a perfect CW-A-dg-module of amplitude contained in* $[a, b]$, *and* $A \longrightarrow A'$ *is a morphism in* $k - cdga$, *then* $A' \otimes^{\mathbb{L}}_A M$ *is a perfect CW-A-dg-module of amplitude contained in* $[a, b]$.

(2) *Let* $A \in k - cdga_{\leq 0}$ *be a* $(-1)$*-connected* $k - cdga$. *Then, any perfect A-dg-module is a perfect CW-A-dg-module of amplitude* $[a, b]$ *for some integer* $a \leq b$.

(3) *Let* $A \in k - cdga_{\leq 0}$ *be a* $(-1)$*-connected* $k - cdga$, *and* $M \in A - Mod_{dg}$. *If there exists a s-et covering* $A \longrightarrow A'$ *such that* $A' \otimes^{\mathbb{L}}_A M$ *is a perfect CW-A-dg-module of amplitude contained in* $[a, b]$, *then so is* $M$.

PROOF. Only (2) and (3) requires a proof. Furthermore, (3) clearly follows from (2) and the local nature of perfect modules (see Cor. 1.3.7.4), and thus it only remains to prove (2). But this is proved in [**EKMM**, III.7]. $\square$

We define a sub-$D$-stack $\mathbf{Perf}^{CW}_{[a,b]} \subset \mathbf{Perf}$, consisting of all perfect modules locally equivalent to some CW-dg-modules of amplitude contained in $[a, b]$. Precisely, for $A \in k - cdga$, $\mathbf{Perf}^{CW}_{[a,b]}(A)$ is the sub-simplicial set of $\mathbf{Perf}(A)$ which is the union of all connected components corresponding to $A$-dg-modules $M$ such that there is an s-ét covering $A \longrightarrow A'$ with $A' \otimes^{\mathbb{L}}_A M$ a perfect CW-$A'$-dg-module of amplitude contained in $[a, b]$.

PROPOSITION 2.3.5.4. *The D-stack* $\mathbf{Perf}^{CW}_{[a,b]}$ *is* 1*-geometric.*

PROOF. We proceed by induction on $n = b - a$. For $n = 0$, the $D$-stack $\mathbf{Perf}^{CW}_{[a,a]}$ is simply the stack of vector bundles **Vect**, which is 1-geometric by our general result Cor. 1.3.7.12. Assume that $\mathbf{Perf}^{CW}_{[a,b]}$ is known to be 1-geometric for $b - a < n$. Shifting if necessary, we can clearly assume that $a = 0$, and thus that $b = n - 1$.

That the diagonal of $\mathbf{Perf}^{CW}_{[a,b]}$ is $(-1)$-representable comes from the fact that $\mathbf{Perf}^{CW}_{[a,b]} \longrightarrow \mathbf{Perf}$ is a monomorphisms and from Prop. 2.3.3.1. It remains to construct a 1-atlas for $\mathbf{Perf}^{CW}_{[a,b]}$.

Let $A \in k - cdga$. We consider the model category $Mor(A - Mod_{dg})$, whose objects are morphisms in $A - Mod_{dg}$ (and with the usual model category structure, see e.g. [**Ho1**]). We consider the subcategory $Mor(A - Mod_{dg})'$ consisting cofibrant objects $u : M \rightarrow N$ in $Mor(A - Mod_{dg})$ (i.e. $u$ is a cofibration between cofibrant $A$-dg-modules) such that $M$ is projective of finite presentation and $N$ is locally a perfect CW-$A$-dg-module of amplidtude contained in $[0, n-1]$. Morphisms in $Mor(A - Mod_{dg})'$ are taken to be equivalences in $Mor(A - Mod_{dg})$. The correspondence $A \mapsto Mor(A - Mod_{dg})'$ defines a pseudo-functor on $k - cdga$, and after strictification and passing to the nerve we get this way a simplicial presheaf

$$\begin{array}{rccc} F: & k - cdga & \longrightarrow & SSet_{\mathbb{V}} \\ & A & \mapsto & N(Mor(A - Mod_{dg})'). \end{array}$$

There exists a morphism of $D$-stacks

$$F \longrightarrow \mathbf{Vect} \times^h \mathbf{Perf}^{CW}_{[0,n-1]}$$

that sends an object $M \to N$ to $(M, N)$. This morphism is easily seen to be $(-1)$-representable, as its fiber at an $A$-point $(M, N)$ is the $D$-stack of morphisms from $M$ to $N$, or equivalently is $\mathbb{R}\underline{Spec}\, B$ where $B$ is the derived free $A$-cdga over $M^\vee \otimes^{\mathbb{L}}_A N$. By induction hypothesis we deduce that the $D$-stack $F$ itself if 1-geometric.

Finally, there exists a natural morphism of $D$-stacks

$$p : F \longrightarrow \mathbf{Perf}^{CW}_{[-1,n-1]}$$

sending a morphism $u : M \to N$ in $Mor(A - Mod_{dg})'$ to its homotopy cofiber. The morphism $p$ is an epimorphism of $D$-stacks by definition of being locally a perfect CW-dg-module, thus it only remains to show that $p$ is fip-smooth. Indeed, this would imply the existence of a 1-atlas for $\mathbf{Perf}^{CW}_{[-1,n-1]}$, and thus that it is 1-geometric. Translating we get that $\mathbf{Perf}^{CW}_{[a,b]}$ is 1-geometric for $b - a = n$.

In order to prove that $p$ is fip-smooth, let $A \in k - cdga$, and $K : \mathbb{R}\underline{Spec}\, A \longrightarrow \mathbf{Perf}^{CW}_{[-1,n-1]}$ be an $A$-point, corresponding to a perfect CW-$A$-dg-module of amplitude contained in $[-1, n-1]$. We consider the homotopy cartesian square

$$\begin{array}{ccc} F & \xrightarrow{p} & \mathbf{Perf}^{CW}_{[-1,n-1]} \\ \uparrow & & \uparrow \\ F' & \xrightarrow[p']{} & \mathbb{R}\underline{Spec}\, A. \end{array}$$

We need to prove that $p$ is a fip-smooth morphism. The $D$-stack $F'$ has natural projection $F' \longrightarrow \mathbf{Vect}$ given by $F' \longrightarrow F \longrightarrow \mathbf{Vect}$, where the second morphism sends a morphism $M \to N$ to the vector bundle $M$. We therefore get a morphism of $D$-stacks

$$F' \longrightarrow \mathbf{Vect} \times^h \mathbb{R}\underline{Spec}\, A \longrightarrow \mathbb{R}\underline{Spec}\, A.$$

As the morphism $\mathbf{Vect} \longrightarrow *$ is smooth, it is enough to show that the morphism

$$F' \longrightarrow \mathbf{Vect} \times^h \mathbb{R}\underline{Spec}\, A$$

is fip-smooth. For this we consider the homotopy cartesian square

$$\begin{array}{ccc} F' & \longrightarrow & \mathbf{Vect} \times^h \mathbb{R}\underline{Spec}\, A \\ \uparrow & & \uparrow \\ F'_0 & \longrightarrow & \mathbb{R}\underline{Spec}\, A, \end{array}$$

where the section $\mathbb{R}\underline{Spec}\, A \longrightarrow \mathbf{Vect}$ correspond to a trivial rank $r$ vector bundle $A^r$. It only remains to show that the morphism $F'_0 \longrightarrow \mathbb{R}\underline{Spec}\, A$ is fip-smooth. The $D$-stack $F'_0$ over $A$ can then be easily described (using for example our general Cor. B.0.8) as

$$\begin{array}{cccc} F'_0 : & A - cdga & \longrightarrow & SSet_{\mathbb{V}} \\ & (A \to A') & \mapsto & Map_{A-Mod}(K[-1], A^r). \end{array}$$

In other words, we can write $F'_0 \simeq \mathbb{R}\underline{Spec}\, B$, where $B$ is the derived free $A - cdga$ over $(K[-1])^r$. But, as $K$ is a perfect CW-$A$-dg-module of amplitude contained in $[-1, n-1]$, $(K[-1])^r$ is a perfect CW $A$-dg-module of amplitude contained in $[0, n]$. The proposition will then follow from the general lemma.

Let $A \in k - cdga$, and recall that the forgetful functor $A - cdga \longrightarrow A - Mod_{dg}$ is right Quillen, and that its derived left adjoint

$$\mathrm{Ho}(A - Mod_{dg}) \longrightarrow \mathrm{Ho}(A - cdga)$$

sends, by definition, an $A$-module $E$ to the *derived free $A$-cdga over $E$*.

LEMMA 2.3.5.5. *Let $A \in k - cdga$, and $K$ be a perfect CW-A-dg-module of amplitude contained in $[0, n]$. Let $B$ be the derived free $A - cdga$ over $K$, and*

$$p : Y := \mathbb{R}\underline{Spec}\, B \longrightarrow X := \mathbb{R}\underline{Spec}\, A$$

*be the natural projection. Then $p$ is fip-smooth.*

PROOF. Let $k'$ be any commutative $k$-algebra, and $B \longrightarrow k$ be any morphism in $\mathrm{Ho}(k - cdga)$, corresponding to a point $x : Spec\, k \longrightarrow X$. Then, we have an isomorphism in $D(k')$

$$\mathbb{L}_{Y/X,x} \simeq K \otimes^{\mathbb{L}}_A k'.$$

We thus see that $\mathbb{L}_{Y/X,x}$ is a perfect complex of $k'$-modules of Tor amplitude concentrated in degrees $[0, n]$. In particular, it is clear that for any complex $M$ of $k'$-modules such that $\pi_i(M) = 0$ for any $i > 0$, then we have

$$[\mathbb{L}_{Y/X,x}, M]_{D(k')} \simeq \pi_0(\mathbb{L}^{\vee}_{Y/X,x} \otimes^{\mathbb{L}}_{k'} M) \simeq 0.$$

This shows that the $A$-algebra $B$ is fip-smooth, and thus shows the lemma. □

The previous lemma finishes the proof of the proposition. □

A direct consequence of Prop. 2.3.5.4 and Thm. 1.4.3.2 is the following.

COROLLARY 2.3.5.6. *The $D$-stack $\mathbf{Perf}^{CW}_{[a,b]}$ has an obstruction theory (relative to the HA context $(C(k), C(k)_{\leq 0}, k - cdga_0)$. For any $A \in k - cdga_0$ and any point $E : X := \mathbb{R}\underline{Spec}\, A \longrightarrow \mathbf{Perf}^{CW}_{[a,b]}$ corresponding to a perfect CW-A-dg-module $E$, there are natural isomorphisms in $\mathrm{Ho}(A - Mod_{dg})$*

$$\mathbb{L}_{\mathbf{Perf}^{CW}_{[a,b]},E} \simeq E^{\vee} \otimes^{\mathbb{L}}_A E[-1]$$

$$\mathbb{T}_{\mathbf{Perf}^{CW}_{[a,b]},E} \simeq E^{\vee} \otimes^{\mathbb{L}}_A E[1].$$

PROOF. The first part of the corollary follows from our Thm. 1.4.3.2 and Prop. 2.3.5.4. Let $A$ and $E : X := \mathbb{R}\underline{Spec}\, A \longrightarrow \mathbf{Perf}^{CW}_{[a,b]}$ as in the statement. We have

$$\mathbb{L}_{\mathbf{Perf}^{CW}_{[a,b]},E} \simeq \mathbb{L}_{\Omega_E \mathbf{Perf}^{CW}_{[a,b]},E}[-1].$$

Moreover, $\Omega_E \mathbf{Perf}^{CW}_{[a,b]} \simeq \mathbb{R}\underline{Aut}(E)$, where $\mathbb{R}\underline{Aut}(E)$ is the $D$-stack of self-equivalences of the perfect module $E$ as defined in §1.3.7. By Prop. 1.3.7.14 we know that $\mathbb{R}\underline{Aut}(E)$ is representable, and furthermore that the natural inclusion morphism

$$\mathbb{R}\underline{Aut}(E) \longrightarrow \mathbb{R}\underline{End}(E)$$

is a formally étale morphism of representable $D$-stacks. Therefore, we have

$$\mathbb{L}_{\Omega_E \mathbf{Perf}^{CW}_{[a,b]},E} \simeq \mathbb{L}_{\mathbb{R}\underline{End}(E),Id}.$$

Finally, we have

$$\mathbb{R}\underline{End}(E) \simeq \mathbb{R}\underline{Spec}\, B,$$

where $B$ is the derived free $A$-cdga over $E^{\vee} \otimes^{\mathbb{L}}_A E$. This implies that

$$\mathbb{L}_{\mathbb{R}\underline{End}(E),Id} \simeq E^{\vee} \otimes^{\mathbb{L}}_A E,$$

and by what we have seen that

$$\mathbb{L}_{\mathbf{Perf}^{CW}_{[a,b]},E} \simeq E^{\vee} \otimes^{\mathbb{L}}_A E[-1].$$

□

REMARK 2.3.5.7. (1) It is important to note that, if

$$N^* : D\mathrm{St}(k) \longrightarrow D^-\mathrm{St}(k)$$

denotes the restriction functor, we have

$$N^*(\mathbf{Perf}^{CW}_{[a,b]}) \simeq \mathbf{Perf}_{[a,b]},$$

where $\mathbf{Perf}_{[a,b]}$ is the sub-stack of $\mathbf{Perf} \in D^-\mathrm{St}(k)$ consisting of perfect modules of Tor-amplitude contained in $[a,b]$. However, the two $D$-stacks $\mathbf{Perf}^{CW}_{[a,b]}$ and $j(\mathbf{Perf}_{[a,b]})$ are not the same. Indeed, for any $A \in k - cdga$ with $A'$ as $(-1)$-connective cover, we have

$$j(\mathbf{Perf}_{[a,b]})(A) \simeq \mathbf{Perf}_{[a,b]}(A').$$

In general the natural morphism

$$- \otimes^{\mathbb{L}}_A A' : \mathbf{Perf}_{[a,b]}(A') \longrightarrow \mathbf{Perf}^{CW}_{[a,b]}(A)$$

is not an equivalence. For example, let us suppose that there exist a non zero element $x \in \pi_{-1}(A) \simeq [A, A[1]]$, then the matrix

$$\begin{pmatrix} Id_A & x \\ 0 & Id_{A[1]} \end{pmatrix}$$

defines an equivalence $A \bigoplus A[1] \simeq A \bigoplus A[1]$ of $A$-dg-modules which is not induced by an equivalence $A' \bigoplus A'[1] \simeq A' \bigoplus A'[1]$ of $A'$-dg-modules.

(2) We can also show that the $D^-$-stack $\mathbf{Perf}_{[a,b]} \in D^-\mathrm{St}(k)$ is $(n+1)$-geometric for $n = b - a$. The proof is essentially the same as for Prop. 2.3.5.4 and will not be reproduced here. Of course, the formula for the cotangent complex remains the same. See [**To-Va1**] for more details.

**2.3.5.3. CW-dg-algebras.** Recall the existence of the following diagram of $D$-stacks

$$\begin{array}{ccc} & & \mathbf{Ass} \\ & & \downarrow \\ \mathbf{Perf}^{CW}_{[a,b]} & \longrightarrow & \mathbf{Perf}. \end{array}$$

DEFINITION 2.3.5.8. *The $D$-stack of CW-dg-algebras of amplitude contained in $[a,b]$ is defined by the following homotopy cartesian square*

$$\begin{array}{ccc} \mathbf{Ass}^{CW}_{[a,b]} & \longrightarrow & \mathbf{Ass} \\ \downarrow & & \downarrow \\ \mathbf{Perf}^{CW}_{[a,b]} & \longrightarrow & \mathbf{Perf}. \end{array}$$

Let $B$ be an associative $k$-dga, and let $M$ be a $B$-bi-dg-module (i.e. a $B \otimes^{\mathbb{L}}_k B^{op}$-dg-module). We can form the square zero extension $B \oplus M$, which is another associative $k$-dga together with a natural projection $B \oplus M \longrightarrow B$. The simplicial set of (derived) derivations from $B$ to $M$ is then defined as

$$\mathbb{R}Der_k(B,M) := Map_{k-dga/B}(B, B \oplus M).$$

The same kind of proof as for Prop. 1.2.1.2 shows that there exists an object $\mathbb{L}^{Ass}_B \in \mathrm{Ho}(B \otimes^{\mathbb{L}}_k B^{op} - Mod_{dg})$ an natural isomorphisms in $\mathrm{Ho}(SSet_{\mathbb{V}})$

$$\mathbb{R}Der_k(B,M) \simeq Map_{B \otimes^{\mathbb{L}}_k B^{op} - Mod}(\mathbb{L}^{Ass}_B, M).$$

We then set

$$\mathbb{R}\underline{Der}_k(B,M) := \mathbb{R}\underline{Hom}_{B\otimes_k^{\mathbb{L}}B^{op}-Mod}(\mathbb{L}_B^{Ass},M) \in \mathrm{Ho}(C(k)),$$

where $\mathbb{R}\underline{Hom}_{B\otimes_k^{\mathbb{L}}B^{op}-Mod}$ denotes the $\mathrm{Ho}(C(k))$-enriched derived Hom's of the $C(k)$-model category $B\otimes_k^{\mathbb{L}}B^{op}-Mod_{dg}$.

COROLLARY 2.3.5.9. *The D-stack* $\mathbf{Ass}_{[a,b]}^{CW}$ *is 1-geometric. For any global point* $B : Spec\, k \longrightarrow \mathbf{Ass}_{[a,b]}^{CW}$, *corresponding to an associative k-dga B, one has a natural isomorphism in* $\mathrm{Ho}(C(k))$

$$\mathbb{T}_{\mathbf{Ass}_{[a,b]}^{CW},B} \simeq \mathbb{R}\underline{Der}_k(B,B)[1].$$

PROOF. Using Prop. 2.3.3.2 we see that the natural projection

$$\mathbf{Ass}_{[a,b]}^{CW} \longrightarrow \mathbf{Perf}_{[a,b]}^{CW}$$

is $(-1)$-representable. Therefore, by Prop. 2.3.5.4 we know that $\mathbf{Ass}_{[a,b]}^{CW}$ is 1-geometric. In patricular, Thm. 1.4.3.2 implies that it has an obstruction theory relative to the HA context $(C(k), C(k)_{\leq 0}, k-cdga_0)$. We thus have

$$\mathbb{T}_{\mathbf{Ass}_{[a,b]}^{CW},B} \simeq \mathbb{T}_{\Omega_B\mathbf{Ass}_{[a,b]}^{CW},B}[1].$$

The identification

$$\mathbb{T}_{\Omega_B\mathbf{Ass}_{[a,b]}^{CW},B} \simeq \mathbb{R}\underline{Der}_k(B,B)$$

follows from the exact same argument as Prop. 2.2.6.9, using $\mathbb{L}_B^{Ass}$ instead of $\mathbb{L}_B^{\mathcal{O}}$. □

**2.3.5.4. The *D*-stack of negative CW-dg-categories.** Recall from §2.3.3.2 the existence of the morphism of $D$-stacks

$$B : \mathbf{Ass} \longrightarrow \mathbf{Cat}_*,$$

sending an associative $k$-algebra $C$ to the dg-category $BC$ having a unique object and $C$ as endomorphisms of this object.

DEFINITION 2.3.5.10. *Let $n \leq 0$ be an integer. The D-*stack of CW-dg-categories of amplitude contained in $[n,0]$ *is defined as the full sub-D-stack* $\mathbf{Cat}_{*,[n,0]}^{CW}$ *of* $\mathbf{Cat}_*$ *consisting of the essential image of the morphism*

$$\mathbf{Ass}_{[n,0]}^{CW} \longrightarrow \mathbf{Ass} \longrightarrow \mathbf{Cat}_*.$$

More precisely, for $A \in k-cdga$, one sets $\mathbf{Cat}_{*,[n,0]}^{CW}(A)$ to be the sub-simplicial set of $\mathbf{Cat}_*(A)$ consisting of $A$-dg-categories $\mathcal{D}$ such that for any $* \in Ob(\mathcal{D})$ the $A$-dg-module $\mathcal{D}(*,*)$ is (locally) a perfect CW-$A$-dg-module of amplitude contained in $[n,0]$.

Recall that for any associative $k$-dga $B$, one has a model category $B\otimes_k^{\mathbb{L}}B^{op}-Mod_{dg}$ of $B$-bi-dg-modules. This model category is naturally tensored and co-tensored over the symmetric monoidal model category $C(k)$, making it into a $C(k)$-model category in the sense of [**Ho1**]. The derived $\mathrm{Ho}(C(k))$-enriched Hom's of $B\otimes_k^{\mathbb{L}}B^{op}-Mod_{dg}$ will then be denoted by $\mathbb{R}\underline{Hom}_{B\otimes_k^{\mathbb{L}}B^{op}}$.

Finally, for any associative $k$-dga $B$, one sets

$$\mathbb{HH}_k(B,B) := \mathbb{R}\underline{Hom}_{B\otimes_k^{\mathbb{L}}B^{op}}(B,B) \in \mathrm{Ho}(C(k)),$$

where $B$ is considered as $B$-bi-dg-module in the obvious way.

THEOREM 2.3.5.11. (1) *The morphism*

$$B : \mathbf{Ass}_{[n,0]}^{CW} \longrightarrow \mathbf{Cat}_{*,[n,0]}^{CW}$$

*is a 1-representable fip-smooth covering of D-stacks.*

(2) *The associated D-stack to* $\mathbf{Cat}^{CW}_{*,[n,0]}$ *is a 2-geometric D-stack.*

(3) *If C is an associative k-dg-algebra, corresponding to a point* $C : Spec\, k \longrightarrow \mathbf{Ass}^{CW}_{[n,0]}$, *then one has a natural isomorphism in* $\mathrm{Ho}(C(k))$

$$\mathbb{T}_{\mathbf{Cat}^{CW}_{*,[n,0]},BC} \simeq \mathbb{HH}_k(C,C)[2].$$

PROOF. (1) Using Cor. 2.3.3.7, it is enough to show that for any $A \in k-cdga$, and any associative $A$-algebra $C$, which is a perfect CW-$A$-dg-module of amplitude contained in $[n,0]$, the morphism of $D$-stack $K(C^*,1) \longrightarrow \mathbb{R}\underline{Spec}\, A$ is 1-representable and fip-smooth. For this it is clearly enough to show that the $D$-stack $C^*$ is representable and that the morphism $C^* \longrightarrow \mathbb{R}\underline{Spec}\, A$ is fip-smooth. We already have seen during the proof of Prop. 2.3.3.5 that $\overline{C^*}$ is representable and a Zariski open sub-$D$-stack of $\mathbb{R}\underline{Spec}\, F$, where $F$ is the free $A$-cdga over $C^\vee$. It is therefore enough to see that $A \longrightarrow F$ is fip-smooth, which follows from Lem. 2.3.5.5 as by assumption $C^\vee$ is a perfect CW-$A$-dg-module of amplitude contained in $[0,n]$.

(2) Follows from Cor. 2.3.5.9 and Cor. 1.3.4.5.

(3) We consider a point $C : Spec\, k \longrightarrow \mathbf{Ass}^{CW}[n,0]$, and the homotopy cartesian square

$$\begin{array}{ccc}
\mathbf{Ass}^{CW}_{[n,0]} & \longrightarrow & \mathbf{Cat}^{CW}_{*,[n,0]} \\
\uparrow & & \uparrow{\scriptstyle C} \\
K(C^*,1) & \longrightarrow & Spec\, k.
\end{array}$$

By (2), Cor. 2.3.5.9 and Thm. 1.4.3.2 we know that all the stacks in the previous square have an obstruction theory, and thus a cotangent complex (relative to the HA context $(C(k), C(k)_{\leq 0}, k - cdga_0)$). Therefore, one finds a homotopy fibration sequence of complexes of $k$-modules

$$\mathbb{T}_{K(C^*,1),*} \longrightarrow \mathbb{T}_{\mathbf{Ass}^{CW}_{[n,0]},C} \longrightarrow \mathbb{T}_{\mathbf{Cat}^{CW}_{*,[n,0]},BC}$$

that can also be rewritten as

$$C[1] \longrightarrow \mathbb{R}\underline{Der}_k(C,C)[1] \longrightarrow \mathbb{T}_{\mathbf{Cat}^{CW}_{*,[n,0]},BC}.$$

The morphism $C \longrightarrow \mathbb{R}\underline{Der}_k(C,C)$ can be described in the following way. The $C$-bi-dg-module $\mathbb{L}^{Ass}_C$ can be easily identified with the homotopy fiber (in the model category of $C \otimes^{\mathbb{L}}_k C^{op}$-dg-modules) of the multiplication morphism

$$C \otimes^{\mathbb{L}}_k C^{op} \longrightarrow C.$$

The natural morphism $\mathbb{L}^{Ass}_C \longrightarrow C \otimes^{\mathbb{L}}_k C^{op}$ then induces our morphism on the level of derivations

$$C \simeq \mathbb{R}\underline{Hom}_{C\otimes^{\mathbb{L}}_k C^{op}-Mod}(C\otimes^{\mathbb{L}}_k C^{op}, C) \longrightarrow \mathbb{R}\underline{Hom}_{C\otimes^{\mathbb{L}}_k C^{op}-Mod}(\mathbb{L}^{Ass}_C, B) \simeq \mathbb{R}\underline{Der}_k(C,C).$$

In particular, we see that there exists a natural homotopy fiber sequence in $C(k)$

$$\mathbb{HH}_k(C,C) = \mathbb{R}\underline{Hom}_{C\otimes^{\mathbb{L}}_k C^{op}-Mod}(C,C) \longrightarrow C \longrightarrow \mathbb{R}\underline{Der}_k(C,C).$$

We deduce that there exists a natural isomorphism in $\mathrm{Ho}(C(k))$

$$\mathbb{T}_{\mathbf{Cat}^{CW}_{*,[n,0]},BC}[-2] \simeq \mathbb{HH}_k(C,C).$$

□

An important corollary of Thm. 2.3.5.11 is given by the following fact. It appears in many places in the literature but we know of no references including a proof of it. For this, we recall that for any commutative $k$-algebra $k'$ we denote by $Cat_*(k')$ the category of equivalences between $k'$-dg-categories satisfying the following two conditions.

- For any two objects $x$ and $y$ in $\mathcal{D}$ the complex of $k'$-module $\mathcal{D}(x,y)$ is perfect (and cofibrant in $C(k')$).
- The category $\pi_0(\mathcal{D})$ possesses a unique object up to isomorphism.

We finally let $Cat_*^{[n,0]}(k')$ be the full subcategory of $Cat_*(k')$ consisting of objects $\mathcal{D}$ such that for any two objects $x$ and $y$ the perfect complex $\mathcal{D}(x,y)$ has Tor amplitude contained in $[n,0]$ for some $n \leq 0$.

COROLLARY 2.3.5.12. *Let $\mathcal{D} \in Cat_*^{[n,0]}(k')$. Then, the homotopy fiber $\mathbb{D}ef_{\mathcal{D}}$, taken at the point $\mathcal{D}$, of the morphism of simplicial sets*

$$N(Cat_*^{[n,0]}(k[\epsilon])) \longrightarrow N(Cat_*^{[n,0]}(k))$$

*is given by*

$$\mathbb{D}ef_{\mathcal{D}} \simeq Map_{C(k)}(k, \mathbb{HH}_k(\mathcal{D},\mathcal{D})[2]).$$

*In particular, we have*

$$\pi_i(\mathbb{D}ef_{\mathcal{D}}) \simeq \mathbb{HH}_k^{2-i}(\mathcal{D},\mathcal{D}).$$

In the above corollary we have used $\mathbb{HH}_k(\mathcal{D},\mathcal{D})$, the Hochschild complex of a dg-category $\mathcal{D}$. It is defined the same way as for associative dg-algebras, and when $\mathcal{D}$ is equivalent to $BC$ for an associative dg-alegbra $C$ we have

$$\mathbb{HH}_k(\mathcal{D},\mathcal{D}) \simeq \mathbb{HH}_k(C,C).$$

Finally, we would like to mention that restricting to negatively graded dg-categories seems difficult to avoid if we want to keep the existence of a cotangent complex.

COROLLARY 2.3.5.13. *Assume that $k$ is a field. Let $C \in \mathbf{Ass}_{[a,b]}^{CW}(k)$ be a $k$-point corresponding to an associative dg-algebra $C$. If we have $H^i(C) \neq 0$ for some $i > 0$ then the $D$-stack $\mathbf{Cat}_*$ does not have a cotangent complex at the point $BC$.*

PROOF. Suppose that $\mathbf{Cat}_*$ does have a cotangent complex at the point $BC$. Then, as so does the $D$-stack $\mathbf{Ass}_{[a,b]}^{CW}$ (by Cor. 2.3.5.9), we see that the homotopy fiber, taken at the point $BC$, of the morphism $B : \mathbf{Ass} \longrightarrow \mathbf{Cat}_*$ has a cotangent complex at $C$. This homotopy fiber is $K(C^*,1)$ (see Cor. 2.3.3.7), and thus we would have

$$\mathbb{L}_{K(C^*,1),C} \simeq C^\vee[-1].$$

Let $k[\epsilon_{i-1}]$ be the square zero extension of $k$ by $k[i-1]$ (i.e. $\epsilon_{i-1}$ is in degree $-i+1$), for some $i > 0$ as in the statement. We now consider the homotopy fiber sequence

$$Map_{C(k)}(C^\vee[-1], k[i-1]) \longrightarrow K(C^*,1)(k[\epsilon_{i-1}]) \longrightarrow K(C^*,1)(k).$$

Considering the long exact sequence in homotopy we find

$$\pi_1(K(C^*,1)(k[\epsilon_{i-1}])) = H^0(C)^* \oplus H^{i-1}(C) \longrightarrow \pi_1(K(C^*,1)(k)) = H^0(C)^* \longrightarrow$$

$$\pi_0(Map_{C(k)}(C^\vee[-1], k[i-1])) = H^i(C) \longrightarrow \pi_0(K(C^*,1)(k[\epsilon_{i-1}])) = *.$$

This shows that the last morphism must be injective, which can not be the case as soon as $H^i(C) \neq 0$. □

REMARK 2.3.5.14. (1) Some of the results in Thm. 2.3.5.11 were announced in [**To-Ve2**, Thm. 5.6]. We need to warn the reader that [**To-Ve2**, Thm. 5.6] is not correct for the description of $\widetilde{\mathbb{R}Cat_{\mathcal{O}}}$ briefly given before that theorem (the same mistake appears in [**To2**, Thm. 4.4]). Indeed, $\widetilde{\mathbb{R}Cat_{\mathcal{O}}}$ would correspond to *isotrivial* deformations of dg-categories, for which the underlying complexes of morphisms stays locally constant. Therefore, the tangent complex of $\widetilde{\mathbb{R}Cat_{\mathcal{O}}}$ can not be the full Hochschild complex as stated in [**To-Ve2**, Thm. 5.6]. Our theorem 2.3.5.11 corrects this mistake.

(2) We like to consider our Thm. 2.3.5.11 (3) and Cor. 2.3.5.12 as a possible explanation of the following sentence in [**Ko-So**, p. 266]:
*"In some sense, the full Hochschild complex controls deformations of the $A_\infty$-category with one object, such that its endomorphism space is equal to $A$."*

Furthermore, our homotopy fibration sequence

$$K(C^*,1) \longrightarrow \mathbf{Ass}^{CW}_{[n,0]} \longrightarrow \mathbf{Cat}^{CW}_{*,[n,0]}$$

is the geometric global counter-part of the well known exact triangles of complexes (see e.g. [**Ko**, p. 59])

$$C[1] \longrightarrow \mathbb{R}\underline{Der}_k(C,C)[2] \longrightarrow \mathbb{HH}_k(C,C)[2] \xrightarrow{+1}$$

as we pass from the former to the latter by taking tangent complexes at the point $C$.

(3) We saw in Cor. 2.3.5.13 that the full $D$-stack $\mathbf{Cat}_*$ can not have a reasonable infinitesimal theory. We think it is important to mention that even Cor. 2.3.5.12 cannot reasonably be true if we remove the assumption that $\mathcal{D}(x,y)$ is of Tor-amplitude contained in $[n,0]$ for some $n \leq 0$. Indeed, for any commutative $k$-algebra $k'$, the morphism

$$B : \mathbf{Ass}(k') \longrightarrow \mathbf{Cat}_*(k')$$

is easily seen to induce an isomorphism on $\pi_0$ and a surjection on $\pi_1$. From this and the fact that $\mathbb{T}_{\mathbf{Ass},C} = \mathbb{R}\underline{Der}_k(C,C)[1]$, we easily deduce that the natural morphism

$$\pi_0(\mathbb{R}\underline{Der}_k(C,C)[1]) \longrightarrow \pi_0(\mathbb{D}ef_{BC})$$

is surjective. Therefore, if Cor. 2.3.5.12 were true for $\mathcal{D} = BC$, we would have that the morphism

$$H^1(\mathbb{R}\underline{Der}_k(C,C)) \longrightarrow \mathbb{HH}^2(C,C)$$

is surjective, or equivalently that the natural morphism

$$H^2(C) \longrightarrow H^2(\mathbb{R}\underline{Der}_k(C,C))$$

is injective. But this is not the case in general, as the morphism $C \longrightarrow \mathbb{R}\underline{Der}_k(C,C)$ can be zero (for example when $C$ is commutative). It is therefore not strictly correct to state that the Hochschild cohomology of an associative dg-algebra controls its deformation as a dg-category, contrary to what appears in several references (including some of the authors !).

(4) For $n=0$, we can easily show that the restriction $N^*(\mathbf{Cat}^{CW}_{*,[0,0]}) \in D^-\mathrm{St}(k)$ is a 2-geometric $D^-$-stack in the sense of §2.2. Furthermore, its truncation $t_0 N^*(\mathbf{Cat}^{CW}_{*,[0,0]}) \in \mathrm{Ho}(k - Aff^{\sim,et})$ is naturally equivalent to the Artin 2-stack of $k$-linear categories with one object. The tangent complex of

$N^*(\mathbf{Cat}^{CW}_{*,[0,0]})$ at a point $C$ corresponding to a $k$-alegbra projective of finite type over $k$ is then the usual Hochschild complex $HH(C,C)$ computing the Hochschild cohomology of $C$.

Of course $\mathbf{Cat}^{CW}_{*,[n,0]}$ is only a rough approximation to what should be the stack of dg-categories, and in particular we think that $\mathbf{Cat}^{CW}_{*,[n,0]}$ is not suited for dealing with dg-categories coming from algebraic geometry. Let for example $X$ be a smooth and projective variety over a field $k$; it is known that the the derived category $D_{qcoh}(X)$ is of the form $\mathrm{Ho}(B - Mod_{dg})$ for some associative $dg$-algebra $B$ which is perfect as a complex of $k$-modules. It is however very unlikely that $B$ can be chosen to be concentrated in degrees $[n,0]$ for some $n \leq 0$ (by construction $H^i(B) = Ext^i(E,E)$ for some compact generator $E \in D_{qcoh}(X)$). So the dg-algebra $B$ when considered as a dg-category will *not* define a point in $\mathbf{Cat}^{CW}_{*,[n,0]}$. Another, more serious problem comes from the fact that the dg-algebra $B$ is not uniquely determined, but is only unique up to Morita equivalence. As a consequence, the variety $X$ can deform and $B$ might not follow this deformation (though another, Morita equivalent, dg-algebra will follow the deformation), and thus $X \mapsto B$ will not be a morphism of stacks (even locally around $X$). Therefore, it seems very important to consider the stack $\mathbf{Cat}_*$ modulo Morita equivalences. We also think that passing to Morita equivalences will solve the problem of the non geometricity of $\mathbf{Cat}_*$ mentioned above. This direction is currently being investigated by M. Anel.

CHAPTER 2.4

# Brave new algebraic geometry

In this final chapter we briefly present brave new algebraic geometry[1], i.e. algebraic geometry over ring spectra. We will emphasize the main differences with derived algebraic geometry, and the subject will be studied in more details in future works.

As in the case of complicial algebraic geometry, we will present two distinct HAG contexts (see Cor. 2.4.1.11) which essentially differs in the choice of the class **P**. The first one, where **P** is chosen to be the class of strongly étale morphisms (see Def. 2.4.1.3), is suited for defining *brave new* Deligne-Mumford stacks, and, as all the contexts based on "strong" morphisms, it is geometrically very close to usual algebraic geometry. The second HAG context, where **P** is chosen to be the class of fip-smooth morphisms (i.e. formally perfect and formally i-smooth morphisms), is weaker and is similar to the corresponding weak context already presented in complicial algebraic geometry: it allows to define *brave new* geometric stacks which are not Deligne-Mumford. Since the notion of fip-smooth morphism for brave new rings behaves differently from commutative rings (see Prop. 2.4.1.5), the geometric intuition in this context is once again a bit far from standard algebraic geometry (e.g. smooth morphisms are not necessary flat). Nonetheless we think that this context is very interesting, as it is not only geometrically reasonable (e.g. it satisfies Artin's conditions), but it is also able to "see" some of the interesting new phenomena arising in the theory of structured ring spectra.

## 2.4.1. Two HAG contexts

We let $\mathcal{C} := Sp^{\Sigma}$, the category of symmetric spectra in $\mathbb{U}$. The model structure we are going to use on $Sp\Sigma$ is the so called positive stable model structure described in [**Shi**]. This model structure is Quillen equivalent to the usual model structure, but is much better behaved with respect to homotopy theory of monoids and modules objects. The model category $Sp^{\Sigma}$ is a symmetric monoidal model category for the smash product of symmetric spectra. Furthermore, all our assumptions 1.1.0.1, 1.1.0.3, 1.1.0.2 and 1.1.0.4 are satisfied thanks to [**MMSS**, Theorem 14.5], [**Shi**, Thm. 3.1, Thm. 3.2], and [**Shi**, Cor. 4.3] in conjunction with Lemma [**HSS**, 5.4.4].

The category $Comm(Sp^{\Sigma})$ is the category of commutative symmetric ring spectra, together with the positive stable model structure. The category $Comm(Sp^{\Sigma})$ will be denoted by $S - Alg$, and its objects will simply be called *commutative $S$-algebras* or also *bn rings* (where *bn* stands for *brave new*). For any $E \in Sp^{\Sigma}$, we will set

$$\pi_i(E) := \pi_i^{stab}(RE),$$

where $RE$ is a fibrant replacement of $E$ in $Sp^{\Sigma}$, and $\pi_i^{stab}(RE)$ are the naive stable homotopy groups of the $\Omega$-spectrum $RE$. Note that if $E \in Sp^{\Sigma}$ is fibrant, then there is a natural isomorphism $\pi_*^{stab}E \simeq \pi_* E$, and that a map $f : E' \to E''$ is a weak equivalence in $Sp^{\Sigma}$ if and only if $\pi_* f$ is an isomorphism.

[1]The term "brave new rings" was invented by F. Waldhausen to describe structured ring spectra; we have only adapted it to our situation.

When $A$ is a commutative $S$-algebra, the $\mathbb{Z}$-graded abelian group $\pi_*(A)$ has a natural structure of a commutative graded algebra. In the same way, when $M$ an $A$-module, $\pi_*(M)$ becomes a graded $\pi_*(A)$-module. An object $E$ will be called connective, or $(-1)$-connected, if $\pi_i(E) = 0$ for all $i < 0$. We let $\mathcal{C}_0$ be $Sp^{\Sigma}_c$, be full subcategory of connective objects in $Sp^{\Sigma}$. We let $\mathcal{A}$ be $S - Alg_0$, the full subcategory of $Comm(Sp^{\Sigma})$ consisting of commutative $S$-algebras $A$ with $\pi_i(A) = 0$ for any $i \neq 0$. If we denote by $H : CommRings \longrightarrow Comm(\mathcal{C})$ the Eilenberg-MacLane functor, then $S - Alg_0$ is the subcategory of $S - Alg$ formed by all the commutative $S$-algebras equivalent to some $Hk$ for some commutative ring $k$.

LEMMA 2.4.1.1. *The triplet* $(Sp^{\Sigma}, Sp^{\Sigma}_c, S - Alg_0)$ *is a HA context.*

PROOF. The only thing to check is that any object $A \in S - Alg_0$ is $Sp^{\Sigma}_c$-good in the sense of Def. 1.1.0.10. Thanks to the equivalence between the homotopy theory of $Hk$-modules and of complexes of $k$-modules (see [**EKMM**, Thm. IV.2.4]), this has already been proved during the proof of Lem. 2.3.1.1 (2) . □

The following example lists some classes of formally étale maps in brave new algebraic geometry, according to our general definitions in Chapter 1.1.

EXAMPLE 2.4.1.2.
(1) If $A$ and $B$ are connective $S$-algebras, a morphism $A \to B$ is formally *thh*-étale if and only if it is formally étale ([**Min**, Cor. 2.8]).
(2) A morphism of (discrete) commutative rings $R \to R'$ is formally étale if and only if the associated morphism $HR \to HR'$ of bn rings is formally étale if and only if the associated morphism $HR \to HR'$ of bn rings is formally *thh*-étale ([**HAGI**, §5.2]).
(3) the complexification map $KO \to KU$ is *thh*-formally étale (by [**Ro**, p. 3]) hence formally étale. More generally, the same argument shows that any *Galois extension* of bn rings, according to J. Rognes [**Ro**], is formally *thh*-étale, hence formally étale.
(4) There exist examples of formally étale morphisms of bn-rings which are not *thh*-étale (see [**Min**] or [**HAGI**, §5.2]).

As in the case of complicial algebraic geometry (Ch. 2.3) we find it useful to introduce also *strong* versions for properties of morphisms between commutative $S$-algebras.

DEFINITION 2.4.1.3. (1) *Let* $A \in S - Alg$, *and* $M$ *be an* $A$*-module. The* $A$*-module* $M$ *is* strong *if the natural morphism*
$$\pi_*(A) \otimes_{\pi_0(A)} \pi_0(M) \longrightarrow \pi_*(M)$$
*is an isomorphism.*
(2) *A morphism* $A \longrightarrow B$ *in* $S - Alg$ *is* strongly flat *(resp.* strongly (formally) smooth, *resp.* strongly (formally) étale, *resp.* a strong Zariski open immersion*) if* $B$ *is strong as an* $A$*-module, and if the morphism of affine schemes*
$$Spec\, \pi_0(B) \longrightarrow Spec\, \pi_0(A)$$
*is flat (resp. (formally) smooth, resp. (formally) étale, resp. a Zariski open immersion).*

One of the main difference between derived algebraic geometry and unbounded derived algebraic geometry was that the strong notions of flat, smooth, étale and Zariski open immersion are not as easily related to the corresponding general notions presented in §1.2. In the present situation, the comparison is even more loose as typical phenomena arising from the existence of Steenrod operations in characteristic $p$, make the notion of smooth morphisms of $S$-algebras rather subtle, and definitely different from the above notion of strongly smooth morphisms. We do not think this is a problem of the theory, but rather we think of this as an interesting new feature of brave new algebraic geometry, as compared to derived algebraic geometry, and well worthy of investigation.

PROPOSITION 2.4.1.4. *Let $f : A \longrightarrow B$ be a morphism in $S - Alg$.*

1. *If $A$ and $B$ are connective, the morphism $f$ is étale (resp. a Zariski open immersion) in the sense of Def. 1.2.6.1, if and only if $f$ is strongly étale (resp. a strong Zariski open immersion).*
2. *If the morphism $f$ is strongly flat ( resp. strongly étale, resp. a strong Zariski open immersion), then it is flat (resp. étale, resp. a Zariski open immersion) in the sense of Def. 1.2.6.1.*

PROOF. (1) The proof is the same as for Thm. 2.2.2.6.

(2) The proof is the same as for Prop. 2.3.1.4. □

The reader will notice that the proof of Thm. 2.2.2.6 (2) does not apply to the present context as a smooth morphism of commutative rings is in general not smooth when considered as a morphism of commutative $S$-algebras. The typical example of this phenomenon is the following.

PROPOSITION 2.4.1.5.

- *The canonical map $H\mathbb{Q} \to H(\mathbb{Q}[T])$ is smooth, i-smooth and perfect.*
- *The canonical map $H\mathbb{F}_p \to H(\mathbb{F}_p[T])$ is strongly smooth but* not *formally smooth, nor formally i-smooth (Def. 1.2.7.1).*

PROOF. For any discrete commutative ring $k$, we have a canonical map $a_k : Hk[T] := F_{Hk}(Hk) \to H(k[T])$ of commutative $Hk$-algebras, corresponding to the map $H(k \to k[T])$ pointing the element $T$. Now

$$\pi_*(Hk[T]) \simeq \bigoplus_{r\geq 0} \mathrm{H}_*(\Sigma_r, k)$$

where in the group homology $\mathrm{H}_*(\Sigma_r, k)$, $k$ is a trivial $\Sigma_r$-module. Since $\mathrm{H}_n(\Sigma_r, \mathbb{Q}) = 0$, for $n \neq 0$, and $\mathrm{H}_0(\Sigma_r, \mathbb{Q}) \simeq \mathbb{Q}$, for any $r \geq 0$ we see that $a_{\mathbb{Q}}$ is a stable homotopy equivalence, and therefore a weak equivalence ([**HSS**, Thm. 3.1.11]). In other words $H(\mathbb{Q}[T])$ "is" the free commutative $H\mathbb{Q}$-algebra on one generator; therefore it is finitely presented over $H\mathbb{Q}$ and, since for any (discrete) commutative ring $k$ we have

$$\mathbb{L}_{Hk[T]/Hk} \simeq Hk[T],$$

the cotangent complex $\mathbb{L}_{H(\mathbb{Q}[T])/H\mathbb{Q}}$ is free of rank one over $H(\mathbb{Q}[T])$, hence projective and perfect. So $H\mathbb{Q} \to H(\mathbb{Q}[T])$ is smooth and perfect.

Let's move to the char $p > 0$ case. It is clear that $H\mathbb{F}_p \to H(\mathbb{F}_p[T])$ is strongly smooth; let's suppose that it is formally smooth. In particular $\mathbb{L}_{H(\mathbb{F}_p[T])/H\mathbb{F}_p}$ is a projective $H(\mathbb{F}_p[T])$-module. Therefore $\pi_*\mathbb{L}_{H(\mathbb{F}_p[T])/H\mathbb{F}_p}$ injects into

$$\pi_* \coprod_E H(\mathbb{F}_p[T]) \simeq \prod_E \mathbb{F}_p[T]$$

(concentrated in degree 0). But, by [**Ba-McC**, Thm. 4.2] and [**Ri-Rob**, Thm. 4.1],

$$\pi_*\mathbb{L}_{H(\mathbb{F}_p[T])/H\mathbb{F}_p} \simeq (H(\mathbb{F}_p[T]))_*(H\mathbb{Z})$$

and the last ring has $(H\mathbb{F}_p)_*(H\mathbb{Z})$ as a direct summand (using the augmentation $H(\mathbb{F}_p[T]) \to H\mathbb{F}_p$). Now, it is known that $(H\mathbb{F}_p)_*(H\mathbb{Z})$ is not concentrated in degree 0 (it is a polynomial $\mathbb{F}_p$-algebra in positive degrees generators, for $p = 2$, and the tensor product of such an algebra with an exterior $\mathbb{F}_p$-algebra for odd $p$). Therefore $H\mathbb{F}_p \to H(\mathbb{F}_p[T])$ cannot be formally smooth. □

We remark again that, as now made clear by the proof above, the conceptual reason for the non-smoothness of $H\mathbb{F}_p \to H(\mathbb{F}_p[T])$ is essentially the existence of (non-trivial) Steenrod operations in characteristic $p > 0$. Since formal smoothness is stable under base-change, we also conclude that $H\mathbb{Z} \to H(\mathbb{Z}[T])$ is not formally smooth. The same argument also shows that this morphism is not formally i-smooth.

The following example shows that the converse of Prop. 2.4.1.4 (2) is false in general.

EXAMPLE 2.4.1.6. The complexification map $m : KO \to KU$ is formally étale but not strongly formally étale. In fact, we have

$$\pi_*m : \pi_*(KO) = \mathbb{Z}[\eta, \beta, \lambda^{\pm 1}]/(\eta^3, 2\eta, \eta\beta, \beta^2 - 4\lambda) \longrightarrow \pi_*(KO) = \mathbb{Z}[\nu^{\pm 1}],$$

with $deg(\eta) = 1, deg(\beta) = 4, deg(\lambda) = 8, deg(\nu) = 2$, $\pi_*m(\eta) = 0$, $\pi_*m(\beta) = 2\nu^2$ and $\pi_*m(\lambda) = \nu^4$. In particular $\pi_0 m$ is an isomorphism (hence étale) but $m$ is not strong. We address the reader to [**HAGI**, Rmk. 5.2.9] for an example, due to M. Mandell, of a non connective formally étale extension of $H\mathbb{F}_p$, which is therefore not strongly formally étale. There also exist examples of Zariski open immersion $HR \longrightarrow A$, here $R$ is a smooth commutative $k$-algebra, such that $A$ possesses non trivial negative homotopy groups (see [**HAGI**, §5.2] for more details).

The opposite model category $S - Alg$ will be denoted by $SAff$. We will endow it with the following *strong étale model topology.*

DEFINITION 2.4.1.7. *A family of morphisms* $\{Spec\, A_i \longrightarrow Spec\, A\}_{i \in I}$ *in* $SAff$ *is a* strong étale covering family *(or simply* s-ét covering family*) if it satisfies the following two conditions.*

1. *Each morphism* $A \longrightarrow A_i$ *is strongly étale.*
2. *There exists a finite sub-set* $J \subset I$ *such that the family* $\{A \longrightarrow A_i\}_{i \in J}$ *is a formal covering family in the sense of 1.2.5.1.*

Using the definition of strong étale morphisms, we immediately check that a family of morphisms $\{Spec\, A_i \longrightarrow Spec\, A\}_{i \in I}$ in $SAff$ is a s-ét covering family if and only if there exists a finite sub-set $J \subset I$ satisfying the following two conditions.

- For all $i \in I$, the natural morphism
$$\pi_*(A) \otimes_{\pi_0(A)} \pi_0(A_i) \longrightarrow \pi_*(A_i)$$
is an isomorphism.
- The morphism of affine schemes
$$\coprod_{i \in J} Spec\, \pi_0(A_i) \longrightarrow Spec\, \pi_0(A)$$
is étale and surjective.

LEMMA 2.4.1.8. *The s-ét covering families define a model topology on* $SAff$*, that satisfies assumption 1.3.2.2.*

PROOF. The same as for Lem. 2.2.2.13. □

The model topology s-ét gives rise to a model category of stacks $SAff^{\sim, \text{s-ét}}$.

DEFINITION 2.4.1.9. (1) *An* $S$-stack *is an object* $F \in SAff^{\sim,s\text{-}ét}$ *which is a stack in the sense of Def. 1.3.2.1.*
(2) *The* model category of $S$-stacks *is* $SAff^{\sim,s\text{-}ét}$, *and its homotopy category will be simply denoted by* $\mathrm{St}(S)$.

We now set $\mathbf{P}$ to be the class of fip-smooth morphisms (i.e. formally perfect and formally i-smooth morphisms) in $SAlg$, and $\mathbf{P}_{\text{s-ét}}$ be the class of strongly étale morphisms.

LEMMA 2.4.1.10. (1) *The class* $\boldsymbol{P}_{s\text{-}ét}$ *of strongly étale morphisms and the s-ét model topology satisfy assumptions 1.3.2.11.*
(2) *The class* $\boldsymbol{P}$ *of fip-smooth morphisms and the s-ét model topology satisfy assumptions 1.3.2.11.*

PROOF. It is essentially the same as for Lem. 2.3.2.1. □

COROLLARY 2.4.1.11. (1) *The 5-tuple* $(Sp^{\Sigma}, Sp^{\Sigma}, S-Alg, s\text{-}ét, \mathbf{P}_{s\text{-}ét})$ *is a HAG context.*
(2) *The 5-tuple* $(Sp^{\Sigma}, Sp^{\Sigma}_{c}, S-Alg_0, s\text{-}ét, \mathbf{P})$ *is a HAG context.*

According to our general theory, the notions of morphisms in $\mathbf{P}$ and $\mathbf{P}_{\text{s-ét}}$ gives two notions of geometric stacks in $SAff^{\sim,\text{s-ét}}$.

DEFINITION 2.4.1.12. (1) A $n$-geometric Deligne-Mumford $S$-stack *is an $n$-geometric $S$-stack with respect to the class* $\boldsymbol{P}_{s\text{-}ét}$ *of strongly étale morphisms.*
(2) An $n$-geometric $S$-stack *is an $n$-geometric $S$-stack with respect to the class* $\boldsymbol{P}$ *of fip-smooth morphisms.*

Of course, as $\mathbf{P}_{\text{s-ét}}$ is included in $\mathbf{P}$ any strong $n$-geometric Deligne-Mumford $S$-stack is an $n$-geometric $S$-stack.

Finally, the reader can easily check the following proposition.

PROPOSITION 2.4.1.13. (1) *The topology s-ét and the class* $\boldsymbol{P}_{s\text{-}ét}$ *satisfy Artin condition (for the HA context* $(Sp^{\Sigma}, Sp^{\Sigma}, S-Alg)$*).*
(2) *The topology s-ét and the class* $\boldsymbol{P}$ *satisfy Artin's condition (for the HA context* $(Sp^{\Sigma}, Sp^{\Sigma}_{c}, S-Alg_0)$*).*

In particular, we obtain as a corollary of Thm. 1.4.3.2 that any $n$-geometric $S$-stack has an obstruction theory relative to the HA context $(Sp^{\Sigma}, Sp^{\Sigma}_{c}, S-Alg_0)$. In the same way, any strong $n$-geometric Deligne-Mumford $S$-stack has an obstruction theory relative to the context $(Sp^{\Sigma}, Sp^{\Sigma}, S-Alg)$.

Without going into details, we mention that all the examples of geometric $D$-stacks given in the previous chapter can be generalized to examples of geometric $S$-stacks. One fundamental example is $\mathbf{Perf}^{CW}_{[a,b]}$ of perfect CW-modules of amplitude contained in $[a,b]$, which by a similar argument as for Prop. 2.3.5.4 is a 1-geometric $S$-stack.

## 2.4.2. Elliptic cohomology as a Deligne-Mumford $S$-stack

In this final section we present the construction of a 1-geometric Deligne-Mumford $S$-stack using the sheaf of spectra of topological modular forms.

The Eilenberg-MacLane spectrum construction (see [**HSS**, Ex. 1.2.5]) gives rise to a fully faithful functor

$$\mathbb{L}H_! : \mathrm{St}(\mathbb{Z}) \longrightarrow \mathrm{St}(S),$$

which starts from the homotopy category of stacks on the usual étale site of affine schemes, i.e.

$$\mathrm{St}(\mathbb{Z}) := \mathrm{Ho}(\mathbb{Z} - Aff^{\sim,\text{ét}}).$$

This functor has a right adjoint, called the *truncation functor*

$$h^0 := H^* : \mathrm{St}(S) \longrightarrow \mathrm{St}(\mathbb{Z}),$$

simply given by composing a simplicial presheaf $F : SAff^{op} \longrightarrow SSet$ with the functor $H : Aff \longrightarrow S - Aff$.

Let us denote by $\overline{\mathcal{E}}$ the moduli stack of generalized elliptic curves with integral geometric fibers, which is the standard compactification of the moduli stack of elliptic curves by adding the nodal curves at infinity (see e.g. [**Del-Rap**, IV], where it is denoted by $\mathcal{M}_{(1)}$); recall that $\overline{\mathcal{E}}$ is a Deligne-Mumford stack, proper and smooth over $Spec\,\mathbb{Z}$ ([**Del-Rap**, Prop. 2.2]).

As shown by recent works of M. Hopkins, H. Miller, P. Goerss, N. Strickland, C. Rezk and M. Ando, there exists a natural presheaf of commutative $S$-algebras on the small étale site $\overline{\mathcal{E}}_{\text{ét}}$ of $\overline{\mathcal{E}}$. We will denote this presheaf by *tmf*. Recall that by construction, if $U = Spec\,A \longrightarrow \overline{\mathcal{E}}$ is an étale morphism, corresponding to an elliptic curve $E$ over the ring $A$, then *tmf*($U$) is the (connective) elliptic cohomology theory associated to the formal group of $E$ (in particular, one has $\pi_0(\mathit{tmf}(U)) = A$). Recall also that the (derived) global sections $\mathbb{R}\Gamma(\overline{\mathcal{E}}, \mathit{tmf})$, form a commutative $S$-algebra, well defined in $\mathrm{Ho}(S - Alg)$, called the *spectrum of topological modular forms*, and denoted by $\mathsf{tmf}$ [2].

Let $U \longrightarrow \overline{\mathcal{E}}$ be a surjective étale morphism with $U$ an affine scheme, and let us consider its nerve

$$\begin{array}{rccc} U_* : & \Delta^{op} & \longrightarrow & Aff \\ & [n] & \mapsto & U_n := \underbrace{U \times_{\overline{\mathcal{E}}} U \times_{\overline{\mathcal{E}}} \cdots \times_{\overline{\mathcal{E}}} U}_{n\ times}. \end{array}$$

This is a simplicial object in $\overline{\mathcal{E}}_{\text{ét}}$, and by applying *tmf* we obtain a co-simplicial object in $S - Alg$

$$\begin{array}{rccc} \mathit{tmf}(U_*) : & \Delta & \longrightarrow & S - Alg \\ & [n] & \mapsto & \mathit{tmf}(U_n). \end{array}$$

Taking $\mathbb{R}\underline{Spec}$ of this diagram we obtain a simplicial object in the model category $S - Aff^{\sim,\text{s-ét}}$

$$\begin{array}{rccc} \mathbb{R}\underline{Spec}\,(\mathit{tmf}(U_*)) : & \Delta^{op} & \longrightarrow & S - Aff^{\sim,\text{s-ét}} \\ & [n] & \mapsto & \mathbb{R}\underline{Spec}\,(\mathit{tmf}(U_n)). \end{array}$$

The homotopy colimit of this diagram will be denoted by

$$\overline{\mathcal{E}}_{\mathbf{S}} := \mathrm{hocolim}_{n \in \Delta^{op}} \mathbb{R}\underline{Spec}\,(\mathit{tmf}(U_*)) \in \mathrm{St}(S).$$

The following result is technically just a remark as there is essentially nothing to prove; however, we prefer to state it as a theorem to emphasize its importance.

THEOREM 2.4.2.1. *The stack $\overline{\mathcal{E}}_{\mathbf{S}}$ defined above is a strong Deligne-Mumford* 1-*geometric $S$-stack. Furthermore $\overline{\mathcal{E}}_{\mathbf{S}}$ is a "brave new derivation" of the moduli stack $\overline{\mathcal{E}}$ of elliptic curves, i.e. there exists a natural isomorphism in* $\mathrm{St}(\mathbb{Z})$

$$h^0(\overline{\mathcal{E}}_{\mathbf{S}}) \simeq \overline{\mathcal{E}}.$$

[2]The notation here is a bit nonstandard: what is usually called the spectrum of topological modular forms is actually the connective cover of the spectrum we have denoted by $\mathsf{tmf}$

PROOF. To prove that $\overline{\mathcal{E}}_{\mathbf{S}}$ is geometric, it is enough to check that the simplicial object $\mathbb{R}\underline{Spec}\,(tmf(U_*))$ is a strongly étale Segal groupoid. For this, recall that for any morphism $U = Spec\,B \to V = Spec\,A$ in $\overline{\mathcal{E}}_{\text{ét}}$, the natural morphism

$$\pi_*(tmf(V)) \otimes_{\pi_0(tmf(V))} \pi_0(tmf(U)) \simeq \pi_*(tmf(V)) \otimes_A B \longrightarrow \pi_*(tmf(U))$$

is an isomorphism. This shows that the functor

$$\mathbb{R}\underline{Spec}\,(tmf(-)) : \overline{\mathcal{E}}_{\text{ét}} \longrightarrow S - Aff^{\sim,\text{s-ét}}$$

preserves homotopy fiber products and therefore sends Segal groupoid objects to Segal groupoid objects. In particular, $\mathbb{R}\underline{Spec}\,(tmf(U_*))$ is a Segal groupoid object. The same fact also shows that for any morphism $U = Spec\,B \to V = Spec\,A$ in $\overline{\mathcal{E}}_{\text{ét}}$, the induced map $tmf(V) \longrightarrow tmf(U)$ is a strong étale morphism. This implies that $\mathbb{R}\underline{Spec}\,(tmf(U_*))$ is a strongly étale Segal groupoid object in representable $S$-stacks, and thus shows that $\overline{\mathcal{E}}_{\mathbf{S}}$ is indeed a strong Deligne-Mumford 1-geometric $S$-stack.

The truncation functor $h^0$ clearly commutes with homotopy colimits, and therefore

$$h^0(\overline{\mathcal{E}}_{\mathbf{S}}) \simeq \text{hocolim}_{n\in\Delta^{op}} h^0(\mathbb{R}\underline{Spec}\,(tmf(U_n))) \in \text{St}(\mathbb{Z}).$$

Furthermore, for any connective representable $S$-stack, $\mathbb{R}\underline{Spec}\,A$, one has a natural isomorphism $h^0(\mathbb{R}\underline{Spec}\,A) \simeq Spec\,\pi_0(A)$. Therefore, one sees immediately that there is a natural isomorphism of simplicial objects in $\mathbb{Z} - Aff^{\sim,\text{ét}}$

$$h^0(\mathbb{R}\underline{Spec}\,(tmf(U_*))) \simeq U_*.$$

Therefore, we get

$$h^0(\overline{\mathcal{E}}_{\mathbf{S}}) \simeq \text{hocolim}_{n\in\Delta^{op}} h^0(\mathbb{R}\underline{Spec}\,(tmf(U_n))) \simeq \text{hocolim}_{n\in\Delta^{op}} U_n \simeq \overline{\mathcal{E}},$$

as $U_*$ is the nerve of an étale covering of $\overline{\mathcal{E}}$. □

Theorem 2.4.2.1 tells us that the presheaf of topological modular forms *tmf* provides a natural geometric $S$-stack $\overline{\mathcal{E}}_{\mathbf{S}}$ whose truncation is the usual stack of elliptic curves $\overline{\mathcal{E}}$. Furthermore, as one can show that the small strong étale topoi of $\overline{\mathcal{E}}_{\mathbf{S}}$ and $\overline{\mathcal{E}}$ coincide (this is a general fact about strong étale model topologies), we see that

$$\mathsf{tmf} := \mathbb{R}\Gamma(\overline{\mathcal{E}}, tmf) \simeq \mathbb{R}\Gamma(\overline{\mathcal{E}}_{\mathbf{S}}, \mathcal{O}),$$

and therefore that topological modular forms can be simply interpreted as *functions on the geometric $S$-stack $\overline{\mathcal{E}}_{\mathbf{S}}$*. Of course, our construction of $\overline{\mathcal{E}}_{\mathbf{S}}$ has essentially been rigged to make this true, so this is not a surprise. However, we have gained a bit from the conceptual point of view: since after all $\overline{\mathcal{E}}$ is a moduli stack, now that we know the existence of the geometric $S$-stack $\overline{\mathcal{E}}_{\mathbf{S}}$ we can ask for a *modular interpretation* of it, or in other words for a direct geometric description of the corresponding simplicial presheaf on $S - Aff$. An answer to this question not only would provide a direct construction of $\mathsf{tmf}$, but would also give a conceptual interpretation of it in a geometric language closer the usual notion of modular forms.

QUESTION 2.4.2.2. *Find a modular interpretation of the $S$-stack $\overline{\mathcal{E}}_{\mathbf{S}}$.*

Essentially, we are asking for the brave new "objects" that the $S$-stack $\overline{\mathcal{E}}_{\mathbf{S}}$ classifies. We could also consider the non-connective version of the $S$-stack $\overline{\mathcal{E}}_{\mathbf{S}}$ (defined through the non-connective version of *tmf*) for which a modular interpretation seems much more accessible.

Very recent work by J. Lurie (see **[Lu2]** for a detailed announcement of his results) answers in fact to Question 2.4.2.2; he shows that such a variant of $\overline{\mathcal{E}}_{\mathbf{S}}$ classifies brave new versions of **[AHS]**'s elliptic spectra plus additional data (called orientations). This moduli-theoretic point of view makes use of some very interesting notions of *brave new abelian varieties*, *brave new formal groups* and their geometry. The complete picture (possibly extended to higher chromatic levels) does not only give an alternative construction of the spectrum $tmf$ (and a better functoriality) but it could be the starting point of a rather new[3] and deep interaction between stable homotopy theory and homotopical algebraic geometry, involving many new questions and objects, and probably also new insights on classical objects of algebraic topology.

[3]J. Lurie's approach has as a byproduct also a natural construction of $G$-*equivariant* versions of elliptic cohomology, for any compact $G$.

APPENDIX A

# Classifying spaces of model categories

The classifying space of a model category $M$ is defined to be $N(M_W)$, the nerve of its subcategory of equivalences. More generally, if $C \subset M_W$ is a full subcategory of the category of equivalences in $M$, which is closed by equivalences in $M$, the classifying space of $C$ is $N(C)$, the nerve of $C$.

We fix a $\mathbb{V}$-small model category $M$ and a full subcategory $C \subset M_W$ closed by equivalences. We consider the model category $(C, C)^{\wedge}$, defined in [**HAGI**, §2.3.2]. Recall that the underlying category of $(C, C)^{\wedge}$ is the category $SSet_{\mathbb{V}}^{C^{op}}$, of $\mathbb{V}$-small simplicial presheaves on $C$. The model structure of $(C, C)^{\wedge}$ is defined as the left Bousfield localization of the levelwise projective model structure on $SSet_{\mathbb{V}}^{C^{op}}$, by inverting all the morphisms in $C$. The important fact is that local objects in $(C, C)^{\wedge}$ are functors $F : C^{op} \longrightarrow SSet_{\mathbb{V}}$ sending all morphisms in $C$ to equivalences.

We define an adjunction

$$N : (C, C)^{\wedge} \longrightarrow SSet/N(C) \qquad (C, C)^{\wedge} \longleftarrow SSet/N(C) : S$$

in the following way. A functor $F : C^{op} \longrightarrow SSet$ is sent to the simplicial set $N(F)$, for which the set of $n$-simplices is the set of parirs

$$N(F)_n := \{(c_0 \to c_1 \to \cdots \to c_n, \alpha)\}$$

where $(c_0 \to c_1 \to \cdots \to c_n)$ is an $n$-simplex in $N(C)$ and $\alpha \in F(c_n)_n$ is an $n$-simplex in $F(c_n)$. Put in an other way, $N(F)$ is the diagonal of the bi-simplicial set

$$(n, m) \mapsto N(C/F_n)_m$$

where $C/F_n$ is the category of objects of the presheaf of $n$-simplices in $F$. The functor $N$ has a right adjoint

$$S : SSet/N(C) \longrightarrow (C, C)^{\wedge}$$

sending $X \longrightarrow N(C)$ to the simplicial presheaf

$$\begin{array}{rccc} S(X) : & C^{op} & \longrightarrow & SSet \\ & x & \mapsto & \underline{Hom}_{SSet/N(C)}(N(h_x), X), \end{array}$$

where $h_x$ is the presheaf of sets represented by $x \in C$ (note that $N(h_x) \to N(C)$ is isomorphic to $N(C/x) \to N(C)$).

PROPOSITION A.0.3. *The adjunction $(N, S)$ is a Quillen equivalence.*

PROOF. First of all we need to check that $(N, S)$ is a Quillen adjunction. For this we use the standard properties of left Bousfield localizations, and we see that it is enough to check that

$$N : SPr(C) \longrightarrow SSet/N(C) \qquad SPr(C) \longleftarrow SSet/N(C) : S$$

is a Quillen adjunction (where $SPr(C)$ is the projective model structure of simplicial presheaves on $C$), and that $S$ preserves fibrant objects. These two facts are clear by definition of $S$ and the description of fibrant objects in $N(C, C)^{\wedge}$.

For any $x \in C$, the morphism $N(h_x) = N(C/x) \longrightarrow N(C)$ is isomorphic in $\mathrm{Ho}(SSet/N(C))$ to $x \longrightarrow N(C)$. Therefore, for $X \in \mathrm{Ho}(SSet/N(C))$ and $x \in C$, the simplicial set

$$\mathbb{R}S(X)(x) \simeq \mathbb{R}\underline{Hom}_{SSet/N(C)}(x, X)$$

is naturally isomorphic in $\mathrm{Ho}(SSet)$ to the homotopy fiber of $X \longrightarrow N(C)$ taken at $x$. This clearly implies that the right derived functor

$$\mathbb{R}S : \mathrm{Ho}(SSet/N(C)) \longrightarrow \mathrm{Ho}((C,C)^\wedge)$$

is conservative. In particular, it only remains to show that the adjunction morphism $Id \longrightarrow \mathbb{R}S\mathbb{L}N$ is an isomorphism. But this last assumption follows from the definition the functor $N$ and from a standard lemma (se for example [**Q3**]), which shows that the homotopy fiber at $x \in C$ of $N(F) \longrightarrow N(C)$ is naturally equivalent to $F(x)$ when $F$ is fibrant in $(C,C)^\wedge$. □

Recall from [**HAGI**, Lem. 4.2.2] that for any object $x \in C$, one can construct a local model for $h_x$ as $\underline{h}_{R(x)}$, sending $y \in C$ to $Hom_C(\Gamma^*(y), R(x))$, where $\Gamma^*$ is a co-simplicial replacement functor in $M$. The natural morphism $h_x \longrightarrow \underline{h}_{R(x)}$ being an equivalence in $(C,C)^\wedge$, one finds using Prop. A.0.3 natural equivalences of simplicial sets

$$\underline{Hom}(h_y, \underline{h}_{R(x)}) \simeq Map_M^{eq}(y,x) \simeq \mathbb{R}\underline{Hom}_{SSet/N(C)}(N(h_y), N(h_x)) \simeq y \times^h_{N(C)} x,$$

where $Map_M^{eq}(y,x)$ is the sub-simplicial set of the mapping space $Map_M(y,x)$ consisting of equivalences. As a corollary of this we find the important result due to Dwyer and Kan. For this, we recall that the simplicial monoid of self equivalences of an object $x \in M$ can be defined as

$$Aut(x) := \underline{Hom}_{(C,C)^\wedge}(\underline{h}_{R(x)}, \underline{h}_{R(x)}).$$

COROLLARY A.0.4. *Let $C \subset M_W$ be a full subcategory of equivalences in a model category $M$, which is stable by equivalences. Then, one has a natural isomorphism in* $\mathrm{Ho}(SSet)$

$$N(C) \simeq \coprod_{x \in \pi_0(N(C)))} BAut(x)$$

*where $Aut(x)$ is the simplicial monoid of self equivalences of $x$ in $M$.*

Another important consequence of Prop. A.0.3 is the following interpretation of mapping spaces in term of homotopy fibers between classifying spaces of certain model categories.

COROLLARY A.0.5. *Let $M$ be a model category and $x, y \in M$ be two fibrant and cofibrant objects in $M$. Then, there exists a natural homotopy fiber sequence of simplicial sets*

$$Map_M(x,y) \longrightarrow N((x/M)_W) \longrightarrow N(M_W),$$

*where the homotopy fiber is taken at $y \in M$.*

PROOF. This follows easily from Cor. A.0.4. □

We will need a slightly more functorial interpretation of Cor. A.0.4 in the particular case where the model category $M$ is simplicial. We assume now that $M$ is a $\mathbb{V}$-small simplicial $\mathbb{U}$-model category (i.e. the model category $M$ is a $SSet_{\mathbb{U}}$-model category in the sense of [**Ho1**]). We still let $C \subset M_W$ be a full subcategory of the category of equivalences in $M$, and still assume that $C \subset M_W$ is stable by equivalences.

We define an $S$-category $\mathcal{G}(C)$ is the following way (recall that an $S$-category is a simplicially enriched category, see for example [**HAGI**, §2.1] for more details and notations). The objects of $\mathcal{G}(C)$ are the objects of $C$ which are furthermore fibrant and cofibrant in $M$. For two objects $x$ and $y$, the simplicial set of morphisms is defined to be

$$\mathcal{G}(C)_{(x,y)} := \underline{Hom}^{eq}_M(x,y),$$

where by definition $\underline{Hom}^{eq}_M(x,y)_n$ is the set of equivalences in $M$ from $\Delta^n \otimes x$ to $y$ (i.e. $\underline{Hom}^{eq}_M(x,y)$ is the sub-simplicial set of $\underline{Hom}_M(x,y)$ consisting of equivalences). Clearly, the $S$-category $\mathcal{G}(C)$ is *groupoid like*, in the sense that its category of connected components $\pi_0(\mathcal{G}(C))$ (also denoted by $\mathrm{Ho}(\mathcal{G}(C))$ is a groupoid (or equivalently every morphism in $\mathcal{G}(C)$ has an inverse up to homotopy). Let $C^{c,f}$ be the full subcategory of $C$ consisting of fibrant and cofibrant objects in $C$. There exist two natural morphisms of $S$-categories

$$C \longleftarrow C^{c,f} \longrightarrow \mathcal{G}(C),$$

where a category is considered as an $S$-category with discrete simplicial sets of morphisms. Passing to the nerves, one gets a diagram of simplicial sets

$$N(C) \longleftarrow N(C^{c,f}) \longrightarrow N(\mathcal{G}(C)),$$

where the nerve functor is extended diagonally to $S$-categories (see e.g. [**D-K**1]). Another interpretation of corollary A.0.4 is the following result.

PROPOSITION A.0.6. *With the above notations, the two morphisms*

$$N(C) \longleftarrow N(C^{c,f}) \longrightarrow N(\mathcal{G}(C)),$$

*are equivalences of simplicial sets.*

PROOF. It is well known that the left arrow is an equivalence as a fibrant-cofibrant replacement functor gives an inverse up to homotopy. For the right arrow, we let $N(\mathcal{G}(C))_n$ be the category of $n$-simplicies in $N(\mathcal{G}(C))$, defined by having the same objects and with

$$(N(\mathcal{G}(C))_n)_{x,y} := (N(\mathcal{G}(C))_{x,y})_n.$$

By definition of the nerve, one has a natural equivalence

$$N(\mathcal{G}(C)) \simeq Hocolim_{n\Delta^{op}}(N(\mathcal{G}(C))_n).$$

Furthermore, it is clear that each functor

$$C^{c,f} = \mathcal{G}(C)_0 \longrightarrow \mathcal{G}(C)_n$$

induces an equivalence of the nerves, as the 0-simplex $[0] \rightarrow [n]$ clearly induces a functor

$$N(\mathcal{G}(C)_n) \longrightarrow N(C^{c,f})$$

which is a homotopy inverse. □

Proposition A.0.6 is another interpretation of Prop. A.0.3, as the $S$-category $\mathcal{G}(C)$ is groupoid-like, the delooping theorem of G. Segal implies that there exists a natural equivalence of simplicial sets

$$x \times^h_{N(C)} y \longleftarrow x \times^h_{N(C^{c,f})} y \longrightarrow x \times^h_{N(\mathcal{G}(C))} y \longleftarrow \mathcal{G}(C)_{(x,y)} = \underline{Hom}^{eq}_M(x,y).$$

The advantage of Prop. A.0.6 over the more general proposition A.0.3 is that it is more easy to state a functorial property of the equivalences in the following particular context (the equivalence in Prop. A.0.3 can also be made functorial, but it requires some additional work, using for example simplicial localization techniques).

We assume that $G : M \longrightarrow N$ is a simplicial, left Quillen functor between $\mathbb{V}$-small simplicial $\mathbb{U}$-model categories. We let $C \subset M_W$ and $D \subset N_W$ be two full sub-categories stable by equivalences, and we suppose that all objects in $M$ and $N$ are fibrant. Finally, we assume that the functor $G$ restricted to cofibrant objects sends $C^c := M^c \cap C$ to $D^c := N^c \cap D$. In this situation, we define an $S$-functor

$$G : \mathcal{G}(C) \longrightarrow \mathcal{G}(D)$$

simply by using the simplicial enrichment of $G$. Then, one has a commutative diagram of $S$-categories

$$\begin{array}{ccccc} C & \longleftarrow & C^c & \longrightarrow & \mathcal{G}(C) \\ & & \downarrow{\scriptstyle G} & & \downarrow{\scriptstyle G} \\ D & \longleftarrow & D^c & \longrightarrow & \mathcal{G}(D), \end{array}$$

and thus a commutative diagram of simplicial sets

$$\begin{array}{ccccc} N(C) & \longleftarrow & N(C^c) & \longrightarrow & N(\mathcal{G}(C)) \\ & & \downarrow{\scriptstyle G} & & \downarrow{\scriptstyle G} \\ N(D) & \longleftarrow & N(D^c) & \longrightarrow & N(\mathcal{G}(D)). \end{array}$$

The important fact here is that this construction is associative with respect to composition of the simplicial left Quillen functor $G$. In other words, if one has a diagram of simplicial model categories $M_i$ and simplicial left Quillen functors, together with sub-categories $C_i \subset (M_i)_W$ satisfying the required properties, then one obtain a commutative diagram of diagrams of simplicial sets

$$\begin{array}{ccccc} N(C_i) & \longleftarrow & N(C_i^c) & \longrightarrow & N(\mathcal{G}(C_i)) \\ & & \downarrow{\scriptstyle G_i} & & \downarrow{\scriptstyle G_i} \\ N(D_i) & \longleftarrow & N(D_i^c) & \longrightarrow & N(\mathcal{G}(D_i)). \end{array}$$

APPENDIX B

# Strictification

Let $I$ be a $\mathbb{U}$-small category. For any $i \in I$ we let $M_i$ be a $\mathbb{U}$-cofibrantly generated model category, and for any $u : i \to j$ morphism in $I$ we let

$$u^* : M_j \longrightarrow M_i \qquad M_j \longleftarrow M_i : u_*$$

be a Quillen adjunction. We suppose furthermore that for any composition

$$i \xrightarrow{u} j \xrightarrow{v} k$$

one has an equality of functors

$$u^* \circ v^* = (v \circ u)^*$$

Such a data $(\{M_i\}_i, \{u^*\}_u)$ will be called, according to [**H-S**], a *($\mathbb{U}$-)cofibrantly generated left Quillen presheaf over $I$*, and will be denoted simply by the letter $M$.

For any cofibrantly generated left Quillen presheaf $M$ on $I$, we consider the category $M^I$, of $I$-diagrams in $M$ in the following way. Objects in $M^I$ are given by the data of objects $x_i \in M_i$ for any $i \in I$, together with morphisms $\phi_u : u^*(x_j) \longrightarrow x_i$ for any $u : i \to j$ in $I$, making the following diagram commutative

$$\begin{array}{ccc} u^*v^*(x_k) & \xrightarrow{u^*(\phi_v)} & u^*(x_j) \\ {\scriptstyle Id}\downarrow & & \downarrow{\scriptstyle \phi_u} \\ (v \circ u)^*(x_k) & \xrightarrow[\phi_{v \circ u}]{} & x_i \end{array}$$

for any $i \xrightarrow{u} j \xrightarrow{v} k$ in $I$. Morphisms in $M^I$ are simply given by families morphisms $f_i : x_i \longrightarrow y_i$, such that $f_i \circ \phi_u = \phi_u \circ f_j$ for any $i \to j$ in $I$.

The category $M^I$ is endowed with a model structure for which the fibrations or equivalences are the morphisms $f$ such that for any $i \in I$ the induced morphism $f_i$ is a fibration or an equivalence in $M_i$. As all model categories $M_i$ are $\mathbb{U}$-cofibrantly generated, it is not hard to adapt the general argument of [**Hi**, 11.6] in order to prove that $M^I$ is also a $\mathbb{U}$-cofibrantly generated model category.

We define an object $x \in M^I$ to be *homotopy cartesian* if for any $u : i \to j$ in $I$ the induced morphism

$$\mathbb{L}u^*(x_i) \longrightarrow x_j$$

is an isomorphism in $\mathrm{Ho}(M_j)$. The full subcategory of cartesian objects in $M^I$ will be denoted by $M^I_{cart}$.

There exists a presheaf of categories $(-/I)^{op}$ over $I$, having the opposite comma category $(i/I)^{op}$ as value over the object $i$, and the natural functor $(i/I)^{op} \longrightarrow (j/I)^{op}$ for any morphism $j \longrightarrow i$ in $I$. For any $\mathbb{U}$-cofibrantly left Quillen presheaf $M$ over $I$, we define a morphism of presheaves of categories over $I$

$$M^I \times (-/I)^{op} \longrightarrow M,$$

where $M^I$ is seen as a constant presheaf of categories, in the following way. For an object $i \in I$, the functor

$$M^I \times (i/I)^{op} \longrightarrow M_i$$

sends an object $(x, u : i \to j)$ to $u^*(x_j) \in M_i$, and a morphism $(x, u : i \to j) \longrightarrow (y, v : i \to k)$, given by a morphism $x \to y$ in $M^I$ and a commutative diagram in $I$

$$\begin{array}{ccc} i & & \\ {\scriptstyle v}\downarrow & \searrow^{u} & \\ k & \xrightarrow{w} & j, \end{array}$$

is sent to the morphism in $M_i$

$$u^*(x_j) \simeq v^*(w^*(x_j)) \longrightarrow v^*(x_k).$$

For a diagram $u : i \to j$ in $I$, the following diagram

$$\begin{array}{ccc} M^I \times (j/I)^{op} & \longrightarrow & M_j \\ \downarrow & & \downarrow {\scriptstyle u^*} \\ M^I \times (i/I)^{op} & \longrightarrow & M_i \end{array}$$

clearly commutes, showing that the above definition actually defines a morphism

$$M^I \times (-/I)^{op} \longrightarrow M.$$

We let $(M^I_{cart})^{cof}_W$ be the subcategory of $M^I$ consisting of homotopy cartesian and cofibrant objects in $M^I$ and equivalences between them. In the same way we consider the sub-presheaf $M^c_W$ whose value at $i \in I$ is the subcategory of $M_i$ consisting of cofibrant objects in $M_i$ and equivalences between them. Note that the functors $u^* : M_j \longrightarrow M_i$ being left Quillen for any $u : i \longrightarrow j$, preserves the sub-categories of equivalences between cofibrant objects.

We have thus defined a morphism of presheaves of categories

$$(M^I_{cart})^c_W \times (-/I)^{op} \longrightarrow M^c_W,$$

and we now consider the corresponding morphism of simplicial presheaves obtained by applying the nerve functor

$$N((M^I_{cart})^c_W) \times N((-/I)^{op}) \longrightarrow N(M^c_W),$$

that is considered as a morphism in the homotopy category $\mathrm{Ho}(SPr(I))$, of simplicial presheaves over $I$. As for any $i$ the category $(i/I)^{op}$ has a final object, its nerve $N((i/I)^{op})$ is contractible, and therefore the natural projection

$$N((M^I_{cart})^c_W) \times N((-/I)^{op}) \longrightarrow N((M^I_{cart})^c_W)$$

is an isomorphism in $\mathrm{Ho}(SPr(I))$. We therefore have constructed a well defined morphism i $\mathrm{Ho}(SPr(I))$, from the constant simplicial presheaf $N((M^I_{cart})^c_W)$ to the simplicial presheaf $N(M^c_W)$. By adjunction this gives a well defined morphism in $\mathrm{Ho}(SSet)$

$$N((M^I_{cart})^c_W) \longrightarrow Holim_{i \in I} N(M^c_W).$$

The strictification theorem asserts that this last morphism is an isomorphism in $\mathrm{Ho}(SSet)$. As this seems to be a folklore result we will not include a proof.

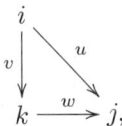

THEOREM B.0.7. *For any $\mathbb{U}$-small category $I$ and any $\mathbb{U}$-cofibrantly generated left Quillen presheaf $M$ on $I$, the natural morphism*

$$N((M^I_{cart})^c_W) \longrightarrow Holim_{i\in I^{op}} N(M^c_W)$$

*is an isomorphism in* $\mathrm{Ho}(SSet)$.

PROOF. When $M$ is the constant Quillen presheaf of simplicial sets this is proved in [**D-K**3]. The general case can treated in a similar way. See also [**H-S**, Thm. 18.6] for a stronger result. □

Let $I$ and $M$ be as in the statement of Thm. B.0.7, and let $M_0$ be a $\mathbb{U}$-cofibrantly model category. We consider $M_0$ as a constant left Quillen presheaf on $I$, for which all values are equal to $M_0$, and all transition functors are identities. We assume that there exists a left Quillen natural transformation $\phi : M_0 \longrightarrow M$. By this, we mean the data of left Quillen functors $\phi_i : M_0 \longrightarrow M_i$ for any $i \in I$, such that for any $u : i \to j$ one has $u^* \circ \phi_j = \phi_i$. In this case, we define a functor

$$\phi : M_0 \longrightarrow M^I,$$

by the obvious formula $\phi(x)_i := \phi_i(x)$ for $x \in M_0$, and the transition morphisms of $\phi(x)$ all being identities. The functor $\phi$ is not a left Quillen functor, but preserves equivalences between cofibrant objects, and thus possesses a left derived functor

$$\mathbb{L}\phi : \mathrm{Ho}(M_0) \longrightarrow \mathrm{Ho}(M^I).$$

One can even show that this functor possesses a right adjoint, sending an object $x \in M^I$ to the homotopy limit of the diagram in $M_0$, $i \mapsto \mathbb{R}\psi_i(x_i)$, where $\psi_i$ is the right adjoint to $\phi_i$.

One also has a natural transformation of presheaves of categories

$$(M_0)^c_W \longrightarrow M^c_W$$

inducing a natural morphism of simplicial presheaves on $I$

$$N((M_0)^c_W) \longrightarrow N(M^c_W),$$

and thus a natural morphism in $\mathrm{Ho}(SSet)$

$$N((M_0)^c_W) \longrightarrow Holim_{i\in I^{op}} N(M^c_W).$$

COROLLARY B.0.8. *Let $I$, $M$ and $M_0$ be as above, and assume that the functor*

$$\mathbb{L}\phi : \mathrm{Ho}(M_0) \longrightarrow \mathrm{Ho}(M^I)$$

*is fully faithful and that its image consists of all homotopy cartesian objects in* $\mathrm{Ho}(M^I)$. *Then the induced morphism*

$$N((M_0)^c_W) \longrightarrow Holim_{i\in I^{op}} N(M^c_W)$$

*is an isomorphism in* $\mathrm{Ho}(SSet)$.

PROOF. Indeed, we consider the functor

$$\begin{array}{cccc} G: & (M_0)^c_W & \longrightarrow & (M^I_{cart})^c_W \\ & x & \mapsto & \phi(Qx), \end{array}$$

where $Qx$ is a functorial cofibrant replacement of $x$. By hypothesis, the induced morphism on the nerves

$$N((M_0)^c_W) \longrightarrow N(M^I_{cart})^c_W$$

is an isomorphism in $\mathrm{Ho}(SSet)$. We now consider the commutative diagram in $\mathrm{Ho}(SSet)$

$$\begin{array}{ccc} N((M_0)^c_W) & \longrightarrow & N(M^I_{cart})^c_W \\ & \searrow & \downarrow \\ & & Holim_{i \in I^{op}} N(M^c_W). \end{array}$$

The right vertical arrow being an isomorphism by Thm. B.0.7 we deduce the corollary. $\square$

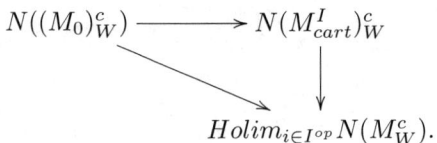

APPENDIX C

# Representability criterion (after J. Lurie)

The purpose of this appendix is to give a sketch of a proof of the following special case of J. Lurie's representability theorem. Lurie's theorem is much deeper and out of the range of this work. We will not need it in its full generality and will content ourselves with this special case, largely enough for our applications.

THEOREM C.0.9. *(J. Lurie, see* [**Lu1**]*) Let $F$ be a $D^-$-stack. The following conditions are equivalent.*

(1) *$F$ is an $n$-geometric $D^-$-stack.*
(2) *$F$ satisfies the following three conditions.*
  (a) *The truncation $t_0(F)$ is an Artin $(n+1)$-stack.*
  (b) *$F$ has an obstruction theory relative to $sk - Mod_1$.*
  (c) *For any $A \in sk - Alg$, the natural morphism*
$$\mathbb{R}F(A) \longrightarrow Holim_k \mathbb{R}F(A_{\leq k})$$
  *is an isomorphism in* $\mathrm{Ho}(SSet)$.

SKETCH OF PROOF. The only if part is the easy part. $(a)$ is true by Prop. 2.2.4.4 and $(b)$ by Cor. 2.2.3.3. For $(c)$ one proves the following more general lemma.

LEMMA C.0.10. *Let $f : F \longrightarrow G$ be an $n$-representable morphism. Then for any $A \in sk - Alg$, the natural square*

$$\begin{array}{ccc}
\mathbb{R}F(A) & \longrightarrow & Holim_k \mathbb{R}F(A_{\leq k}) \\
\downarrow & & \downarrow \\
\mathbb{R}G(A) & \longrightarrow & Holim_k \mathbb{R}G(A_{\leq k})
\end{array}$$

*is homotopy cartesian.*

PROOF. We prove this by induction on $n$. For $n = -1$, one reduces easily to the case of a morphism between representable $D^-$-stacks, for which the result simply follows from the fact that $A \simeq Holim_k A_{\leq k}$. Let us now assume that $n \geq 0$ and the result prove for $m < n$. Let $A \in sk - Alg$ and

$$x \in \pi_0(Holim_k \mathbb{R}G(A) \times^h_{\mathbb{R}G(A_{\leq k})} \mathbb{R}F(A_{\leq k}))$$

with projections

$$x_k \in \pi_0(\mathbb{R}G(A) \times^h_{\mathbb{R}G(A_{\leq k})} \mathbb{R}F(A_{\leq k})).$$

We need to prove that the homotopy fiber $H$ of

$$\mathbb{R}F(A) \longrightarrow Holim_k \mathbb{R}G(A) \times^h_{\mathbb{R}G(A_{\leq k})} \mathbb{R}F(A_{\leq k})$$

at $x$ is contractible. Replacing $F$ by $F \times^h_G X$ where $X := \mathbb{R}\underline{Spec}\, A$, and $G$ by $X$, one can assume that $G$ is a representable stack and $F$ is an $n$-geometric stack. As $G$ is representable, the morphism

$$\mathbb{R}G(A) \longrightarrow Holim_k \mathbb{R}G(A_{\leq k})$$

is an equivalence. Therefore, we are reduced to the case where $G = *$. The point $x$ is then a point in $\pi_0(Holim_k \mathbb{R}F(A_{\leq k}))$, and we need to prove that the homotopy fiber $H$, taken at $x$, of the morphism

$$\mathbb{R}F(A) \longrightarrow Holim_k \mathbb{R}F(A_{\leq k})$$

is contractible. Using Cor. 2.2.2.9 one sees easily that this last statement is local on the small étale site of $A$. By a localization argument we can therefore assume that each projection $x_k \in \pi_0(\mathbb{R}F(A_{\leq k}))$ of $x$ is the image of a point $y_k \in \pi_0(\mathbb{R}U(A_{\leq k}))$, for some representable $D^-$-stack $U$ and a smooth morphism $U \longrightarrow F$.

Using Prop. 1.4.2.6 and Lem. 2.2.1.1 we see that the homotopy fiber of the morphism

$$\mathbb{R}U(A_{\leq k+1}) \longrightarrow \mathbb{R}U(A_{\leq k}) \times^h_{\mathbb{R}F(A_{\leq k})} \mathbb{R}F(A_{\leq k+1})$$

taken at $y_k$ is equivalent to $Map_{A_{\leq k}-Mod}(\mathbb{L}_{U/F,y_k}, \pi_{k+1}(A)[k+1])$. Cor. 2.2.5.3 then implies that when $k$ is big enough, the homotopy fibers of the morphisms

$$\mathbb{R}U(A_{\leq k+1}) \longrightarrow \mathbb{R}U(A_{\leq k}) \times^h_{\mathbb{R}F(A_{\leq k})} \mathbb{R}F(A_{\leq k+1})$$

are simply connected, and thus this morphism is surjective on connected components. This easily implies that the points $y_k$ can be thought as a point $y \in \pi_0(Holim_k \mathbb{R}U(A_{\leq k}))$ whose image in $\pi_0(Holim_k \mathbb{R}D(A_{\leq k}))$ is equal to $x$.

We then consider the diagram

$$\begin{array}{ccc} \mathbb{R}U(A) & \longrightarrow & Holim_k \mathbb{R}U(A_{\leq k}) \\ \downarrow & & \downarrow \\ \mathbb{R}F(A) & \longrightarrow & Holim_k \mathbb{R}F(A_{\leq k}). \end{array}$$

By induction on $n$ we see that this diagram is homotopy cartesian, and that the top horizontal morphism is an equivalence. There, the morphism induced on the homotopy fibers of the horizontal morphisms is an equivalence, showing that $H$ is contractible as required. □

Conversely, let $F$ be a $D^-$-stack satisfying conditions $(a)-(c)$ of C.0.9. The proof goes by induction on $n$. Let us first $n = -1$. We start by a lifting lemma.

LEMMA C.0.11. *Let $F$ be a $D^-$-stack satisfying the conditions $(a) - (c)$ of Thm. C.0.9. Then, for any affine scheme $U_0$, and any étale morphism $U_0 \longrightarrow t_0(F)$, there exists a representable $D^-$-stack $U$, a morphism $u : U \longrightarrow F$, with $\mathbb{L}_{F,u} \simeq 0$, and a homotopy cartesian square in $k - D^-Aff^{\sim,\text{ét}}$*

$$\begin{array}{ccc} U_0 & \longrightarrow & t_0(F) \\ \downarrow & & \downarrow \\ U & \longrightarrow & F. \end{array}$$

PROOF. We are going to construct by induction a sequence of representable $D^-$-stacks

$$U_0 \longrightarrow U_1 \ldots \longrightarrow U_k \longrightarrow U_{k+1} \ldots \longrightarrow F$$

in $k - D^-Aff^{\sim,\text{ét}}/F$ satisfying the following properties.

- One has $U_k = \mathbb{R}\underline{Spec}\, A_k$ with $\pi_i(A_k) = 0$ for all $i > k$.
- The corresponding morphism $A_{k+1} \longrightarrow A_k$ induces isomorphisms on $\pi_i$ for all $i \leq k$.

- The morphism $u_k : U_k \longrightarrow F$ are such that

$$\pi_i(\mathbb{L}_{U_k/F,u_k}) = 0 \; \forall \; i \leq k+1.$$

Assume for the moment that this sequence is constructed, and let $A := Holim_{A_k}$, and $U := \mathbb{R}\underline{Spec}\, A$. The points $u_k$, defines a well defined point in $\pi_0(Holim_k \mathbb{R}F(A_k))$, which by condition $(d)$ induces a well defined morphism of stacks $u : U \longrightarrow F$. Clearly, one has a homotopy cartesian square

$$\begin{array}{ccc} U_0 & \longrightarrow & t_0(F) \\ \downarrow & & \downarrow \\ U & \longrightarrow & F. \end{array}$$

Let $M$ be any $A$-module. Again, using condition $(d)$, one sees that

$$\mathbb{D}er_F(U, M) \simeq Holim_k \mathbb{D}er_F(U_k, M_{\leq k}) \simeq Holim_k Map_{A_k - Mod_s}(\mathbb{L}_{U_k/F,u_k}, M_k) \simeq 0.$$

This implies that $\mathbb{L}_{U/F,u} = 0$.

It remains to explain how to construct the sequence of $U_k$. This is done by induction. For $k = 0$, the only thing to check is that

$$\pi_i(\mathbb{L}_{U_0/F,u_0}) = 0 \; \forall \; i \leq 1.$$

This follows easily from Prop. 1.4.2.6 and the fact that $u_0 : U_0 \longrightarrow t_0(F)$ is étale.

Assume now that all the $U_i$ for $i \leq k$ have been constructed. We consider $u_k : U_k \longrightarrow F$, and the natural morphism

$$\mathbb{L}_{U_k} \longrightarrow \mathbb{L}_{U_k/F,u_k} \longrightarrow (\mathbb{L}_{U_k/F,u_k})_{\leq k+2} = N_{k+1}[k+2],$$

where $N_{k+1} := \pi_{k+2}(\mathbb{L}_{U_k/F,u_k})$. The morphism

$$\mathbb{L}_{U_k} \longrightarrow N_{k+1}[k+2]$$

defines a square zero extension of $A_k$ by $N_{k+1}[k+1]$, $A_{k+1} \longrightarrow A_k$. By construction, and using Prop. 1.4.2.5 there exists a well defined point in

$$\mathbb{R}\underline{Hom}_{U_k/k-D^-Aff^{\sim,\text{ét}}/F}(U_{k+1}, F),$$

corresponding to a morphism in $\mathrm{Ho}(k - D^-Aff^{\sim,\text{ét}}/F)$

$$\begin{array}{ccc} U_k & \longrightarrow & U_{k+1} \\ & \searrow & \downarrow \\ & & F. \end{array}$$

It is not hard to check that the corresponding morphism $\mathbb{R}\underline{Spec}\, A_{k+1} = U_{k+1} \longrightarrow F$ has the required properties. □

Let us come back to the case $n = -1$. Lemma C.0.11 implies that there exists a representable $D^-$-stack $U$, a morphism $U \longrightarrow F$ such that $\mathbb{L}_{F,u} \simeq 0$, and inducing an isomorphism $t_0(U) \simeq t_0(F)$. Using Prop. 1.4.2.6 and Lem. 2.2.1.1, it is not hard to show by induction on $k$, that for any $A \in sk - Alg$ with $\pi_i(A) = 0$ for $i > k$, the induced morphism

$$\mathbb{R}U(A) \longrightarrow \mathbb{R}F(A)$$

is an isomorphism. Using condition $(c)$ for $F$ and $U$ one sees that this is also true for any $A \in sk - Alg$. Therefore, $U \simeq F$, and $F$ is a representable $D^-$-stack.

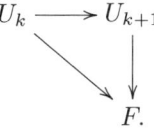

We now finish the proof of Thm. C.0.9 by induction on $n$. Let us suppose we know Thm. C.0.9 for $m < n$ and let $F$ be a $D^-$-stack satisfying conditions $(a) - (c)$ for rank $n$. By induction we see that the diagonal of $F$ is $(n-1)$-representable and the hard point is to prove that $F$ has an $n$-atlas. For this we lift an $n$-atlas of $it_0(F)$ in the following way. Starting from a smooth morphism $U_0 \longrightarrow it_0(F)$, we to construct by induction a sequence of representable $D^-$-stacks

$$U_0 \longrightarrow U_1 \ldots \longrightarrow U_k \longrightarrow U_{k+1} \cdots \longrightarrow F$$

in $k - D^-Aff^{\sim,\text{ét}}/F$ satisfying the following properties.

- One has $U_k = \mathbb{R}\underline{Spec}\, A_k$ with $\pi_i(A_k) = 0$ for all $i > k$.
- The corresponding morphism $A_{k+1} \longrightarrow A_k$ induces isomorphisms on $\pi_i$ for all $i \leq k$.
- The morphism $u_k : U_k \longrightarrow F$ is such that for any $M \in A_k - Mod_s$ with $\pi_i(M) = 0$ for all $i > k+1$ and $\pi_0(M) = 0$, one has
$$[\mathbb{L}_{U_k/F,u_k}, M]_{Sp(A_k - Mod_s)} = 0.$$

Such a sequence can be constructed by induction on $k$ using obstruction theory as in Lem. C.0.11. We then let $A = Holim_k A_k$, and $U = \mathbb{R}\underline{Spec}\, A$. Then, condition $(c)$ implies the existence of a well defined morphism $U \longrightarrow F$, which is seen to be smooth using Cor. 2.2.5.3. Using this lifting of smooth morphisms from $it_0$ to $F$, one produces an $n$-atlas for $F$ by lifting an $n$-atlas of $it_0(F)$. $\square$

# Bibliography


[EGAI] A. Grothendieck, J. Dieudonné, *Eléments de Géométrie Algébrique*, I, Springer-Verlag, New York 1971.

[AHS] M. Ando, M. J. Hopkins, N. P. Strickland, *Elliptic spectra, the Witten genus, and the theorem of the cube*, Inv. Math. **146**, (2001), 595-687.

[Ar] M. Artin, *Algebraization of formal moduli I*, in "Global Analysis" (papers in honor of K. Kodaira), University of Tokyo Press, 1969, p. 21-71.

[SGA4-I] M. Artin, A. Grothendieck, J. L. Verdier, *Théorie des topos et cohomologie étale des schémas- Tome 1*, Lecture Notes in Math **269**, Springer Verlag, Berlin, 1972.

[SGA4-II] M. Artin, A. Grothendieck, J. L. Verdier, *Théorie des topos et cohomologie étale des schémas- Tome 2*, Lecture Notes in Math **270**, Springer Verlag, Berlin, 1972.

[Ba] M. Basterra, *André-Quillen cohomology of commutative $S$-algebras*, J. Pure Appl. Algebra **144**, 1999, 111-143.

[Ba-McC] M. Basterra, R. McCarthy, *Gamma Homology, Topological André-Quillen Homology and Stabilization*, Topology and its Applications **121**/3, 2002, 551-566.

[Be1] K. Behrend, *Differential Graded Schemes I: Perfect Resolving Algebras*, Preprint math.AG/0212225.

[Be2] K. Behrend, *Differential Graded Schemes II: The 2-category of Differential Graded Schemes*, Preprint math.AG/0212226.

[Bl] B. Blander, *Local projective model structure on simpicial presheaves*, $K$-theory **24** (2001) No. 3, 283-301.

[Ci-Ka1] I. Ciocan-Fontanine, M. Kapranov, *Derived Quot schemes*, Ann. Sci. Ecole Norm. Sup. (4) **34** (2001), 403-440.

[Ci-Ka2] I. Ciocan-Fontanine, M. Kapranov, *Derived Hilbert Schemes*, preprint available at math.AG/0005155.

[Del1] P. Deligne, *Catégories Tannakiennes*, in *Grothendieck Festschrift Vol. II*, Progress in Math. **87**, Birkhauser, Boston 1990.

[Del2] P. Deligne, *Le groupe fondamental de la droite projective moins trois points*, in *Galois groups over* $\mathbb{Q}$, Math. Sci. Res. Inst. Publ., 16, Springer Verlag, New York, 1989.

[Del-Rap] P. Deligne, M. Rapoport, *Les schémas de modules de courbes elliptiques*, 143-317 in *Modular functions of one variable II*, LNM 349, Springer, Berlin 1973.

[Dem-Gab] M. Demazure, P. Gabriel, *Groupes algébriques, Tome I*, Masson & Cie. Paris, North-Holland publishing company, 1970.

[Du] D. Dugger *Universal homotopy theories*, Adv. Math. **164** (2001), 144-176.

[Du2] D. Dugger *Combinatorial model categories have presentations*, Adv. in Math. **164** (2001), 177-201.

[D-K1] W. Dwyer, D. Kan, *Simplicial localization of categories*, J. Pure and Appl. Algebra **17** (1980), 267-284.

[D-K2] W. Dwyer, D. Kan, *Equivalences between homotopy theories of diagrams*, in *Algebraic topology and algebraic $K$-theory, Annals of Math. Studies* **113**, Princeton University Press, Princeton, 1987, 180-205.

[D-K3] W. Dwyer, D. Kan, *Homotopy commutative diagrams and their realizations*, J. Pure Appl. Algebra **57** (1989) No. 1, 5-24.

[DHK] W. Dwyer, P. Hirschhorn, D. Kan, *Model categories and more general abstract homotopy theory*, Book in preparation, available at http://www-math.mit.edu/~psh.

[DS] W. Dwyer, J. Spalinski, *Homotopy theories and model categories*, Handbook of Algebraic Topology, edited by I. M. James, Elsevier, 1995, 73-126.

[EKMM] A.D. Elmendorf, I. Kriz, M.A. Mandell, J.P. May, *Rings, modules, and algebras in stable homotopy theory*, Mathematical Surveys and Monographs, vol. 47, American Mathematical Society, Providence, *RI*, 1997.

[Goe-Ja] P. Goerss, J.F. Jardine, *Simplicial homotopy theory*, Progress in Mathematics, Vol. **174**, Birkhauser Verlag 1999.

[Goe-Hop] P. Goerss, M. Hopkins, *André-Quillen (co)homology for simplicial algebras over simplicial operads*, Une Dégustation Topologique [Topological Morsels]: Homotopy Theory in the Swiss Alps (D. Arlettaz and K. Hess, eds.), Contemp. Math. 265, Amer. Math. Soc., Providence, RI, 2000, pp. 41-85.

[Go] J. Gorski, *Representability of the derived Quot functor*, in preparation.

[Gr] A.Grothendieck, *Catégories cofibrées additives et complexe cotangent relatif*, Lecture Note in Mathematics 79, Springer-Verlag, Berlin, 1968.

[EGAI] A. Grothendieck, J. Dieudonné, *Eléments de Géométrie Algébrique I*, Springer Verlag, Berlin, 1971.

[EGAIV] A. Grothendieck, *Eléments de Géométrie Algébrique IV. Etude locale des schémas et des morphismes de schémas*, Publ. Math. I.H.E.S., **20**, **24**, **28**, **32** (1967).

[Ha] M. Hakim, *Topos annelés et schémas relatifs*, Ergebnisse der Mathematik und ihrer Grenzgebiete, Band 64. Springer-Verlag Berlin-New York, 1972.

[Hin1] V. Hinich, *Homological algebra of homotopical algebras*, Comm. in Algebra **25** (1997), 3291-3323.

[Hin2] V. Hinich, *Formal stacks as dg-coalgebras*, J. Pure Appl. Algebra **162** (2001), No. 2-3, 209-250.

[Hi] P. S. Hirschhorn, *Model Categories and Their Localizations*, Math. Surveys and Monographs Series 99, AMS, Providence, 2003.

[H-S] A. Hirschowitz, C. Simpson, *Descente pour les n-champs*, preprint available at math.AG/9807049.

[Hol] S. Hollander, *A homotopy theory for stacks*, preprint available at math.AT/0110247.

[Ho1] M. Hovey, *Model categories*, Mathematical surveys and monographs, Vol. **63**, Amer. Math. Soc., Providence 1998.

[Ho2] M. Hovey, *Spectra and symmetric spectra in general model categories*, J. Pure Appl. Alg. **165** (2001), 63-127.

[HSS] M. Hovey, B.E. Shipley, J. Smith, *Symmetric spectra*, J. Amer. Math. Soc. **13** (2000), no. 1, 149-208.

[Ill] L. Illusie, *Complexe cotangent et déformations I*, Lectur Notes in Mathematics **239**, Springer Verlag, Berlin, 1971.

[Ja1] J. F. Jardine, *Simplicial presheaves*, J. Pure and Appl. Algebra **47** (1987), 35-87.

[Ja2] J. F. Jardine, *Stacks and the homotopy theory of simplicial sheaves*, in *Equivariant stable homotopy theory and related areas* (Stanford, CA, 2000). Homology Homotopy Appl. **3** (2001), No. 2, 361-384.

[Jo1] A. Joyal, Letter to Grothendieck.

[Jo2] A. Joyal, unpublished manuscript.

[Jo-Ti] A. Joyal, M. Tierney, *Strong stacks and classifying spaces*, in *Category theory (Como, 1990)*, Lecture Notes in Mathematics **1488**, Springer-Verlag New York, 1991, 213-236.

[Ka2] M. Kapranov, *Injective resolutions of BG and derived moduli spaces of local systems*, J. Pure Appl. Algebra **155** (2001), No. 2-3, 167-179.

[K-P-S] L. Katzarkov, T. Pantev, C. Simspon, *Non-abelian mixed Hodge structures*, preprint math.AG/0006213.

[Ko] M. Kontsevich, *Operads and motives in deformation quantization*, Moshé Flato (1937-1998), Lett. Math. Phys. **48**, (1999), No. 1, 35-72.

[Ko-So] M. Kontsevich, Y. Soibelman, *Deformations of algebras over operads and the Deligne conjecture*, Conférence Moshé Flato 1999, Vol. 1 (Dijon), 255-307, Math. Phys. Stud. **21**, Kluwer Acad. Publ, Dordrecht, 2000.

[Kr-Ma] I. Kriz, J. P. May, *Operads, algebras, modules and motives*, Astérisque **233**, 1995.

[La-Mo] G. Laumon and L. Moret-Bailly, *Champs algébriques*, A Series of Modern Surveys in Mathematics vol. **39**, Springer-Verlag 2000.

[Lu1] J.Lurie, PhD Thesis, MIT, Boston, 2004.

[Lu2] J. Lurie, *A survey of elliptic cohomology*, Preprint December 2005 (available at http://www.math.harvard.edu/~lurie/papers/survey.pdf).

[MMSS] M. Mandell, J. P. May, S. Schwede, B. Shipley *Model categories of diagram spectra*, Proc. London Math. Soc. **82** (2001), 441–512.

[May] J.P. May, *Pairings of categories and spectra*, JPAA **19** (1980), 299-346.

[May2] J.P. May, *Picard groups, Grothendieck rings, and Burnside rings of categories*, Adv. in Math. **163**, 2001, 1-16.

[Mil] J. S. Milne, *Étale cohomology*, Princeton University Press, 1980.

[Min] V. Minasian, *André-Quillen spectral sequence for THH*, Topology and Its Applications, **129**, (2003) 273-280.

[MCM] R. Mc Carthy, V. Minasian, *HKR theorem for smooth S-algebras*, Journal of Pure and Applied Algebra, Vol **185**, 2003, 239-258, 2003.

[Q1] D. Quillen, *Homotopical algebra*, Lecture Notes in Mathematics **43**, Springer Verlag, Berlin, 1967.

[Q2] D. Quillen, *On the (co-)homology of commutative rings*, Applications of Categorical Algebra (Proc. Sympos. Pure Math., Vol XVII, New York, 1964), 65-87. Amer. Math. Soc., Providence, P.I.

[Q3] D. Quillen, *Higher algebraic K-theory I*, in *Algebraic K-theory I-Higher K-theories*, Lecture Notes in Mathematics **341**, Springer Verlag, Berlin.

[Re] C. Rezk, *Spaces of algebra structures and cohomology of operads*, Thesis 1996, available at http://www.math.uiuc.edu/ rezk.

[Ri-Rob] B. Richter, A. Robinson, *Gamma-homology of group algebras and of polynomial algebras*, To appear in the "Proceedings of the Northwestern conference" 2002.

[Ro] J. Rognes, *Galois extensions of structured ring spectra*, Preprint math.AT/0502183.

[Schw-Shi] S. Schwede, B. Shipley, *Stable model categories are categories of modules*, Topology **42** (2003), 103 − −153.

[Shi] B. Shipley, *A convenient model category for commutative ring spectra*, Preprint 2002.

[S1] C. Simpson, *Homotopy over the complex numbers and generalized cohomology theory*, in *Moduli of vector bundles (Taniguchi Symposium, December 1994)*, M. Maruyama ed., Dekker Publ. (1996), 229-263.

[S2] C. Simpson, *A Giraud-type characterization of the simplicial categories associated to closed model categories as $\infty$-pretopoi*, Preprint math.AT/9903167.

[S3] C. Simpson, *Algebraic (geometric) n-stacks*, Preprint math.AG/9609014.

[S4] C. Simpson, *The Hodge filtration on non-abelian cohomology*, Preprint math.AG/9604005.

[Sm] J. Smith, *Combinatorial model categories*, unpublished.

[Sp] M. Spitzweck, *Operads, algebras and modules in model categories and motives*, Ph.D. Thesis, Mathematisches Institüt, Friedrich-Wilhelms-Universität Bonn (2001), available at http://www.uni-math.gwdg.de/spitz/.

[Tab] G. Tabuada, *Une structure de catégorie de modèles de Quillen sur la catégorie des dg-catégories*, Preprint math.KT/0407338.

[To1] B. Toën, *Champs affines*, Preprint math.AG/0012219.

[To2] B. Toën, *Homotopical and higher categorical structures in algebraic geometry*, Hablitation Thesis available at math.AG/0312262

[To3] B. Toën, *Vers une interprétation Galoisienne de la théorie de l'homotopie*, Cahiers de topologie et geometrie differentielle categoriques, Volume XLIII (2002), 257-312.

[To-Va1] B. Toën, M. Vaquié, *Moduli of objects in dg-categories*, Preprint math.AG/0503269.

[To-Va2] B. Toën, M. Vaquié, *Au-dessous de Spec $\mathbb{Z}$*, Preprint math.AG/0509684.

[HAGI] B. Toën, G. Vezzosi, *Homotopical algebraic geometry I: Topos theory*, Advances in Mathematics, **193**, Issue 2 (2005), p. 257-372.

[To-Ve1] B. Toën, G. Vezzosi, *Segal topoi and stacks over Segal categories*, December 25, 2002, to appear in Proceedings of the Program *"Stacks, Intersection theory and Non-abelian Hodge Theory"*, MSRI, Berkeley, January-May 2002 (also available as Preprint math.AG/0212330).

[To-Ve2] B. Toën, G. Vezzosi, *From HAG to DAG: derived moduli spaces*, p. 175-218, in *"Axiomatic, Enriched and Motivic Homotopy Theory"*, Proceedings of the NATO Advanced Study Institute, Cambridge, UK, (9-20 September 2002), Ed. J.P.C. Greenlees, NATO Science Series II, Volume 131 Kluwer, 2004.

[To-Ve3] B. Toën, G. Vezzosi, *"Brave New" algebraic geometry and global derived moduli spaces of ring spectra*, to appear in Proceedings of the Euroworkshop *"Elliptic Cohomology and Higher Chromatic Phenomena"* (9 - 20 December 2002), Isaac Newton Institute for Mathematical Sciences (Cambridge, UK), H. Miller, D. Ravenel eds. (also available as Preprint math.AT/0309145).

[To-Ve4] B. Toën, G. Vezzosi, *Algebraic geometry over model categories. A general approach to Derived Algebraic Geometry*, Preprint math.AG/0110109.

[We] C. Weibel, *An introduction to homological algebra*, Cambridge Univ. Press, Cambridge, 1995.

# Index

## Editorial Information

To be published in the *Memoirs*, a paper must be correct, new, nontrivial, and significant. Further, it must be well written and of interest to a substantial number of mathematicians. Piecemeal results, such as an inconclusive step toward an unproved major theorem or a minor variation on a known result, are in general not acceptable for publication.

Papers appearing in *Memoirs* are generally at least 80 and not more than 200 published pages in length. Papers less than 80 or more than 200 published pages require the approval of the Managing Editor of the Transactions/Memoirs Editorial Board.

As of January 31, 2008, the backlog for this journal was approximately 17 volumes. This estimate is the result of dividing the number of manuscripts for this journal in the Providence office that have not yet gone to the printer on the above date by the average number of monographs per volume over the previous twelve months, reduced by the number of volumes published in four months (the time necessary for preparing a volume for the printer). (There are 6 volumes per year, each usually containing at least 4 numbers.)

A Consent to Publish and Copyright Agreement is required before a paper will be published in the *Memoirs*. After a paper is accepted for publication, the Providence office will send a Consent to Publish and Copyright Agreement to all authors of the paper. By submitting a paper to the *Memoirs*, authors certify that the results have not been submitted to nor are they under consideration for publication by another journal, conference proceedings, or similar publication.

## Information for Authors

*Memoirs* are printed from camera copy fully prepared by the author. This means that the finished book will look exactly like the copy submitted.

**Initial submission.** The AMS uses Centralized Manuscript Processing for initial submissions. Authors should submit a PDF file using the Initial Manuscript Submission form found at `www.ams.org/cgi-bin/peertrack/submission.pl`, or send one copy of the manuscript to the following address: Centralized Manuscript Processing, MEMOIRS OF THE AMS, 201 Charles Street, Providence, RI 02904-2294 USA. If a paper copy is being forwarded to the AMS, indicate that it is for it Memoirs and include the name of the corresponding author, contact information such as email address or mailing address, and the name of an appropriate Editor to review the paper (see the list of Editors below).

The paper must contain a *descriptive title* and an *abstract* that summarizes the article in language suitable for workers in the general field (algebra, analysis, etc.). The *descriptive title* should be short, but informative; useless or vague phrases such as "some remarks about" or "concerning" should be avoided. The *abstract* should be at least one complete sentence, and at most 300 words. Included with the footnotes to the paper should be the 2000 *Mathematics Subject Classification* representing the primary and secondary subjects of the article. The classifications are accessible from `www.ams.org/msc/`. The list of classifications is also available in print starting with the 1999 annual index of *Mathematical Reviews*. The Mathematics Subject Classification footnote may be followed by a list of *key words and phrases* describing the subject matter of the article and taken from it. Journal abbreviations used in bibliographies are listed in the latest *Mathematical Reviews* annual index. The series abbreviations are also accessible from `www.ams.org/publications/`. To help in preparing and verifying references, the AMS offers MR Lookup, a Reference Tool for Linking, at `www.ams.org/mrlookup/`.

**Electronically prepared manuscripts.** The AMS encourages electronically prepared manuscripts, with a strong preference for $\mathcal{A}_{\mathcal{M}}\mathcal{S}$-LaTeX. To this end, the Society has prepared $\mathcal{A}_{\mathcal{M}}\mathcal{S}$-LaTeX author packages for each AMS publication. Author packages include instructions for preparing electronic manuscripts, samples, and a style file that generates the particular design specifications of that publication series. Though $\mathcal{A}_{\mathcal{M}}\mathcal{S}$-LaTeX is the highly preferred format of TeX, author packages are also available in $\mathcal{A}_{\mathcal{M}}\mathcal{S}$-TeX.

Authors may retrieve an author package from the AMS website starting from `www.ams.org/tex/` or via FTP to `ftp.ams.org` (login as `anonymous`, enter username as password, and type `cd pub/author-info`). The *AMS Author Handbook* and the *Instruction Manual* are available in PDF format following the author packages link from `www.ams.org/tex/`. The author package can also be obtained free of charge by sending email to `tech-support@ams.org` (Internet) or from the Publication Division, American Mathematical Society, 201 Charles St., Providence, RI 02904-2294, USA. When requesting an author package, please specify $\mathcal{A}_{\mathcal{M}}\mathcal{S}$-LaTeX or $\mathcal{A}_{\mathcal{M}}\mathcal{S}$-TeX and the publication in which your paper will appear. Please be sure to include your complete mailing address.

**After acceptance.** The final version of the electronic file should be sent to the Providence office (this includes any TeX source file, any graphics files, and the DVI or PostScript file) immediately after the paper has been accepted for publication.

Before sending the source file, be sure you have proofread your paper carefully. The files you send must be the EXACT files used to generate the proof copy that was accepted for publication. For all publications, authors are required to send a printed copy of their paper, which exactly matches the copy approved for publication, along with any graphics that will appear in the paper.

Accepted electronically prepared files can be submitted via the web at `www.ams.org/submit-book-journal/`, sent via FTP, or sent on CD-Rom or diskette to the Electronic Prepress Department, American Mathematical Society, 201 Charles Street, Providence, RI 02904-2294 USA. TeX source files, DVI files, and PostScript files can be transferred over the Internet by FTP to the Internet node `ftp.ams.org` (130.44.1.100). When sending a manuscript electronically via CD-Rom or diskette, please be sure to include a message identifying the paper as a Memoir.

Electronically prepared manuscripts can also be sent via email to `pub-submit@ams.org` (Internet). In order to send files via email, they must be encoded properly. (DVI files are binary and PostScript files tend to be very large.)

**Electronic graphics.** Comprehensive instructions on preparing graphics are available at `www.ams.org/jourhtml/`. A few of the major requirements are given here.

Submit files for graphics as EPS (Encapsulated PostScript) files. This includes graphics originated via a graphics application as well as scanned photographs or other computer-generated images. If this is not possible, TIFF files are acceptable as long as they can be opened in Adobe Photoshop or Illustrator. No matter what method was used to produce the graphic, it is necessary to provide a paper copy to the AMS.

Authors using graphics packages for the creation of electronic art should also avoid the use of any lines thinner than 0.5 points in width. Many graphics packages allow the user to specify a "hairline" for a very thin line. Hairlines often look acceptable when proofed on a typical laser printer. However, when produced on a high-resolution laser imagesetter, hairlines become nearly invisible and will be lost entirely in the final printing process.

Screens should be set to values between 15% and 85%. Screens which fall outside of this range are too light or too dark to print correctly. Variations of screens within a graphic should be no less than 10%.

**Inquiries.** Any inquiries concerning a paper that has been accepted for publication should be sent to `memo-query@ams.org` or directly to the Electronic Prepress Department, American Mathematical Society, 201 Charles St., Providence, RI 02904-2294 USA.

## Editors

This journal is designed particularly for long research papers, normally at least 80 pages in length, and groups of cognate papers in pure and applied mathematics. Papers intended for publication in the *Memoirs* should be addressed to one of the following editors. The AMS uses Centralized Manuscript Processing for initial submissions to AMS journals. Authors should follow instructions listed on the Initial Submission page found at `www.ams.org/memo/memosubmit.html`.

**Algebra** to ALEXANDER KLESHCHEV, Department of Mathematics, University of Oregon, Eugene, OR 97403-1222; email: `ams@noether.uoregon.edu`

**Algebraic geometry and its application** to MINA TEICHER, Emmy Noether Research Institute for Mathematics, Bar-Ilan University, Ramat-Gan 52900, Israel; email: `teicher@macs.biu.ac.il`

**Algebraic geometry** to DAN ABRAMOVICH, Department of Mathematics, Brown University, Box 1917, Providence, RI 02912; email: `amsedit@math.brown.edu`

**Algebraic number theory** to V. KUMAR MURTY, Department of Mathematics, University of Toronto, 100 St. George Street, Toronto, ON M5S 1A1, Canada; email: `murty@math.toronto.edu`

**Algebraic topology** to ALEJANDRO ADEM, Department of Mathematics, University of British Columbia, Room 121, 1984 Mathematics Road, Vancouver, British Columbia, Canada V6T 1Z2; email: `adem@math.ubc.ca`

**Combinatorics** to JOHN R. STEMBRIDGE, Department of Mathematics, University of Michigan, Ann Arbor, Michigan 48109-1109; email: `JRS@umich.edu`

**Complex analysis and harmonic analysis** to ALEXANDER NAGEL, Department of Mathematics, University of Wisconsin, 480 Lincoln Drive, Madison, WI 53706-1313; email: `nagel@math.wisc.edu`

**Differential geometry and global analysis** to LISA C. JEFFREY, Department of Mathematics, University of Toronto, 100 St. George St., Toronto, ON Canada M5S 3G3; email: `jeffrey@math.toronto.edu`

**Functional analysis and operator algebras** to DIMITRI SHLYAKHTENKO, Department of Mathematics, University of California, Los Angeles, CA 90095; email: `shlyakht@math.ucla.edu`

**Geometric analysis** to WILLIAM P. MINICOZZI II, Department of Mathematics, Johns Hopkins University, 3400 N. Charles St., Baltimore, MD 21218; email: `trans@math.jhu.edu`

**Geometric analysis** to MARK FEIGHN, Math Department, Rutgers University, Newark, NJ 07102; email: `feighn@andromeda.rutgers.edu`

**Harmonic analysis, representation theory, and Lie theory** to ROBERT J. STANTON, Department of Mathematics, The Ohio State University, 231 West 18th Avenue, Columbus, OH 43210-1174; email: `stanton@math.ohio-state.edu`

**Logic** to STEFFEN LEMPP, Department of Mathematics, University of Wisconsin, 480 Lincoln Drive, Madison, Wisconsin 53706-1388; email: `lempp@math.wisc.edu`

**Number theory** to JONATHAN ROGAWSKI, Department of Mathematics, University of California, Los Angeles, CA 90095; email: `jonr@math.ucla.edu`

**Partial differential equations** to GUSTAVO PONCE, Department of Mathematics, South Hall, Room 6607, University of California, Santa Barbara, CA 93106; email: `ponce@math.ucsb.edu`

**Partial differential equations and dynamical systems** to PETER POLACIK, School of Mathematics, University of Minnesota, Minneapolis, MN 55455; email: `polacik@math.umn.edu`

**Probability and statistics** to RICHARD BASS, Department of Mathematics, University of Connecticut, Storrs, CT 06269-3009; email: `bass@math.uconn.edu`

**Real analysis and partial differential equations** to DANIEL TATARU, Department of Mathematics, University of California, Berkeley, Berkeley, CA 94720; email: `tataru@math.berkeley.edu`

**All other communications to the editors** should be addressed to the Managing Editor, ROBERT GURALNICK, Department of Mathematics, University of Southern California, Los Angeles, CA 90089-1113; email: `guralnic@math.usc.edu`.

# Titles in This Series

TITLES IN THIS SERIES

876 **Gabriel Debs and Jean Saint Raymond,** Borel liftings of Borel sets: Some decidable and undecidable statements, 2007

875 **C. Krattenthaler and T. Rivoal,** Hypergéométrie et fonction zêta de Riemann, 2007

874 **Sonia Natale,** Semisolvability of semisimple Hopf algebras of low dimension, 2007

873 **A. J. Duncan,** Exponential genus problems in one-relator products of groups, 2007

872 **Anthony V. Geramita, Tadahito Harima, Juan C. Migliore, and Yong Su Shin,** The Hilbert function of a level algebra, 2007

871 **Pascal Auscher,** On necessary and sufficient conditions for $L^p$-estimates of Riesz transforms associated to elliptic operators on $\mathbb{R}^n$ and related estimates, 2007

870 **Takuro Mochizuki,** Asymptotic behaviour of tame harmonic bundles and an application to pure twistor $D$-modules, Part 2, 2007

869 **Takuro Mochizuki,** Asymptotic behaviour of tame harmonic bundles and an application to pure twistor $D$-modules, Part 1, 2007

868 **Gelu Popescu,** Entropy and multivariable interpolation, 2006

867 **Vilmos Totik,** Metric properties of harmonic measures, 2006

866 **William Craig,** Semigroups underlying first-order logic, 2006

865 **Nathanial P. Brown,** Invariant means and finite representation theory of $C*$-algebras, 2006

864 **John M. Lee,** Fredholm operators and Einstein metrics on conformally compact manifolds, 2006

863 **M. Lübke and A. Teleman,** The Universal Kobayashi-Hitchin correspondence on Hermitian manifolds, 2006

862 **Alberto Canonaco,** The Beilinson complex and canonical rings of irregular surfaces, 2006

861 **Leon A. Takhtajan and Lee-Peng Teo,** Weil-Petersson metric on the universal Teichmüller space, 2006

860 **Thomas M. Fiore,** Pseudo limits, biadjoints and pseudo algebras: Categorical foundations of conformal field theory, 2006

859 **N. Arcozzi, R. Rochberg, and E. Sawyer,** Carleson measures and interpolating sequences for Besov spaces on complex balls, 2006

858 **Enrico Valdinoci, Berardino Sciunzi, and Vasile Ovidiu Savin,** Flat level set regularity of $p$-Laplace phase transitions, 2006

857 **Donatella Danielli, Nocola Garofalo, and Duy-Minh Nhieu,** Non-doubling Ahlfors measures, perimeter measures, and the characterization of the trace spaces of Sobolev functions in Carnot-Carathéodory spaces, 2006

856 **Vladimir Bolotnikov and Harry Dym,** On boundary interpolation for matrix valued Schur functions, 2006

855 **Yevgenia Kashina, Yorck Sommerhäuser, and Yongchang Zhu,** On higher Frobenius-Schur indicators, 2006

854 **Noam Greenberg,** The role of true finiteness in the admissible recursively enumerable degrees, 2006

853 **Joachim Krieger,** Stability of spherically symmetric wave maps, 2006

852 **Viorel Barbu, Irena Lasiecka, and Roberto Triggiani,** Tangential boundary stabilization of Navier-Stokes equations, 2006

851 **Jie Wu,** On maps from loop suspensions to loop spaces and the shuffle relations on the Cohen groups, 2006